Human Subjects Research after the Holocaust

Sheldon Rubenfeld • Susan Benedict

Editors

Human Subjects Research after the Holocaust

Foreword by Arthur L. Caplan

 Springer

Editors
Sheldon Rubenfeld, MD
Clinical Professor of Medicine,
  Baylor College of Medicine
Clinical Professor of Nursing,
  The University of Texas Health
  Science Center at Houston School
  of Nursing
President, Center for Medicine
  after the Holocaust
Houston, Texas
USA

Susan Benedict, CRNA, PhD, FAAN
Assistant Dean and Department Chair:
  Acute and Continuing Care
Professor and Director of Global Health
Co-director Campus-wide Program in
  Interprofessional Ethics
The University of Texas Health Science
  Center at Houston School of Nursing
Houston, Texas
USA

ISBN 978-3-319-05701-9   ISBN 978-3-319-05702-6 (eBook)
DOI 10.1007/978-3-319-05702-6
Springer Cham Heidelberg New York Dordrecht London

Library of Congress Control Number: 2014941704

Printed on acid-free paper

Springer is part of Springer Science+Business Media (www.springer.com)

We dedicate this book to all victims, known and unknown, of human subjects research during and after the Holocaust.

Preface

In December 2012 and April 2013, the Center for Medicine after the Holocaust (CMATH) convened meetings on human subjects research after the Holocaust. The first was held in Houston and was cosponsored by The Methodist Hospital Research Institute (now Houston Methodist Research Institute) with additional support provided by The University of Texas Health Science Center at Houston School of Nursing. The second meeting was held in Berlin. CMATH asked the speakers to submit manuscripts for publication on their topic of choice.

A remarkable transformation took place, particularly in Houston, where the material was relatively new to some of the distinguished speakers. Subsequently, moved and challenged by the day's events and presentations, especially those of Eva Kor and Traute Lafrenz Page, some authors asked if they could submit manuscripts that reflected the impact of the day on them. As a result, the papers in this volume are not necessarily the ones that were presented at the meetings. For example, bioethicist Joseph Fins, whose original lecture was "The Rights of Patients with Severe Brain Injury, Specifically Patients in the Minimally Conscious State," chose instead to write "Teaching the Holocaust to Medical Students: A Reflection on Pedagogy and Medical Ethics." Research scientist Mauro Ferrari, who presented "A Portrait of Nanomedicine and Its Bioethical Implications," chose to write a personal essay entitled "No Exceptions, No Excuses: A Testimonial." Most remarkably, Linda Emanuel, who was to speak on "Human Subjects Research at the End of Life," was so moved by the day's program that she spontaneously gave a completely different lecture, "The Psychophysiology of Attribution: Why Appreciative Respect Can Keep us Safe."

CMATH's mission is to challenge doctors, nurses, and bioscientists to personally confront the medical ethics of the Holocaust and to apply that knowledge to contemporary practice and research. We were delighted to discover that, right before our eyes, many speakers accepted our challenge and were transformed by the experience. It is our hope that readers of this book will be similarly affected.

Houston, TX   Sheldon Rubenfeld
                                                                    Susan Benedict

January 27, 2014   International Holocaust Remembrance Day

Foreword

With the publication of this volume, *Human Subjects Research after the Holocaust*, an era has come to an end. Since the conclusion of the Nuremberg trials in 1949, there has not been a thorough, multidisciplinary, international examination of the research conducted by German scientists during the reign of the Nazi party. That lacuna in bioethics and the history of medicine has now been filled. The analyses presented here, ranging from examinations of sulfonamide experiments on prisoners in concentration camps, cruel studies to prevent and treat typhus, and morally polluted anatomical research, to heinous studies on pregnant women and children, are admirably detailed and sophisticated. Future generations of scholars and commentators now have a much sturdier foundation upon which to build their work on what happened and why.

One thing the studies in this book of research conducted by German scientists during the war make clear is that, judged by the standards prevailing at the time, it is impossible and even irresponsible to dismiss what was done as fringe or crackpot science done by fringe or crackpot biomedical researchers. One of the standard responses by some earlier scholars in trying to interpret what was done in the name of biomedical research in the concentration camps and elsewhere was to dismiss such activity in exactly these terms.

Sadly, part of the reason for discounting Nazi scientific research on scientific criteria is that it is then possible to avoid having to examine the ethical rationales and values that led mainstream and prominent researchers to undertake their barbarous studies. To some extent the attempt to marginalize research undertaken during the Nazi era was a result of the fact that the status of the research became swept up in a very different dispute—the acceptability of citing or using findings from those experiments in contemporary research studies or policy applications.

The issue of whether data can be used when acquired by immoral experiments and investigations conducted in camps became the subject of a huge controversy in the early 1990s. Researchers studying the impact of hypothermia on the human body became aware that studies had been done in the Dachau camp both to determine the temperatures at which subjects died and to examine what techniques led to longer survival and revival. The contemporary researchers asked for help in deciding whether to utilize the data from the Dachau hypothermia research.

They received plenty of opinions. Some argued that the data from Dachau was of little use because the research was done on weak and sick subjects. Dr. Robert L. Berger, a heart surgeon, now deceased, at the New England Deaconess Hospital and Harvard Medical School, concluded in a 1990 paper in the *New England Journal of Medicine* that the studies relied on scientifically unsound methods, were carried out erratically, were largely fraudulent, and were led by a known murderer. Robert Pozos, a physiologist and expert in hypothermia who, while working at the University of Minnesota-Duluth in the 1990s and later at San Diego State University, had also served as an expert consultant for the United States Navy, did not agree.

It was his view and remains so that, while some of those involved were not competent, the key researchers involved at Dachau were well-trained researchers in the field of physiology. He further maintained that the findings concerning prolonged exposure to cold temperatures were viewed as legitimate throughout the war by German researchers, that scientists from the victorious allied nations viewed the findings as sound for many years after the end of the war and cited them as such, and that his own work and that of other scientists, when done on volunteers under controlled and far safer conditions, appeared to confirm many of the trends seen in the Dachau findings.

While the validity of the Dachau work continues to be disputed, there is no disputing the fact that it is made far easier to resolve the issue of the morality of continuing to use the data from these and other immoral experiments seven or eight decades after they were conducted if there is nothing of contemporary value in what was reported as valid data. Those who believed that any use of so-called "tainted" data from the Dachau hypothermia studies was and remains unethical spent a great deal of time dismissing the research that was done as worthless and those who conducted it as frauds. This has left the impression among many that nothing of any value done by anyone of any scientific merit or standing occurred in the camps or anywhere else during the Nazi era. The studies in this book show that position to be completely and utterly false. Many competent and even distinguished researchers conducted studies using the modes of inquiry and methodologies that predominated during their day as valid science upon hapless, sick, and coerced subjects, nearly all of whom suffered horribly and many of whom died. Which leaves the outstanding question, also forthrightly addressed by many of the contributors to this volume, how could that happen?

One key factor that permitted what are, on their face, ethically loathsome human studies to be conducted is racism. The subjects of the concentration camp and anatomical studies were dehumanized, viewed as less than human and deserving of no respect due to their religion, ethnicity, gender orientation, political views, status as prisoners, or Nazi race theory. Author after author in the papers collected here notes how Nazi racial views distorted the ethics of those who conducted research so that they did not feel they were violating the interests or rights of any "human" subject.

Also important is the fact that we are viewing a sample of research conducted during a titanic war where the stakes were nothing less than world domination. As the war progressed and began to turn against Germany and her allies, the military became increasingly desperate to find answers quickly to pressing challenges such as battling typhus, finding answers to healing burns and wounds, keeping pilots and seamen alive in cold waters, and in carrying out the "Final Solution"—the extermination of the Jews—as a key goal of the Reich. Many of the leading scientists were involved in research to advance German national security, the cause of the war, and to save the German people from the wrath of her enemies.

In the Cold War years and as the War on Terror continues to evolve, there are many questions, thoughtfully presented in this book, raised by the work done by German scientists who were motivated by patriotism in an atmosphere of racism with little sympathy for "enemy prisoners." The desire to do what is necessary to secure the security of one's family and nation is often praised as noble. What that norm does to distort the ethical duties owed by doctors and scientists to human subjects in biomedical research may not be.

Arthur L. Caplan, PhD
Drs. William F. and Virginia Connolly Mitty Chair
Director, Division of Medical Ethics
NYU Langone Medical Center
New York, NY

Acknowledgments

We gratefully acknowledge the numerous people who have supported the Center for Medicine after the Holocaust (CMATH) and have made this book a possibility and, now, a reality. We are especially grateful to Miriam Weiner and Tony Rodriguez of The Conference on Jewish Material Claims Against Germany (Claims Conference) for their support of CMATH from its earliest days through the present. Their support has enabled us to bring together scholars from the United States, Germany, Austria, England, and Israel to contribute to this book as well as to share scholarship related to medicine after the Holocaust.

This book would not have been possible without the enthusiastic support of the Methodist Hospital Research Institute (MHRI) of Houston, Texas and, especially its president and CEO, Mauro Ferrari, PhD. Ron Girotto, the former president and CEO of the Methodist Hospital System, initially brought Dr. Ferrari and CMATH together, and Marc Boom, MD, the current president and CEO of Houston Methodist, and Roberta Schwartz, Houston Methodist Hospital Executive Vice President, continue to enthusiastically support our joint efforts. Dr. Mary Schwartz provided invaluable liaison between parties and sage advice all along. MHRI staff, including Tong Sun, Rebecca Hall, Kari Stein, and Homer Quintana, enabled our scholars to gather together in Houston to share ideas that became the intellectual impetus for this book.

We are appreciative of the support of The University of Texas Health Science Center at Houston School of Nursing and, in particular, Dean Patricia Starck, for their longstanding support of CMATH. We are also grateful to Jessica Sussman of Frosch Travel who facilitated the gathering of our European scholars to exchange ideas that contributed to the development of this book. Several scholars were unable to contribute chapters but provided us with important scholarship. Among them are Prof. Dr. Gerhard Baader, Dr. Eva Marie Ulmer, Prof. Dr. med Andreas Frewer, Vivien Spitz, and Prof. Dr. med Wolfgang U. Eckart.

We are grateful to David Packer, our editor at Springer who guided this book from proposal to product, to Sara Kate Heukerott, his editorial assistant, and to their staff at Springer for providing a steady hand through the editing and publishing processes. We are very thankful for Kathy Kobos, our personal and indispensable editorial assistant, for her unflagging energy and tireless devotion to the book and for getting everything just right.

Finally, we'd like to thank Linda and Jeff, our loving spouses, who fully supported us from the very start of this entire project and who gracefully and genially managed without our company many nights and weekends. They have been the rocks on which the foundations of this book were built.

Houston, Texas   Sheldon Rubenfeld, MD
                                                    Susan Benedict, CRNA, PhD, FAAN
January 27, 2014   International Holocaust Remembrance Day

ועידת התביעות
Claims Conference
The Conference on Jewish Material
Claims Against Germany
www.claimscon.org

Contents

Chapter 1
Introduction: How Did It Go So Wrong?

Sheldon Rubenfeld

> Human beings are a species splendid in their array of
> moral equipment, tragic in their propensity to misuse it,
> and pathetic in their constitutional ignorance of the misuse.
> From *The Moral Animal* by Robert Wright (1994).

If we think about medicine and the Holocaust, the name Josef Mengele is likely to come to mind. Dr. Mengele making "selections" on the train platform at Auschwitz is an iconic image of the Holocaust. He may be the most notorious Nazi physician and human subjects researcher, primarily because of his twin experiments. Yet, compared to the number of victims he sent to the gas chambers, Mengele killed a relatively small number of people in his experiments. Also, the number of murders he committed is small compared to other lesser-known Nazi physicians.

Consider Irmfried Eberl who was in charge of the Brandenburg euthanasia center where the first gas chamber was designed and deployed. Eberl extinguished approximately 10,000 German "lives not worth living" at Brandenburg and then killed almost the same number at Bernburg, another euthanasia center. Eberl was subsequently rewarded with command of the Treblinka death camp where he murdered 250,000 Warsaw Jews before he was fired for his inefficient disposal of bodies (Strous 2009). If the sobriquet "Angel of Death" is synonymous with Josef Mengele (United States Holocaust Memorial Museum [USHMM], "Josef Mengele"), what possible epithet can describe Irmfried Eberl?

Although accurate numbers are elusive, there were at least 23,000 victims of human subjects research (Weindling, personal communication, July 4, 2013; see also Chap. 11 in this book) compared to 6,000,000 Jews, Gypsies, and other "inferior people" murdered by the Nazis. Why then are the human subjects research violations of the Holocaust so well known despite the relatively small number of victims?

S. Rubenfeld, MD (✉)
Clinical Professor of Medicine, Baylor College of Medicine; Clinical Professor of Nursing, The University of Texas Health Science Center at Houston School of Nursing; President, Center for Medicine after the Holocaust, 3122 Robinhood, Houston, TX 77005, USA
e-mail: sheldonrubenfeld@mac.com

S. Rubenfeld and S. Benedict (eds.), *Human Subjects Research after the Holocaust*,
DOI 10.1007/978-3-319-05702-6_1, © Springer International Publishing Switzerland 2014

Several answers come to mind. The Nuremberg Doctors' Trial focused on Nazi human subjects research as opposed to the medical profession's other more serious violations of medical ethics and basic human rights. Brigadier General Telford Taylor, chief counsel for the trials of war criminals in Nuremberg, delivered the opening statement highlighting the researchers' brutal treatment of the victims (Taylor 1992). The Nuremberg Code, a powerful statement about human subjects research and informed consent, was a major and the most universally recognized outcome of this trial (Freyhofer 2004, pp. 103–104). Although it was virtually ignored for 30 years, the code finally received the attention it deserved after Jean Heller's 1972 exposé of the Tuskegee syphilis study once again raised the specter of and forced the world to think about the Nazi medical experiments (see Chaps. 13 and 14 in this book).

The German medical profession did all it could to cover up its enthusiastic, voluntary role in the design and implementation of the Nazi policies of eugenics, euthanasia, and extermination. The victorious Allies also had ulterior motives for not publicizing the crimes of Nazi bioscientists other than the medical experiments. The United States and the Soviet Union were beginning the Cold War, and they each wanted to acquire Nazi scientists with potentially valuable skills and knowledge even though they may have been guilty of war crimes. At one time or another, the US military employed 4 of the 23 defendants at the Nuremberg Doctors' Trial (Mitscherlich and Mielke 1992, pp. 106–107) and more than 700 German scientists were brought to America as part of Operation Paperclip, including physician Hubertus Strughold, the father of both German and American space medicine ("Operation Paperclip" n.d.; see Chap. 17 in this book). Prosecuting Nazi physicians and bioscientists for medical war crimes may have been a lower priority than rebuilding Germany as a stalwart ally against the Soviet Union (see Chap. 13 in this book).

The Nuremberg Doctors' Trial brought to light another reason that America and its allies downplayed Nazi medical crimes other than human subjects research violations. The defendants' lawyers disputed prosecutors' claims that the Nazi researchers were violating well-established regulations for human subjects research by presenting embarrassing evidence of allied medical misconduct (Schmidt 2004, p. 161). Prosecuting Nazi physicians for involuntary sterilization, anti-miscegenation laws, and euthanasia could have proven equally embarrassing because America had similar laws, policies, and proposals antedating those of Germany, as will be shown.

Another reason for the notoriety of the Nazi human subjects research is the intimate nature of experimentation on humans. The physicians on the ramp at Auschwitz who selected nameless Jews for mass murder had no relationships with their victims. On the other hand, researchers typically were required to know more about their victims as part of their experimental protocols. The relationship between the researchers and their victims resembled a "doctor-patient" relationship. While most people cannot imagine traveling by cattle car to a death camp, they are familiar with the doctor-patient relationship. Even though there is nothing but contempt for the medical murderers in the concentration camps, some survivors of the experiments have ambivalent feelings about their physician captors (Posner and Ware 2000, p. 35). For example, the twins in Mengele's studies knew Mengele's name, he often knew their names, and at least one twin believed Mengele saved them from the gas chambers by selecting them for his experiments (Posner and Ware 2000, p. 29). Needless to say, theirs was a doctor-patient relationship fraught with mortal danger, but it was a relationship nonetheless (Posner and Ware 2000, p. 38). Because of the personal nature of the doctor-patient relationship and human subjects research, and of the impersonal nature of mass murder, people may be more likely to recall the inhumane medical experiments.

German Physicians

Consider the biography of the most famous medical miscreant, Josef Mengele. He earned doctorates in physical anthropology from the University of Munich in 1935 and in medicine from the University of Frankfurt in 1938 ("Auschwitz II–Birkenau" n.d.). His dissertation in the Anthropological Institute at the University of Munich was entitled "Racial-Morphological Examination of the Anterior Portion of the Lower Jaw in Four Racial Groups." Mengele's medical dissertation, published in 1938, was entitled "Genealogical Studies in the Cases of Cleft Lip-Jaw-Palate." In January 1937, he became an assistant to Dr. Otmar von Verschuer at the Institute for Hereditary Biology and Racial Hygiene in Frankfurt. Mengele would later send specimens from his twin studies at Auschwitz to von Verschuer, who was well known for his twin research. (USHMM, "Josef Mengele").

In 1931, at the age of 20, Mengele joined the Steel Helmet (*Stahlhelm*). He joined the SA (*Sturmabteilung*) in 1923 and applied for Nazi party membership in 1937. Upon being accepted into the party, he applied for membership in the SS (*Schutzstaffel*). Mengele joined the military in 1938, the reserve medical corps in 1940, and then served three years with an Armed SS (*Waffen* SS) unit, including service as a medical expert for the Race and Settlement Main Office in the summer of 1940 at the Central Immigration Office North-East in Posen (today Poznan). Ultimately, he was wounded and declared medically unfit for combat, and in January 1943, he began work at the Kaiser Wilhelm Institute for Anthropology, Heredity, and Eugenics, directed by von Verschuer, his former mentor. In April of 1943, after being promoted to SS captain, Mengele asked to be transferred to Auschwitz because of its potential for human subjects research, which was funded at least in part by von Verschuer through a grant from the German Research Society (Lifton 1986, p. 341). He arrived at Auschwitz at the end of May 1943 (Posner and Ware 2000, p. 18) and resumed his research, including experiments on identical twins like Eva and Miriam Mozes (see Chap. 2 in this book).

Although Mengele was in US custody immediately after WWII, he was released and worked in Germany under false papers from 1945 to 1949 before escaping to Argentina (Posner and Ware 2000, pp. 59–93). West German courts issued a warrant for his arrest, but despite periodic sensational reports of Mengele sightings, he eluded capture and died of a stroke in 1979 (Posner and Ware 2000, pp. 287–288). By that time, he had become one of the most notorious Nazi war criminals and the most infamous Nazi doctor and bioscientist. Few people know of the involvement of the medical profession in Nazi war crimes and the Holocaust beyond what they know of Josef Mengele.

A final reason for emphasizing the human subjects research violations over the Nazi physicians' other criminal behavior: American and other Western physicians were not ready to deal with the implications of doctors and biomedical researchers committing atrocious medical crimes. To inoculate themselves against the necessity to study medicine during the Holocaust and its implications for today's human subjects research, American physicians and bioscientists accepted myths about the Nazi medical profession (Proctor 2000). The following myths enable them to ignore or dismiss the more serious and numerous crimes committed by the Nazi medical profession:

- Physicians were forced to cooperate.
- Few physicians were involved.
- Those involved were either incompetent, mad, or sadistic.
- Their ideas were beyond the pale, not mainstream ideas.
- There was no scientific rationale for the military, genetic, and physiology experiments.
- Because today's medical professionals are not Nazis, we are not capable of doing such evil.
- Because our government is not a Nazi dictatorship, liberal democracies are not capable of doing such evil.
- The Nazi medical policies were legally, morally, and ethically indefensible— physicians performing evil acts cannot possibly be motivated by ethical beliefs.

In fact, German medical professionals and bioscientists—the best in the world at the time—were deeply, voluntarily, and enthusiastically involved in eugenics, the Nazi discriminatory and murderous public health care policies, and the Holocaust.

German Medicine: World Leader

Mengele was heir to a great German medical tradition that was the envy of the world in the 1930s. William Osler studied in Germany after receiving his medical degree from McGill Medical College in Montreal in 1872 and ultimately combined the German system of postgraduate medical training with the English system of bedside teaching to revolutionize the American medical curriculum ("Celebrating Contributions of William Osler" 1993; Bliss 1999, pp. 125, 175, 221). The Carnegie Foundation funded Abraham Flexner's survey of German and other European medical schools to improve medical education in the United States and Canada. In the early 1900s, the Flexner report transformed 155 North American medical schools into a much smaller number of "Germanized" university-affiliated medical schools that employed the scientific method (Flexner 1912, pp. x, 7–10).

American bioscientists pursuing careers in medical research traveled to Germany to learn their methods. For example, famed cardiovascular surgical pioneer Michael E. DeBakey journeyed to Heidelberg during the Third Reich to study with surgeon Martin Kirschner at about the same time that Josef Mengele was studying for his two degrees. Later in life, reflecting on this experience, DeBakey said: "I was looking forward to a career in surgery and research in a medical institution, and the training available in the United States was mostly pretty mediocre . . . there was simply nothing in the United States comparable to the prestigious universities in Germany" (DeBakey 2010, p. 221).

Many countries emulated Germany's national social insurance innovations. Chancellor Otto von Bismarck initiated national social welfare programs in the 1880s, including health insurance, and other countries, such as France, Switzerland, Belgium, the Netherlands, Luxembourg, Austria, and Japan, quickly did the same (Starr 1982, p. 237). The United States did not adopt Germany's social welfare programs at the time, but in 1912 Theodore Roosevelt's Progressive Party included a national health insurance plank in his failed presidential bid (Starr 1982, pp. 243–254). One hundred years later the United States has a burgeoning social welfare program and is closer than ever to a national health care policy.

Volker Roelcke has established the scientific validity of the objectives of Nazi human subjects research (2010, pp. 17–28). In Chaps. 5 and 6 of this book, Roelcke and Sabine Hildebrandt demonstrate that German scientific research was methodologically sound, albeit ethically corrupt.

Once these myths about the incompetence and irrationality of German bioscientists have been dispelled, a serious examination of the foundations of the unethical Nazi human subjects research can begin (Lindert et al. 2013). Given that Nazi medicine was the best in the world at the time, how did the German doctors go so wrong?

Eugenics and *Gleichschaltung*: The Slippery Slope

While there may be many explanations for the origins of the confrontation between Eva Mozes and Josef Mengele, two stand out: eugenics and *Gleichschaltung*. The *Merriam-Webster Dictionary* defines *eugenics* as "a science that deals with the improvement (as by control of human mating) of hereditary qualities of a race or breed." Positive eugenics promotes the transmission of desirable genetic traits by procreation and the provision of medical care for superior races. Negative eugenics inhibits the transmission of undesirable genetic traits by discouraging procreation and medical care for inferior races. Although eugenics was popular in many countries prior to WWII, only Germany carried eugenics to its murderous but logical conclusion of extermination and genocide (see Chap. 3 in this book). *Gleichschaltung* (synchronization), the enforced unification of all societal forces, enabled the Nazis to harness all institutions in Germany to Nazi policies through the leadership principle (*Führerprinzip*) (Weikart 2009, pp. 111–113).

Eugenics is not a modern concept. According to the Hebrew Bible, "Now there arose a new king over Egypt, who knew not Joseph." The King attempted to kill all Jewish males but was thwarted by God through the actions of two genuine medical heroines, the midwives Shifra and Puah, who defied the King (Exodus 1: 8–20). The ancient Greeks and Romans also believed in eugenics. While Jews valued human life above almost all other values, Greeks valued quality of life, as Plato stated in *The Republic* about Asclepius, the Greek god of medicine and healing:

> But he makes no attempt to cure those whose constitution is basically diseased by treating them with a series of evacuations and doses which can only lead to an unhappy prolongation of life, and the production of children as unhealthy as themselves. No, he thought no treatment should be given to the man who cannot survive the routine of his ordinary job, and who is therefore of no use either to himself or society. (Plato 2003, p. 105)

In Jerusalem, the holy was beautiful, whereas in Athens and Rome, the beautiful was holy.

Hippocrates, regarded as the father of modern medicine, and his disciples rebelled against the suicide, infanticide, and euthanasia that were common in ancient Greece and proposed a life-centered oath for physicians (Dowbiggin 2003, pp. 2–3). The Hippocratic Oath, although it required a pledge to pagan deities, was nonetheless endorsed by the Catholic Church and remained unchanged and of vital importance until it became irrelevant in the twentieth century.

Charles Darwin's theory of natural selection changed our perception of the human species. His cousin, Francis Galton, coined the term "eugenics" from the Greek *eugenēs*, or "wellborn," and classified human "races" as either superior or inferior (Gillham 2001, pp. 1, 207). The rediscovery in 1900 of Gregor Mendel's 1866 publication about the transmission of traits in peas gave intellectual respectability to eugenics, which became the conventional wisdom among American academics, such as David Starr Jordan and Charles Eliot, presidents of Stanford and Harvard; scientists, such as Alexander Graham Bell and Nobel Laureate Alexis Carrel; presidents, such as Theodore Roosevelt and Woodrow Wilson; attorneys, such as Chief Justice Oliver Wendell Holmes, Jr. and Madison Grant (1916); social activists, such as Margaret Sanger; and philanthropic organizations, such as the Carnegie and Rockefeller Foundations (Bruinius 2006, p. 165).

Eugenics was strongly embraced by Germany, where it was known as racial hygiene. Many universities had endowed chairs and departments in this new academic discipline. As racial hygiene became widely accepted and admired, the prestige of German academic physicians and human subjects researchers, which was already higher than in most other countries, steadily rose; during the Third Reich, 30 percent of all university rectors were physicians (Proctor 1988, p. 94). A controversy erupted in the late 1920s within the German racial hygiene movement: Was eugenics for the betterment of all superior races, including the Jewish race (Galton considered Jews a superior race), or were Jews to be counted among the

Applied Biology
    Sterilization Law
        Nuremberg Laws
            Child Euthanasia
                Adult T4 Euthanasia
                    Human Subjects Research
                        Wannsee and the Final Solution

Fig. 1.1 The slippery slope

inferior races? The Nazis won this debate and fatally entwined their anti-Semitism with the eugenics movement (Proctor 1988, pp. 46–47).

Historian Robert Proctor (1988) pointed out that Hitler needed, counted on, and received the support of physicians much more than any other professional group to implement his public health care policy (p. 62). In 1929, four years before Hitler was democratically elected as chancellor of Germany, more than 2,700 physicians, approximately 6 percent of all German physicians, had joined the Nazi Physicians League to provide support for his policy (p. 65). At one meeting of the Nazi Physician's League, Hitler said, "… you, you National Socialist doctors, I cannot do without you for a single day, not for a single hour. If not for you, if you fail me, then all is lost. For what good are our struggles, if the health of our people is in danger?" (p. 62).

Bavarian Cabinet Minister Hans Schemm called Hitler's policy applied biology while Rudolf Hess, Hitler's deputy, called it applied racial science (p. 62). Robert Jay Lifton describes the Nazi state as a biocracy whose policy was:

> … absolute control over the evolutionary process, over the biological future. Making widespread use of the Darwinian term "selection," the Nazis sought to take over the natural functions of nature (natural selection) and God (the Lord giveth and the Lord taketh away) in orchestrating their own "selections," their own version of human evolution. (1986, p. 17)

Eugenics, *Gleichschaltung*, and applied biology proved to be a very potent combination, so much so that Germany became the foremost eugenic nation, quickly descending a slippery ethical slope (Fig. 1.1).

If the Nazis could sterilize and murder their own German citizens, they would certainly have no qualms about using Jews and other "lives not worth living" in inhumane and murderous experiments (Pasternak 2006; Spitz 2005).

Sterilization Law

The Law for the Prevention of Genetically Diseased Offspring, commonly called the Sterilization Law, was passed on July 14, 1933. The law permitted involuntary sterilization of anyone suffering from the following presumably genetically determined illnesses: feeblemindedness (the most common reason people were sterilized), blindness, deafness, severe alcoholism, Huntington's disease, schizophrenia, severe malformations, and insanity. Doctors were required to register every case of genetic illness, with fines of up to 150 Reichsmark for failure to register such a person. The law established 181 Genetic Health Courts and Appellate Genetic Health Courts in 1934 to administer the Sterilization Law. Each court was presided over by a lawyer and two doctors. Approximately 400,000 Germans, primarily patients in neuropsychiatric facilities, were approved for sterilization (Proctor 1988, pp. 95–117). While this law did not specifically target Jews, it encouraged experiments to discover efficient methods of sterilization to develop a work force that could not reproduce (see Chap. 7 in this book). A 1935 amendment to the law allowed forced abortions, setting the stage for research on pregnant women (Proctor 1988, p. 123; see also Chaps. 8 and 18 in this book).

Nuremberg Laws

Three "Nuremberg Laws" were passed between September and November 1935: the Reich Citizenship Law, the Law for the Protection of German Blood and German Honor, and the Law for the Protection of the Genetic Health of the German People. According to Robert Proctor (1988), the Reich Citizenship Law differentiated between "citizens" (males and married women of German blood with full privileges) and "residents" (unmarried women and Jews for example, whose privileges were proscribed). The Law for the Protection of German Blood and German Honor defined "full-Jews" and "half-breeds" and criminalized marriage and sexual relations between Jews and non-Jews. Finally, the Law for the Protection of the Genetic Health of the German People mandated medical examinations before marriage to see if "racial damage" might occur. Because it was feared many Jews had converted to Christianity, religion alone was no longer a reliable criterion of one's race. Only physicians trained in genetics could ensure that racial policy was grounded in "a sound scientific basis." Therefore, physicians were responsible for issuing certificates that one was "fit to marry," certificates that were required to obtain a marriage license (Proctor 1988, pp. 131–142).

Child Euthanasia

According to Henry Friedlander (1995), child euthanasia began in the winter of 1938–1939 when a family successfully petitioned Hitler to euthanize their handicapped infant, which was done at the University of Leipzig pediatric clinic by lethal injection. On August 18, 1939, the Reich Committee for the Scientific Registering of Serious Hereditary and Congenital Illnesses introduced compulsory registration of all malformed newborns. Doctors and midwives were paid 28 Reichsmark for the obligatory reporting of children with idiocy, Down syndrome, microcephaly, hydrocephaly, physical deformities, or forms of spastic paralysis. Information was forwarded to the Reich committee where three expert referees, all physicians, made the selections without actually seeing the children. Each child's registration form was marked either with a "+" if selected to die, with a "−" if selected to live, or with a "?" in cases requiring further consideration. Approximately 5,000 children were killed in the children's euthanasia program (Friedlander 1995, pp. 39–61).

Some parents were convinced to part with their child by their family doctor or public health nurses; other parents, misled to believe that the Reich committee offered improved treatment, requested admission of their children into 1 of 22 specialized facilities (Friedlander 1995, p. 47). At these extermination facilities, children were killed by injections of morphine, gassing with cyanide or chemical warfare agents, starved to death, or left without heat until they died of hypothermia. Parents received a standardized letter stating that their child had died from one of several fabricated causes and, due to the danger of an epidemic, the body had been cremated. On July 12, 1941, the child euthanasia program was expanded to include not just infants but all handicapped minors; medical personnel failing to register such children could be fined or imprisoned. Prior to 1943, when the child euthanasia program was expanded to include healthy children of inferior races, Jewish children were excluded because they did not deserve the "merciful act" of euthanasia (Proctor 1988, pp. 185–188; see Chaps. 9 and 10 in this book).

In addition to promoting the merciful and benevolent aspects of its euthanasia and other eugenic programs, the Nazis touted their economic benefits. Caring for children and adults with disabilities was expensive. Preventing the reproduction of people with presumed genetic disabilities and eliminating nonproductive children and adults was inexpensive, which was a particularly appealing argument during a severe, worldwide depression (Proctor 1988, p. 182).

Adult T4 Euthanasia

Hitler stated in a handwritten note in October of 1939:

> Reich Leader Bouhler and Dr. Med Brandt are charged with the responsibility of enlarging the powers of specific physicians, designated by name, so that patients who, on the basis of human judgement, are considered incurable, can be granted mercy death after the most careful assessment of their condition. (cited in Schmidt 2007, p. 132)

The Committee for the Scientific Treatment of Severe, Genetically Determined Illness, the committee responsible for executing Hitler's order, required physicians at neuropsychiatric institutions to complete forms assessing each patient's ability to work. Physician "experts," who did not examine the patients, evaluated the completed forms. Of the first 283,000 applications, 75,000 were marked for death with a "+" (Proctor 1988, p. 189).

Because the central administrative office of the adult euthanasia program was housed at Tiergartenstrasse 4, the program became known as *Aktion T-4*, or simply T4 (Friedlander 1995, p. 68). Among the bureaucracies supervised by this office, one was responsible for the transportation of patients from one location to another.

Additional covert bureaucracies were charged with other functions related to the mass killing of mental patients: for example, administration, human resources, killing, finances, and communication with relatives. These bureaucracies were established, among other reasons, to obscure the true purpose of the program and to make it difficult for relatives to locate patients (Schmidt 2007, pp. 132–157). Incidentally, this obfuscation relieved individual perpetrators of responsibility for the entire program.

The gas chamber was initially developed to euthanize the mental patient population of Germany; the original numerical target may have been 70,000 patients (Proctor 1988, p. 191). The first large-scale gassing, which used carbon monoxide, occurred in Brandenburg in December 1939 or January 1940 (Friedlander, 1995, p. 87). The essential roles of physicians and nurses are highlighted in this eyewitness report by chemist August Becker:

> The first gassing was administered personally by Dr. Widmann [chief chemist of the German Criminal Police Office]. He operated the controls and regulated the flow of gas. He also instructed the hospital physicians Dr. Eberl and Dr. Baumhardt, who later took over the exterminations in Grafeneck and Hadamar … At this first gassing, approximately 18–20 people were led into the "showers" by the nursing staff. These people were required to undress in another room until they were completely naked. The doors were closed behind them. They entered the room quietly and showed no signs of anxiety. Dr. Widmann operated the gassing apparatus; I could observe through the peep hole that, after a minute, the people either fell down or lay on the benches. There was no great disturbance or commotion. After another 5 minutes, the room was cleared of gas. SS men specially designated for this purpose placed the dead on stretchers and brought them to the ovens … At the end of the experiment Victor Brack [head of T4 program], who was of course also present (and whom I'd previously forgotten), addressed those in attendance. He appeared satisfied by the results of the experiment, and repeated once again that this operation should be carried out only by physicians, according to the motto: "The needle belongs in the hands of the doctor." Karl Brandt [Hitler's personal physician and, ultimately, highest ranking health official in the Third Reich] spoke after Brack, and stressed again that gassings should only be done by physicians. That is how things began in Brandenburg. (cited in Proctor 1988, pp. 189–190)

There were six main killing centers operating between January 1940 and August 1941: Grafeneck, Brandenburg, Bernburg, Hadamar, Hartheim, and Sonnenstein. After the murder of approximately 70,000 mental patients, the gassing stopped (Proctor 1988, p. 191). Public protests, particularly a sermon delivered on August 3, 1941, by Clemens Count von Galen, the bishop of Münster, may have contributed to the end of the official euthanasia program (Schmidt 2007, pp. 162–168). Doctors and nurses killed at least another 70,000 patients by neglect, starvation, and drugs from 1941 to 1945 in the next phase of the euthanasia program, which became known as "Wild Euthanasia" (Lifton 1986, pp. 96–102). Nazi medical personnel were so committed to their mission that, even after the end of the war, Allied troops liberated some surviving patients at gunpoint (Proctor 1988, p. 193). These euthanasia programs provided a wealth of clinical material for human subjects research (see Chaps. 6 and 10 in this book).

The Final Solution

In January 1942, Heinrich Himmler, head of the Gestapo (short for *Geheime Staatspolizei*, or secret state police) and the *Waffen* SS, announced his plan for eliminating Europe's 11 million Jews at the Wannsee Conference on the Final Solution of the Jewish Question. The gas chambers and crematoria were manufactured in larger sizes and quantities and placed in death camps like Auschwitz. Zyklon B, an insecticide invented by Jewish Nobel Laureate Fritz Haber, was ultimately substituted for carbon monoxide (Charles 2005, pp. 245–246). Doctors selected men, women, and children for the gas chambers, supervised their gassing, completed the forms ascertaining the number of deaths, observed the extraction of teeth from the corpses, and selected subjects for research. According to one Auschwitz survivor, "A doctor was not a doctor. A doctor was the selection. That was what the doctor was—the selection" (Lifton 1986, p. 180).

The German medical profession had replaced the doctor-patient relationship with the state-*Volkskörper* (the German people's body) relationship in the Hippocratic Oath—treatment of the individual was subordinated to treatment of the German Aryan population (Koonz 1991, p. 169; Timmerman, 2002; see also Chap. 4 in this book). Jews, Gypsies, and homosexuals among others were considered pathogens, vermin, infectious agents, or genetic impurities that had to be eliminated (Proctor 1988, p. 162). Physicians believed that their oath of loyalty to Hitler was more important than the Hippocratic Oath (Lifton 1986, p. 207). Conveniently, by invoking the transformed Hippocratic Oath, doctors claimed the moral high ground—Nazi bioethics justified genocide.

By the end of the Third Reich, German physicians, nurses, public health officials, and biomedical scientists voluntarily and actively participated in the following:

- Selecting and killing 6,000,000 Jews.
- Selecting and sterilizing 400,000 German patients.
- Selecting and euthanizing 5,000 German children.
- Selecting and euthanizing at least 140,000 German adults.

If doctors were willing to kill German non-Jewish children and adults to protect the *Volkskörper*, they would certainly have no compunction about conducting inhumane and often fatal human subjects research on Jews and others they considered subhuman (see Chap. 16 in this book).

Eugenics in the United States

Given that German medical practice and research were the best in the world at the time of these atrocities, twenty-first century medical practitioners and academic physicians should consider if they, too, are capable of egregiously unethical behavior.

Table 1.1 Eugenics in the Third Reich and in the United States

	Germany	United States
Slave labor	Late 1930s	1641 Massachusetts
Anti-miscegenation laws	1935	1876
Sterilization law	1933	1907 Indiana
		1927 *Buck v. Bell*
Euthanasia	1938	1915 Dr. Haiselden
Immigration restriction	1930s	1924 Johnson Act
Human subjects research	Late 1930s	1906 Bilibid Prison
		1932–1972 Tuskegee

Americans in particular should be concerned because they provided legal, financial, and moral support for Hitler's eugenic policies. A simple table comparing the timing of specific events in the United States and Germany makes this point (Table 1.1).

In Chicago in 1915, Dr. Harry Haiselden euthanized an infant born with multiple congenital abnormalities that could have been surgically corrected. This infant was not the first he let die, but it was the first he publicly acknowledged. Hollywood eventually made a movie, *The Black Stork*, about Dr. Haiselden and his treatment of congenitally handicapped infants (Pernick 1996; Block and Pernick interview n.d.). The movie played for 12 years in American theaters, and Nazi propaganda films promoting euthanasia bore an eerie similarity to it (Gabbard 2010).

Hitler often praised eugenic practices in the United States or used them as propaganda to justify his own policies when they were criticized. As a young corporal in the German army during WWI, he wrote a fan letter to lawyer, eugenicist, and conservationist Madison Grant, author of *The Passing of the Great Race*, a racist tract advocating an Aryan Master Race, saying that "the book was his Bible" (Kühl 2002, p. 85). In 1925, he wrote in *Mein Kampf* about immigration policies:

> There is today one state in which at least weak beginnings toward a better conception are noticeable. Of course, it is not our model German Republic, but the American Union, in which an effort is made to consult reason at least partially. By refusing immigration on principle to elements in poor health, by simply excluding certain races from naturalization, it professes in slow beginnings a view which is peculiar to the folkish state concept. (Hitler 1925/1999, p. 439)

When criticized for his treatment of Jews, Hitler's propagandists pointed out America's long history of mistreatment of blacks and noted that his laws were more lenient in their definition of Jews than American Jim Crow laws were in their definition of blacks (Proctor 1988, p. 174), which often defined a black as anyone having "one drop" of black blood (Davis 1991). Germany's 1933 Sterilization Law emulated Virginia's model sterilization law that was declared constitutional by the US Supreme Court in 1927 in *Buck v. Bell* (Proctor 1988, p. 286).

American scientists also conducted questionable human subjects research. In 1906, US researchers performed a series of cholera experiments on patients sentenced to death in Bilibid Prison in Manila, resulting in 13 deaths (Hornblum 1997). The infamous Tuskegee Syphilis Study of the natural history of untreated syphilis in black males began in 1932, a year before Hitler came to power, and ended

40 years later (see Chap. 15 in this book). And, finally, American philanthropists provided financial support for human subjects research in Germany. For example, the Rockefeller Foundation provided funds to build the Kaiser Wilhelm Institute of Anthropology, Heredity, and Eugenics (Kühl 2002, pp. 20–21; Samaan 2013, p. 425), the same research facility that received specimens taken from Eva Kor and other twins selected for Mengele's human subjects research at Auschwitz (Posner and Ware 2000, p. 41).

One could reasonably conclude that the United States was actually the world's leader in eugenics in the first three decades of the twentieth century. Charles Davenport, director of Cold Spring Harbor Laboratory (CSHL), and Harry Laughlin, superintendent of its Eugenics Record Office (ERO), were especially influential, as demonstrated on CSHL's Web site Image Archives on the American Eugenics Movement at http://www.eugenicsarchive.org/eugenics/ and in publications by Nobel Laureate James Watson, its former president. The Carnegie and Harriman fortunes funded the ERO, just as the Rockefeller fortune funded eugenic research in Germany in the 1920s (Brunius 2006, p. 164). To its credit, CSHL has made its archives available to researchers. Unfortunately, the keepers of the legacy of other great names heavily involved in eugenics, such as Carnegie, Rockefeller, Harriman, Kellogg, Sanger, and the American Museum of Natural History, have not been as forthcoming.

Considering America's strong interest in eugenics, it is reasonable to ask why America did not slide down the same slippery slope as Germany. Friedlander answers the question this way:

> Although the German eugenics movement, led until the Weimar years by Alfred Ploetz and Wilhelm Schallmayer, did not differ radically from the American movement, it was more centralized. Unlike in the United States, where federalism and political heterogeneity encouraged diversity even within a single movement, in Germany one society, the German Society for Race Hygiene (*Deutsche Gesellschaft füe Rassenhygiene*), eventually represented all eugenicists, while one journal, the *Archiv für Rassen- und Gesellschafts-Biologie*, founded by Ploetz in 1904, remained the primary scientific publication of German Eugenics. (1995, pp. 9–10)

While one American state, Indiana, passed the world's first involuntary sterilization law in 1907, it took 20 years for 28 states to adopt similar laws and for the Supreme Court to rule on its constitutionality (Lombardo 2008); the Nazis passed their sterilization law within the first six months of assuming power and applied it to all of Germany. Federalism, as opposed to *Gleichschaltung*, impeded the American federal government from establishing involuntary sterilization as the law of the land and allowed our states to function as independent "laboratories of democracy," a metaphor attributed to Justice Louis D. Brandeis (Greve 2001).

America is the most religious Western country (Pew Research 2002), and its strong Judeo-Christian heritage may have prevented the United States from embracing eugenics to the same extent as Germany. Traditional Western religions believe that life is of supreme value, and under most circumstances, they forbid involuntary sterilization and euthanasia.

Human Subjects Research After the Holocaust

Although the 1947 Nuremberg Code outlined ethical behavior for human subjects researchers, it was widely ignored by medical professionals and legislators. The World Medical Association (WMA) deliberated for 17 years after the publication of the Nuremberg Code before issuing its Declaration of Helsinki, a watered-down version of the code (WMA General Assembly 1964/2013; see also Chap. 14 in this book). The WMA diluted the code because it wanted to save human subjects research from its "overly rigid" provisions (Refshauge 1977, pp. 85–92) and from its "uncompromising language" to protect the subjects of medical research (Katz, 1996). Another reason why the WMA may have had trouble accepting the Nuremberg Code is that its membership included unindicted Nazi physicians who had achieved positions of prominence in international academic medicine (Seidelman 1996).

Even as the documentation of Nazi medical war crimes discredited eugenics, the eugenic impulse persisted. *Eugenics* was transformed into the "fresh, new, and ethically squeaky-clean field" of *medical genetics* "without throwing out the eugenic baby with the Nazi bathwater" (Comfort 2012, p. 132). Some professional journals changed their names; for example, the *Annals of Eugenics* became the *Annals of Human Genetics* in 1954. The name of the man who coined the term *eugenics* was eventually co-opted in the transformation: At University College London, the Francis Galton Laboratory of National Eugenics became the Galton Laboratory of the Department of Human Genetics & Biometry in 1963 ("History of the Dalton Collection," n.d.) and the Galton Chair of Eugenics became the Galton Chair of Genetics (Blaney 2011, p. 108).

Eugenicists, working with the United Nations and other international agencies (Connelly 2008), promoted human subjects reproductive research on an unprecedented scale in undeveloped countries (Franks 2005; Kasun 1999). For example, Margaret Sanger and C. P. Blacker formed the International Planned Parenthood Foundation, and John D. Rockefeller III and Frederick Osborn formed the Population Council, both of which worked with influential obstetrican/gynecologist Alan Guttmacher (Connelly 2008, pp. 201–202). These organizations provided financial support for the United States Agency for International Development (USAID) to promote "population control" in Third World countries (Franks 2005). In one instance, USAID's Office of Population, under the leadership of Dr. Reimert Ravenholt, purchased unsterilized Dalkon Shields in bulk at half-price (Connelly 2008, p. 252). In the United States, the sterilized version of this particular IUD caused 18 deaths, over 200,000 infections, miscarriages, hysterectomies, and other gynecological complications, as well as an unknown number of septic abortions, which resulted in multiple lawsuits (Kluchin 2009, p. 58). By the time the recall order was issued for the Dalkon Shield in 1975, 440,000 women in 42 foreign countries were using them (Connelly 2008, p. 252).

The Hippocratic Oath

Although it does not address human subjects research directly, the original Hippocratic Oath is "the most admired work in Western European medical ethics" (Miles 2004, p. v). As noted above, it was transformed by Nazi physicians from a document about the doctor-patient relationship to one about the state-*Volkskörper* relationship with physicians acting as agents of the state. The traditional oath (Miles 2004, p. xiii–xiv) experienced a revival after the Nuremberg Doctors' Trial, but in a revised and attenuated form in 1948 as the WMA's Declaration of Geneva (WMA General Assembly 1948/2006). A comparison of the opening statement of the traditional oath, the current fifth revision of the Declaration of Geneva, and the Oath of Hippocrates cited by the Association of American Medical Colleges (AAMC) ("The Road to Becoming a Doctor" n.d., p. 6) is instructive:

- *Original Hippocratic Oath*: I swear by Apollo the Physician and by Asclepius and by Health (the god Hygieia) and Panacea and by all the gods as well as goddesses, making them judges (witnesses), to bring the following oath and written covenant to fulfillment, in accordance with my power and my judgment.
- *WMA Declaration*: I solemnly pledge myself to consecrate my life to the service of humanity.
- *AAMC Oath of Hippocrates*: I do solemnly swear, by that which I hold most sacred.

The WMA's modification opened the way for subsequent revisions of the ancient oath by many organizations and, as demonstrated by Orr and his colleagues in 1997, the traditional Hippocratic Oath is administered in very few American medical schools (Orr et al. 1997). In effect, the oath's content has been superseded and greatly expanded by modern biomedical ethics.

In *Principles of Biomedical Ethics*, Beauchamp and Childress (1994) note four clusters of basic principles underlying American biomedical ethics: respect for autonomy (respecting the decision-making capacities of autonomous persons); nonmaleficence (avoiding the causation of harm); beneficence (providing benefits and balancing benefits against risks and costs); and justice (distributing benefits, risks, and costs fairly). Comparing modern secular biomedical ethics to other medical ethics that have a different common morality, such as Jewish medical ethics, Hippocratic medical ethics, and Nazi medical ethics, is revealing. In addition to lack of a common morality, the approach to solving medical ethical dilemmas may differ among the four systems. Judaism, for example, takes a casuistic approach. Nonetheless, it is instructive to compare these four systems of medical ethics as if each had only four clusters of basic principles.

In each ethical system, one principle is usually the dominant principle: human life in Jewish medical ethics; eugenics in Nazi medical ethics; and autonomy in modern secular biomedical ethics. According to Avraham Steinberg, "In Judaism, the value of human life is supreme; therefore, to save a life, nearly all Biblical laws are waived. This approach is in contrast to the secular ethical view, which considers human life to be one of many values and often gives greater weight to 'the quality of life'" (Steinberg 2010, p. 217). While not explicitly stated in the original Hippocratic Oath, it is clear that the Hippocratic school also values human life over quality of life as evidenced by this statement in the oath: "And I will not give a drug that is deadly to anyone if asked (for it), nor will I suggest the way to such counsel. And likewise I will not give a woman a destructive pessary" (Miles 2004, p. xiv). The definitions of nonmaleficence, beneficence, and justice follow from the dominant principle of each of these ethical systems. Nonmaleficence in Jewish and Hippocratic medical ethics means not harming the individual, whereas nonmaleficence in Nazi medical ethics meant not harming the *Volkskörper*, thereby permitting research on (and mass medical murder of) those perceived by the state as harmful to the German people. Beneficence and justice are also defined differently in each system, depending upon its dominant principle.

National socialism valued eugenics over all other principles. Historian and bioethicist Jonathan Moreno warned that a "dominant social philosophy of medicine [such as the 1930s eugenics movement in Germany] . . . can slide down the slope to disaster" (Moreno 2004, p. 2). Lisa Eckenwiler and Felicia Cohn added that establishing the dominance of one value, such as America's "foundational belief in individual autonomy," over other values while ignoring the history of bioethics before the Nuremberg Code is potentially dangerous (2007 p. 5). The danger is especially great if, as opposed to Jewish and Hippocratic medical ethics, autonomy is valued over human life (Gaylin and Jennings 2003, pp. 215–216). For example, valuing individual autonomy over human life can transform a foundational American belief in the unalienable rights to "Life, Liberty and the pursuit of Happiness" (Declaration of Independence 1776) into a belief in the unalienable rights to quality of life, autonomy, and happiness with unforeseen and unintended consequences.

I am not suggesting that modern bioethics is analogous to Nazi bioethics. Rather I take the position attributed to Mark Twain: "History does not necessarily repeat itself, but it does rhyme."

Nazi bioscience cannot be dismissed by accepting the myths that Nazi researchers were incompetent, mad, sadistic, coerced, or few in number, and Nazi bioethics cannot be dismissed by assuming that Nazi physicians did not have ethical justifications for their immoral behavior. Arthur Caplan (2010) presents the issue directly:

> Perhaps the most important reason for the absence of commentary on the ethics of the research done in the camps is that such questions open a door that few bioethicists wish to enter. If moral justifications can be given for why someone deemed mass murder appropriate in the name of public health or thought that it was right to freeze hapless men and women to death or decompress them or infect them with lethal doses of typhus, then to put the question plainly, what good is bioethics? (p. 84)

The intent of this book is to dispel the myths about Nazi human subjects research. Once these myths have been dispelled, physicians, nurses, bioscientists, and public health policymakers can seriously consider the history of human subjects research under National Socialism. Just like Emanuel, Roelcke, and Ferrari, (see Chaps. 20, 21, and 23 in this book), we must summon the courage to personally confront the present-day implications of that history and to study and reflect upon "How Ethics Failed" (Lindert et al. 2013; see also Chaps. 19 and 22 in this book). Because the United States and other Western countries provided moral, legal, and philanthropic support for Nazi medicine, and because American medical education and national health insurance public policies were derived from Germany, intellectual honesty and ethical integrity demand further exploration of human subjects research after the Holocaust.

References

Association of American Medical Colleges (AAMC). n.d. The road to becoming a doctor: The oath of Hippocrates. https://www.aamc.org/download/68806/data/road-doctor.pdf. Accessed 05 Jan 2014.

Auschwitz II–Birkenau: History of a man-made hell. http://www.scrapbookpages.com/AuschwitzScrapbook/History/Articles/Selection.html. Accessed 28 Dec 2013.

Beauchamp, T.L., and J.F. Childress. 1994. *Principles of biomedical ethics*. New York: Oxford University Press.

Blaney, T. 2011. *The chief sea lion's inheritance: Eugenics and the Darwins*. Leicester, UK: Troubador Publishing.

Bliss, M. 1999. *William Osler: A life in medicine*. Toronto, ON: University of Toronto Press.

Block, L. (Interviewer) Pernick, M. S. (interviewee). n.d. Tomorrow's Children [Interview transcript and audio]. Retrieved from National Public Radio website at http://www.npr.org/programs/disability/ba_shows.dir/children.dir/highlights/bsmovsti.html. Accessed 20 Dec 2013.

Bruinius, H. 2006. *Better for all the world: The secret history of forced sterilization and America's quest for racial purity*. New York: Knopf.

Buck v. Bell. 1927. 274 U.S. 200.

Caplan, A.L. 2010. The stain of silence: Nazi ethics and bioethics. In *Medicine after the Holocaust: From the Master Race to the human genome and beyond*, ed. S. Rubenfeld, 83–92. New York: Palgrave Macmillan.

Celebrating the contributions of William Osler: Biography. 1993. Reprinted from McCall, Nancy. ed. *The portrait collection of Johns Hopkins Medicine: A catalog of paintings and photographs at The Johns Hopkins University School of Medicine and The Johns Hopkins Hospital*. Baltimore, MD: The Johns Hopkins University School of Medicine. http://www.medicalarchives.jhmi.edu/osler/biography.htm. Accessed 28 Dec 2013.

Charles, D. 2005. *Mastermind: The rise and fall of Fritz Haber, the Nobel Laureate who launched the age of chemical warfare*. New York: Ecco.

Cold Spring Harbor Laboratory. n.d. Image archive on the American eugenics movement. http://www.eugenicsarchive.org/eugenics/. Accessed 20 Dec 2013.

Comfort, N. 2012. *The science of human perfection: How genes became the heart of American medicine*. New Haven, CT: Yale University Press.

Connelly, M. 2008. *Fatal misconception: The struggle to control world population*. Cambridge, MA: Belknap.

Davis, F.J. 1991. *Who is black? One nation's definition*. University Park, PA: Penn State University Press. http://www.pbs.org/wgbh/pages/frontline/shows/jefferson/mixed/onedrop.html. Accessed 20 Dec 2013.

DeBakey, M.E. 2010. Afterword. In *Medicine after the Holocaust: From the Master Race to the human genome and beyond*, ed. S. Rubenfeld, 221–223. New York: Palgrave Macmillan.

Declaration of Independence. 1776. The charters of freedom. National archives and records administration. http://www.archives.gov/exhibits/charters/declaration_transcript.html. Accessed 08 Jan 2014.

Dowbiggin, I. 2003. *A merciful end: The euthanasia movement in modern America*. New York: Oxford University Press.

Eckenwiler, L.A., and F.G. Cohn. 2007. *The ethics of bioethics: Mapping the moral landscape*. Baltimore, MD: The Johns Hopkins University Press.

Flexner, A. 1912. *Medical education in Europe: A report to the Carnegie Foundation for the advancement of teaching*. Boston, MA: The Merrymount Press.

Franks, A. 2005. *Margaret Sanger's eugenic legacy: The control of female fertility*. Jefferson, NC: McFarland & Co.

Freyhofer, H.H. 2004. *The Nuremberg medical trial: The Holocaust and the origin of the Nuremberg medical code*. New York: Peter Lang Publishing.

Friedlander, H. 1995. *The origins of Nazi genocide: From euthanasia to the final solution*. Chapel Hill: University of North Carolina.

Gabbard, G.O. 2010. In *Medicine after the Holocaust: From the Master Race to the human genome and beyond*, ed. S. Rubenfled, 153–161. New York: Palgrave Macmillan.

Gaylin, W., and B. Jennings. 2003. *The perversion of autonomy: Coercion and constraints in a liberal society*. Washington, DC: Georgetown University Press.

Gillham, N.W. 2001. *A life of Sir Francis Galton: From African exploration to the birth of eugenics*. New York: Oxford University Press.

Grant, M. 1916. *The passing of the great race*. New York: Charles Scribner's Sons.

Greve, M.S. 2001. Laboratories of democracy: Anatomy of a metaphor. AEI (American Enterprise Institute). http://www.aei.org/article/politics-and-public-opinion/elections/laboratories-of-democracy/. Accessed 30 Dec 2013.

Heller, J. 1972. *Syphilis victims in the U.S. study went untreated for 40 years*. New York: New York Times.

History of the Galton Collection. n.d. University College London's Galton collection. http://www.ucl.ac.uk/museums/galton/about/history. Accessed 20 Dec 2013.

Hitler, A. 1925/1999. *Mein Kampf*. New York: First Mariner Books.

Hornblum, A.M. 1997. They were cheap and available: Prisoners as research subjects in twentieth century America. *British Medical Journal* 315(7120): 1437–1441.

Kasun, J. 1999. *War against population: The economics and ideology of population control*. San Francisco, CA: Ignatius Press.

Katz, J. 1996. Human sacrifice and human experimentation: Reflections at Nuremberg. *Occasional Papers*, Paper 5. http://lsr.nellco.org/yale/ylsop/papers/5. Accessed 20 Dec 2013.

Kluchin, R.M. 2009. *Fit to be tied: Sterilization and reproductive rights in America, 1950–1980*. New Brunswick, NJ: Rutgers University Press.

Koonz, C. 1991. Genocide and eugenics: The language of power. In *Lessons and legacies I: The meaning of the Holocaust in a changing world*, vol. 1, ed. P. Hayes, 155–177. Evanston, IL: Northwestern University Press.

Kühl, S. 2002. *The Nazi connection: Eugenics, American racism, and German National Socialism*. New York: Oxford University Press.

Lifton, R.J. 1986. *The Nazi doctors: Medical killing and the psychology of genocide*. New York: Basic Books.

Lindert, J., Y. Stein, H. Guggenheim, J.J.K. Jaakkola, and R.D. Strous (2013). How ethics failed— The role of psychiatrists and physician in Nazi programs from exclusion to extermination, 1933–1945. *Public Health Reviews* 34. http://www.publichealthreviews.eu/show/f/120. Accessed 5 Jan 2014.

Lombardo, P.A. 2008. *Three generations, no imbeciles: Eugenics, the Supreme Court, and Buck v. Bell*. Baltimore, MD: The Johns Hopkins University Press.

Mendel, G. 1866. Versuche über Plflanzenhybriden [Experiments in plant hybridization]. *Verhand- lungen des naturforschenden Vereines in Brünn, Bd. IV für das Jahr 1865, Abhandlungen,* 3–47.

Merriam-Webster Dictionary [Online dictionary]. s.v. "eugenics." http://www.merriam-webster. com/dictionary/eugenics. Accessed 20 Dec 2013.

Miles, S.H. 2004. *The Hippocratic Oath and the ethics of medicine.* New York: Oxford University Press.

Mitscherlich, A., and F. Mielke. 1992. Epilogue: Seven were hanged. In *The Nazi doctors and the Nuremberg Code: Human rights in human experimentation,* ed. G.J. Annas and M.A. Grodin, 105–107. New York: Oxford University Press.

Moreno, J. 2004. Bioethics imperialism. *ASBH Exchange* 7(3): 2.

Operation Paperclip. n.d. http://www.historylearningsite.co.uk/operation_paperclip.htm. Accessed 14 Dec 2013.

Orr, R.D., N. Pang, E.D. Pellegrino, and M. Siegler. 1997. Use of the Hippocratic Oath: A review of twentieth century practice and a content analysis of oaths administered in medical schools in the U.S. and Canada in 1993. *Journal of Clinical Ethics* 8(4): 377–388.

Pasternak, A. 2006. *Inhuman research: Medical experiments in German concentration camps.* Budapest: Akadémiai Kiadó.

Pernick, M.S. 1996. *The black stork: Eugenics and the death of "defective" babies in American medicine and motion pictures since 1915.* New York: Oxford University Press.

Pew Research Global Attitudes Project. 2002. Among wealthy nations U.S. stands alone in its embrace of religion. http://www.pewglobal.org/2002/12/19/among-wealthy-nations/. Accessed 20 Dec 2013.

Plato. 2003. *The republic.* New York: Penguin Books

Posner, G.L., and J. Ware. 2000. *Mengele: The complete story.* New York: Cooper Square Press.

Proctor, R.N. 1988. *Racial hygiene: Medicine under the Nazis.* Cambridge: Harvard University Press.

———. 2000. Nazi science and Nazi medical ethics: Some myths and misconceptions. *Perspectives in Biology and Medicine* 43(3): 335–346.

Refshauge, W. 1977. The place for international standards in conducting research on humans. *Bulletin of the World Health Organization* 55(2): 131–145.

Roelcke, V. 2010. Medicine during the Nazi Period: Historical facts and some implications for teaching medical ethics and professionalism. In *Medicine after the Holocaust: From the Master Race to the human genome and beyond,* ed. S. Rubenfeld, 17–28. New York: Palgrave Macmillan.

Samaan, A.E. 2013. *From a "race of masters" to a "master race": 1948 to 1848.* Charleston, SC: A. E. Samaan/CreateSpace.com.

Schmidt, U. 2004. *Justice at Nuremberg: Leo Alexander and the doctors' trial.* New York: Palgrave Macmillan.

———. 2007. *Karl Brandt: The Nazi doctor: Medicine and power in the Third Reich.* New York: Hambledon Continuum.

Seidelman, W. E. 1996. Nuremberg lamentation: For the forgotten victims of medical science. *British Medical Journal* 313. doi: http://dx.doi.org/10.1136/bmj.313.7070.1463.

Spitz, V. 2005. *Doctors from hell: The horrific account of Nazi experiments on humans.* Boulder, CO: Sentient Publications.

Starr, P. 1982. *The social transformation of medicine: The rise of a sovereign profession and the making of a vast Industry.* New York: Basic Books.

Steinberg, A. 2010. Jewish medical ethics and risky treatments. In *Medicine after the Holocaust: From the master race to the human genome and beyond,* ed. S. Rubenfeld, 213–220. New York: Palgrave Macmillan.

Strous, R.D. 2009. Dr. Irmfried Eberl, 1910–1948: Mass murdering MD. *Israel Medical Association Journal* 11(4): 216–218.

Taylor, T. 1992. Opening statement of the prosecution, December 9, 1946. In *The Nazi doctors and the Nuremberg code: Human rights in human experimentation*, ed. G.J. Annas and M.A. Grodin, 67–93. New York: Oxford University Press.

Timmerman, C. 2002. A model for the new physician: Hippocrates in interwar Germany. In *Reinventing hippocrates*, ed. D. Cantor, 302–324. London: Ashgate.

United States Holocaust Memorial Museum (USHMM). n.d. Josef Mengele. Holocaust encyclopedia. http://www.ushmm.org/wlc/en/article.php?ModuleId=10007060. Accessed 28 Dec 2013.

Weikart, R. 2009. *Hitler's ethic: The Nazi pursuit of evolutionary progress*. New York: Palgrave Macmillan.

World Medical Association (WMA) General Assembly. 1948/2006. WMA declaration of Geneva. http://www.wma.net/en/30publications/10policies/g1/. Accessed 28 Dec 2013.

World Medical Association (WMA) General Assembly. 1964/2013. WMA declaration of Helsinki—Ethical principles for medical research involving human subjects. http://www.aei.org/article/politics-and-public-opinion/elections/laboratories-of-democracy/. Accessed 28 Dec 2013

Wright, R. 1994. *The moral animal: Why we are the way we are: The new science of evolutionary psychology*. New York: Vintage Books.

Chapter 2
Twin Experiments at Auschwitz: A First-Person Account

Eva Mozes Kor

One of the most important and challenging topics confronting today's generation of medical professionals is the topic of human rights in medical research and experimentation. In this book, you will find many academic papers written by experts in the fields of medicine, human behavior, medical research, science, biology, and ethics. I was asked to submit a chapter not because of my great academic credentials, but because it is important to share this perspective: How does it feel to be a human guinea pig used and abused in human experimentation?

This essay is not the result of thorough academic research, but it is the fruit of my soul-searching in trying to make some sense out of being a human guinea pig. I am trying to explain logically something that defies logic because it deals with feelings—it deals with the innermost of my soul, and many people cannot understand it. I am not a religious person, I am not an angel, nor am I a heroine. I am just a human being who survived many deadly experiments, who was treated as a subhuman, yet remained human, and who has been fortunate to discover the secret to self-empowerment and self-healing—namely forgiveness.

Seventy years ago I was merely a disposable human subject. If I would have died, I would have been replaced by twins arriving on the next transport. I hope you grasp the significance of this essay: A disposable human guinea pig from Auschwitz who could not be disposed of by the Auschwitz doctors is writing to you, the medical professionals of today, about ethics in medicine. On behalf of the thousands of guinea pigs who died in the experiments conducted by Dr. Mengele and the other Nazi doctors, as well as on behalf of those of us who lived to talk about it, I submit my thoughts below. This is a most appropriate tribute to the memory of those who died, including my twin sister, Miriam Mozes Zeiger.

E.M. Kor (✉)
Auschwitz Survivor A-7063; Founding Director, CANDLES Holocaust Museum and Education Center, 1532 S. 3rd Street, Terre Haute, IN, USA
e-mail: eva@candlesholocaustmuseum.org

S. Rubenfeld and S. Benedict (eds.), *Human Subjects Research after the Holocaust*, 21
DOI 10.1007/978-3-319-05702-6_2, © Springer International Publishing Switzerland 2014

My thoughts are divided into three parts: (1) How I survived Auschwitz; (2) How it feels to be a human guinea pig and the lessons in ethics for today; and (3) Self-liberation and healing through forgiveness.

How I Survived Auschwitz

In May of 1944, we, the Mozes family—my father Alexander Mozes, age 44; my mother, Jaffa Mozes, age 38; my sister Edit Mozes, age 14; my sister Aliz Mozes, age 12; and Miriam and Eva Mozes, identical twins age 10—were loaded onto cattle cars. We came from a tiny village of 100 families called Portz in Transylvania, Romania. In 1940, our village was occupied by the Hungarian army, so we were part of the Hungarian transport that was sent to the Auschwitz death camp.

It was the dawn of an early spring day in 1944. Our cattle car train came to a sudden stop. I could hear lots of German voices yelling orders outside. There was no room to spare in the packed cattle car. Above the press of bodies, I could see nothing but a patch of gray sky through the barbed wires in the window.

As soon as we stepped down onto the cement platform, my mother grabbed Miriam and me by the hand, hoping that as long as she could hold onto us, she would protect us. I was standing on that cement platform for about ten minutes when, in my childish curiosity, I looked around, trying to figure out what that place was. When I was looking around, I realized that my father and two older sisters were gone. I never saw any of them again.

As Miriam and I were clutching my mother's hand, an SS trooper hurried by shouting, "Zwillinge! Zwillinge!" (Twins! Twins!). He stopped by to look at Miriam and me because we were dressed alike and we looked very much alike.

"Are they twins?" he asked.

"Is it good?" asked my mother.

"Yes," nodded the SS.

"Yes, they are twins," said my mother.

Without any warning or explanation, he grabbed Miriam and me away from our mother. Our screaming and pleading fell on deaf ears. I remember looking back and seeing my mother's arms stretched out in despair as she was pulled away. I never said goodbye to her. I didn't know that this would be the last time we saw her. This was just 30 minutes since we stepped down from the cattle car onto the selection platform, and we were ripped apart from our family forever.

I will never forget that little strip of cement called the selection platform. It measured 85 feet by 35 feet, a strip of land where millions of people were ripped apart from their families. I believe that there is no other strip of land like this in the world.

We became part of a group of little girls, all twins: 13 sets of little girls, aged 2–16 years, and 1 mother who was permitted to stay with her daughters. We were marched to a huge building and made to sit naked for the better part of the day. Late in the afternoon our processing began. We were given short haircuts; the mother's head was shaved. Our clothes were returned with a huge red cross painted on the back because it identified us as human subjects for experiments. To have our own hair and clothes was a privilege we, the twins, were granted.

Then we lined up for registration and tattooing. When my turn came, I decided to give them as much trouble as I possibly could. Four people, two Nazis and two women prisoners, restrained me with all their strength while they heated a pen-like gadget that looked like a writing pen with a metal tip that was heated over the flame of a lamp. When it became red hot, they dipped it into ink and burned into my left arm, dot by dot, the capital letter A-7063; Miriam became A-7064. Auschwitz was the only Nazi camp that tattooed its inmates. Years later Miriam told me that in addition to creating a general confusion, I bit the Nazi holding my arm. I do not remember biting anyone. I must have blocked it out of my mind. After all, I was raised to be a nice girl, and nice girls and boys don't bite.

To look back at my childhood is to remember my experiences as a human guinea pig in the Auschwitz-Birkenau laboratory of Dr. Josef Mengele. To recount such painful memories is to relive the horrors of human experimentation, where people were used as mere objects to research. I envision the chimneys, the smell of burning flesh, the medical injections, the endless blood-taking, the tests, the dead bodies that were all around us, the hunger, and the rats. Nothing that is close to human existence was present in that place.

Nothing on the face of this earth can prepare anyone for a place like Auschwitz. At age 10, Miriam and I became part of a special group of children, twins used as human guinea pigs. One thousand five hundred sets of twins were used by Mengele in his experiments. That first night when Miriam and I went to the latrine, there on the filthy floor were the scattered corpses of three children. This is when I realized that it could happen to Miriam and me unless I did something to prevent it. So I made a silent pledge: *I will do whatever is within my power to make sure that Miriam and I shall not end up on that filthy latrine floor.*

In the barrack, we were children, girls ages 2–16, huddled in our filthy bunk beds crawling with lice and rats. We were starved for food, starved for human kindness, and starved for the love of the mothers and fathers we once had. We had no rights, but we had a fierce determination to live one more day, to survive one more experiment. No one tried to minimize the risks to our lives, and no one showed us any respect or consideration. On the contrary, we knew we were there to be used as guinea pigs at the mercy of Dr. Mengele.

We were used in a variety of experiments. Three times a week we would walk to Auschwitz I, where we would be placed naked in a room for six to eight hours. Every part of my body was measured, compared to charts, and compared to my twin sister. These particular experiments were not dangerous, but they were very demeaning. Even in Auschwitz, I could not cope with the fact that I was treated like a nobody, a nothing. The only way I could cope with it was by blocking it out my mind. Even at that young age, we knew we had to submit to Dr. Mengele's experiments. We were reduced to the lowest form of existence: just a mass of living, breathing cells.

Our families were gone; our lives depended on being a cooperative mass of living cells. There was no escape from it. Our childhood was gone, snatched away by the Nazis. Our bodies were fodder for Mengele's experiments. All we had were our lives, and I concentrated all my efforts, all my talents, all my being on one thing—survival.

Three times a week we were taken to another lab that I called the blood lab. They would tie both of my arms, take a lot of blood from my left arm, and give me a minimum of five injections into my right arm. Those were the deadly ones. After one of those injections, I became very ill with a very high fever, a fact I desperately tried to hide because the rumor was that anyone taken to the hospital never came back.

On the next visit to the blood lab, the doctors measured my fever, and I was taken to the hospital, which was filled with people who looked more dead than alive. The next morning, Dr. Mengele and four other doctors came to see me. They never examined me. They just looked at my fever chart. Dr. Mengele said, laughing sarcastically, "Too bad, she is so young. She has only two weeks to live!"

I was all alone. The doctors I had did not want to save my life; they wanted me dead. Miriam was not with me and I missed her so very much. She was the only kind and loving person I could cuddle up with when I was hungry, cold, and ill. Even though I was very sick and all alone, I refused to die and made a second silent pledge: *I will do everything within my power to prove Dr. Mengele wrong, survive, and be reunited with Miriam.*

For the next two weeks, I was between life and death. I remember waking up on the barrack floor. I was crawling because I no longer could walk—crawling to reach a faucet with water at the other end of the barrack. This barrack was not allocated any food, water, or medicine since people were brought there to die.

As I was crawling, I would fade in and out of consciousness. I kept telling myself, "I must survive, I must survive!" After two weeks, my fever broke and I felt much better. It took me another three weeks before my fever chart showed normal and I was released and reunited with Miriam.

But the happiness of our reunion was short lived. Miriam looked just like the living dead I left in the other barrack. She was sick and lost her desire to fight for her own life. It was very easy to die in Auschwitz—surviving was a full-time job. The will to live made the difference between life and death.

The next day, I volunteered to carry food from the kitchen to the barrack so I could "organize" (which in camp language meant stealing from the Nazis). I wanted to organize raw potatoes. As I entered the kitchen, I saw a long table with sacks of potatoes underneath it. I hesitated for a moment because the rumor was that anyone caught stealing would be hanged. I finally conquered my fear and bent down. I could feel two small potatoes in my hands when somebody grabbed me by my head and pulled me up, yelling into my face, "It's not nice to steal!" I almost burst into laughter because I expected to be dragged to the gallows and hanged. But I was not.

That day I learned a very important lesson: As long as Mengele wanted us alive, no one dared harm us. Next day, I tried my organizing skills again, and I became the happy owner of three raw potatoes. We boiled them after the supervisors went to sleep, and I became a very good organizer. We had potatoes three to four times a week. The steady diet of potatoes gave Miriam enough strength to fight for her own life.

When I was in Auschwitz, I thought the whole world was a big concentration camp. Children lose their point of reference very quickly. One day in July 1944, I saw an airplane fly over Auschwitz. It was flying very low. I could see an American flag on the wings. That gave me the realization that somebody was trying to free us. That knowledge reinforced my determination to live one more day, survive one more experiment. The air raids and fighting by the Allied forces continued and increased greatly until liberation.

On January 27, 1945, just four days before my 11th birthday, Auschwitz was liberated by the Soviet army. We were free, we were alive, and we had triumphed over unbelievable evil. The glory of that day will be forever engraved in my heart, and the echoes from Auschwitz will always be part of my life. My silent pledge that first night in the latrine became a reality, and that was an unbelievable experience.

How It Feels to Be a Human Guinea Pig and the Lessons in Ethics for Today

I, Eva Mozes Kor, a survivor of Mengele's Auschwitz experiments, have learned that human rights in medical experimentation is an issue that needs to be addressed and taught. Those of you who are physicians and research scientists are to be congratulated. You have chosen a wonderful and difficult profession: wonderful because you can save human lives, and difficult because you are walking a very narrow line. You have been trained to use good judgment and clear logic, but you cannot forget that you are dealing with human beings. The moment you forget and cross that narrow line, you are heading in the direction of the Josef Mengeles.

I hope what was done to me will never happen to another human being. Those of you who do research must be compelled to obey the international law. Scientists should continue to do research, and I have benefited from your research. My son Dr. Alex Kor was diagnosed with an advanced case of testicular cancer in 1987 that had metastasized to his lungs and his lymph nodes. I am so grateful that medical research found a cure for my son. I want to thank Dr. Lawrence Einhorn of the Indiana University Medical Center for saving my son's life.

I am personally aware both of the benefits and drawbacks of medical research. It can benefit mankind, but it also can be greatly abused. The scientists must make a moral commitment never to violate a person's human rights and human dignity. The research scientists must respect the wishes of their subjects. Every time you, the scientist, are involved in human experimentation, you should put yourself in the place of your subject and see how you would feel.

You, the scientists of the world, must remember that the research is done for the sake of mankind and not for the sake of science; you must never detach yourselves from the subjects you serve. I hope with all my heart that our sad stories will in some way impel the international community to devise laws and rules to govern human experimentation.

The dignity of all human beings must be respected, preserved, and protected at all costs. Life without it is a mere existence. I experienced such loss of dignity every day as a guinea pig in Dr. Mengele's laboratory. These same doctors had taken an oath to help and to save human life.

Self-Liberation and Healing Through Forgiveness

From 1944 until 1995, for 51 years, I felt deep emotional pain. I was a victim who was angry with the world and who could not be really happy. I was physically liberated from Auschwitz in 1945 by the Soviet army. I liberated myself from the emotional pain in 1995 when I forgave the Nazis, and this is how it happened.

After liberation, Miriam and I were in refugee camps for nine months. We arrived back home to our village to find three crumpled pictures on a bedroom floor, and that was all that was left of our family. We were taken in by an aunt who lost her family, who married a guy who lost his family. We lived in communist Romania from 1945 until 1950.

During that time, I learned that the Communists had beautiful slogans, but freedom was just a dream. We wanted to leave Romania for Israel, but they did not let us leave. For two years we struggled to get an exit visa. When we were finally granted one, the Romanian government took away all our land, property, and everything we had, permitting us to take only what we could wear. I wore three dresses and a winter coat in the middle of the summer.

In 1950, we arrived in Israel. After two years in an agricultural school, we were drafted into the Israeli army. Miriam studied and became a registered nurse. I stayed eight years in the engineering corps, reaching the rank of a sergeant major. In 1960, I met and married an American tourist from Terre Haute, Indiana. He is also a Holocaust survivor who was liberated in Buchenwald by an American colonel from Terre Haute. So I came from Tel-Aviv to Terre Haute, Indiana—quite a jump.

Miriam got married in 1957, and then in 1960, when she was expecting her first child, she developed a severe kidney infection that did not respond to antibiotics. With the second pregnancy in 1963, her kidney infections got worse, and her doctors found out that Miriam's kidneys never grew larger than the kidneys of a ten-year-old child.

After her third pregnancy, her kidneys failed, and in 1987 I donated my left kidney to her. We were a perfect match, but a year later she developed cancerous polyps in her bladder, the only one among 2,000 transplantees to develop such cancer. Doctors in Israel said if we could find our Auschwitz files, it might help them figure out why Miriam was the only one to develop cancer and what substances were injected into her in Auschwitz to combine with the anti-rejection medication to create the cancer. We have never found our Auschwitz files, we have never found out what was injected into our bodies, and I am still hoping for help.

Today as I write this paper, there are still seven vials in Auschwitz, filled with a fluid. They are in a box marked with a "Bayer" logo and the code "Be 1034." I asked the Auschwitz-Birkenau State Museum to test the contents, but so far no action has been taken. These vials have been there for 67 years. Might they give us some answers? Maybe.

Miriam died on June 6, 1993. I was a realtor, and when I came home from an Open House on a Sunday afternoon, there was a message on my answering machine from my brother in-law. He said Miriam had died. I immediately called Israel, telling my brother-in-law that I would catch the first flight to Israel. He told me not to bother. Because of Jewish custom, the funeral would be in ten hours. Israel was eight hours ahead, and it would take me at least 12 hours to get there. I pleaded with him to wait for me. I have never buried any family member—a simple human gesture. I wanted to touch her, I wanted to say goodbye to her and say goodbye to my kidney, but he said no. I was devastated, trying to cope with the death of my only sister.

Three weeks later I received a telephone call from Dr. John Michalczyk who said he heard me speak at a conference on Nazi medicine and he wanted me to come and lecture to some doctors. I said to him that I love to lecture to doctors, and I do. He asked me why, and I explained that when I go to the doctors, they tell me that I am too fat, I don't eat the right food, I don't exercise enough, and then I have to pay them. I would like to tell the good doctors what they do wrong, because they do plenty of things wrong, and I love the fact that then they pay me. He said, "Okay, okay," and we joked about it. Then Dr. Michalczyk asked me, "When you come to the Boston conference, could you bring with you a Nazi Doctor?"

"A Nazi doctor?" I asked stunned. "Where do you think I can find one of those doctors? Last time I looked, they were not advertising in the Yellow Pages," I blurted out. He insisted that I think about it and I promised that I would.

I remembered that the last project Miriam and I worked on together in 1991 and 1992 was a documentary on the Mengele twins done by ZDF, a German television network. In the documentary, a Nazi doctor by the name of Hans Münch appeared, so I figured he might be still alive. I faxed a letter to ZDF in Frankfurt, Germany, telling them that Miriam had just died and asking would they please give me Dr. Münch's telephone number in the memory of Miriam. I had asked for that phone number in 1992, but they refused to give it to me because they said they were not giving out any phone numbers of their guests. An hour or so later, I received a nice sympathy letter and Dr. Münch's phone number.

Next, I called a friend of mine, Tony Van Rentergham, a Dutch resistance fighter who speaks German (my mother tongue is Hungarian and I do not speak German) and asked him to call Dr. Münch and invite him to the Boston conference. Tony called me back the next day and told me that Dr. Münch would not come to Boston, but would meet with me and give me a videotaped interview to take to the conference, and that a Dutch television network would film it and pay for my expenses.

In July of 1993, I was heading to Germany to meet a Nazi doctor. I was very nervous and scared. What I remembered about Nazi doctors I did not want to experience again, but I was curious about what I might learn of our experiments, and understand why this Nazi doctor was willing to meet with me.

We arrived at Dr. Münch's house. He treated me with the utmost respect and kindness. It was overwhelming to be treated that kindly by a Nazi doctor. He said he knew nothing about our experiments because Dr. Mengele said it was top secret and he did not share any details, but he gave us a substantive interview about Nazi medicine.

I felt very much at ease in Dr. Münch's company. I realized that this might be the only time in my life when I could talk to a Nazi face to face, and I wanted to know if he knew anything about the gas chambers in Auschwitz because the revisionists said that there were no gas chambers in Auschwitz. So I asked Dr. Münch if he knew anything about them.

"This is the nightmare I live with every single day of my life," he said. Then he went on to describe the operation of the gas chambers. People would be taken to an undressing room with numbered hangers on the wall. They were told to remember their hanger number. Then they were taken into the gas chamber, which looked clean and even smelled pleasant. Once the gas chambers reached capacity, the doors were closed hermetically. Dr. Münch was stationed outside, looking through a peephole.

A vent-like hatch opened in the ceiling. They used a poisonous gas in pellet form called Zyklon B, which looked like gravel and operated like dry ice. When the pellets dropped through a vent in the ceiling, the gas rose from the floor. People were trying to get away from the rising gas, and the strongest ended up on the top of the pile. When the people on the top stopped moving, Dr. Münch knew that everyone was dead. Then he would sign a certificate stating how many people were just killed in the gas chamber.

This was information I had never heard before. I asked him to come with me to Auschwitz in January 1995, when we would observe 50 years since the liberation of the camp. He immediately said that he would love to, and that he would sign a document testifying to the existence of the gas chambers at Auschwitz. I did not know it was going to be that easy.

When I came home from Germany, I was so excited about having this document signed by a Nazi, which could be used as evidence against those who would deny the reality of the Holocaust or the existence of the gas chambers. I wanted to give him a gift as my way of thanking him, but I did not know how to thank a Nazi doctor. Where does one look for a gift for a Nazi doctor? I went to the local Hallmark card shop and read many cards, but I found no card appropriate for a Nazi doctor. So I went back to lesson number one of my life lessons: "Never ever give up!"

For the next ten months I kept asking myself when I was driving, cooking, cleaning, doing the laundry—when my mind was not busy—"How do I thank this Nazi doctor? What can give this Nazi doctor?" All kinds of ideas popped into my head, until one day the following idea came into my mind: a simple letter of forgiveness, from me to him. I knew he would appreciate it, but what I discovered for myself was much more important: that I had the power to forgive. No one else could give me that power and no one could take it away. It was all mine to use as I pleased. Victims are always hurt, angry, feel hopeless, helpless, and powerless, and I discovered I had a power I did not previously know I possessed.

I began writing my letter to Dr. Münch. I wrote many versions, working through a lot of pain, but I was concerned with my spelling, so I contacted my former English professor to correct it. We met a few times, and then she said, "Eva it's nice that you forgive Dr. Münch, but you really should think about forgiving Dr. Mengele!" I said, "This was just a little letter to Dr. Münch." She asked me to promise her that I would think about forgiving Dr. Mengele, and I promised that I would.

After thinking about it, I realized that I had the power to forgive even the "Angel of Death." I concluded that if I forgave Mengele—the worst of the worst—that I might as well forgive everybody who has ever hurt me. It made me feel really good to have any power in my life. Up to that point, I had always reacted to what was done to me. Now I was initiating action and I was not hurting anybody, so I had no reason not to do it, and it made me feel good.

We arrived in Auschwitz on January 27, 1995. Dr. Münch came with his daughter, son, and granddaughter. I came with my son, Alex, and daughter, Rina. Dr. Münch signed his document about the existence of the gas chambers. I read mine and signed it, and I immediately felt a burden of pain was lifted from my shoulders. I was no longer a victim of Auschwitz, nor a prisoner of my tragic past. I was finally free from Auschwitz and free from Mengele.

Forgiveness is an act of self-healing and self-empowerment that brings serenity, freedom, and peace. Anger is a seed for war; forgiveness is a seed for peace.

In 1984, I founded an organization called CANDLES. In 1995, I founded CANDLES Holocaust Museum and Education Center in Terre Haute, Indiana. I liked the name because candles are used as a memorial, and candles illuminate. CANDLES is an acronym for Children of Auschwitz Nazi Deadly Lab Experiments Survivors. We wanted to shed some light on the darkest chapter of the Holocaust: the chapter of children used as human guinea pigs.

In her diary published in 2007, the Jewish freedom fighter Hannah Senesh recorded the line, "All the darkness can't extinguish a single candle, yet one candle can illuminate all its darkness" (Grossman 2007, p. 152).[1] I am asking you to remember the Nuremberg Code, and as you finish reading this, wherever life takes you, become a glowing light, an example of caring and ethical behavior, and never use a human being in experiments without informed consent.

I am hoping in some small way to send the world a message of hope, a message of healing, a message of forgiveness, and a message of peace. May we spread the

---

[1] The quote appears as a line in Senesh's diary. She attributes the line to a book by the Hebrew novelist Haim Hazaz, the title of which has been translated as *Broken Grindstones* or *Broken Millstones*. Senesh was a Hungarian-born Jew who immigrated to British Mandatory Palestine during the Second World War. There she joined a parachute regiment of the British Army. On a mission to rescue Hungarian Jews who were to be deported to Auschwitz, she was captured and executed by German authorities in Hungary.

light of knowledge throughout the world, to illuminate all the dark corners of human rights and human experimentation across the globe.

Reference

Grossman, R. (ed.). 2007. *Hannah Senesh: Her life and diary, the first complete*. Woodstock, VT: Jewish Lights Publishing.

Chapter 3
Eugenics and Racial Hygiene: Applied Research Strategies before, during, and after National Socialism

Hans-Walter Schmuhl

Eugenics was a field straddling the boundaries among various human sciences: biology, genetics, medicine, psychiatry, hygiene, anthropology, demography, and sociology. However, which of these disciplines was to become the primary science for a eugenics movement depended mainly on the context of each discipline's origins (Adams 1990b; Weingart 1999). Every eugenics movement began with a loose network of men and women who developed ideas, founded the early organizations, and attempted to establish programs in a variety of scientific and social fields—with varying degrees of success.

In Germany the first initiatives came from outsiders at the margins of the scientific community, including Dr. Alfred Ploetz and Dr. Wilhelm Schallmayer who attempted to establish the new concept of racial hygiene, the German term for eugenics (Weiss 1987; Proctor 1988; Weingart et al. 1988; Weindling 1989). The field of racial hygiene attracted scientists from very different disciplines as well as non-scientists and, initially, it was not clear which affiliations would succeed and endure. For instance, at the First Congress of German Sociologists in Frankfurt in 1910, Ploetz attempted to establish racial hygiene as a subdiscipline of sociology, but he lost this debate to Max Weber, a prominent German sociologist (Simmel 1911, pp. 111–165; Peukert 1989). Ploetz was more successful with his speech at the German Association for Public Health Care Convention in 1911 (Ploetz 1911). Similarly, racial hygiene was represented in the International Hygiene Exhibition in Dresden that same year (von Gruber and Rüdin 1911). In 1913, it was accepted into the medical section of the Society of German Natural Scientists and Physicians (Kroll 1983, pp. 117–123).

Racial hygienists further pursued affiliations with other disciplines after the First World War. In the Weimar Republic, racial hygiene enjoyed particularly close connections with anthropology and with psychiatry. By now this was no

H.-W. Schmuhl, apl. Prof. Dr. (✉)
Department of History, Philosophy and Theology, University of Bielefeld, Postfach 10 01 31, 33501 Bielefeld, Germany
e-mail: hschmuhl@uni-bielefeld.de

S. Rubenfeld and S. Benedict (eds.), *Human Subjects Research after the Holocaust*, DOI 10.1007/978-3-319-05702-6_3, © Springer International Publishing Switzerland 2014

coincidence, as racial hygiene was closely associated with the currents predominant in these fields. In Germany, anthropology perceived itself above all as *physical* anthropology, with the *cultural* anthropology of the American variety enjoying only weak influence (Schmuhl 2009). German psychiatry remained bound to its maxim of "mental illnesses are diseases of the brain" (Roelcke 1999, pp. 68–79, 88–95). Because one of its main missions was to investigate the heritability of mental illnesses and mental disabilities, eugenics was seen as a possible deterrent to the spread of mental illnesses as long as there were no somatic therapies for these supposed diseases of the brain. Later on, in National Socialist Germany, new somatic therapies were introduced—electroconvulsive therapy, insulin coma therapy, and cardiazol shock therapy, for example—and combined with eugenics and "euthanasia" (Schmuhl and Roelcke 2013).

In contrast, the connection between eugenics and genetics, a subdiscipline of biology, remained tenuous in Germany. Human genetics was a subdiscipline of genetics but not a very important one. An international comparison makes the differences apparent: In Great Britain, there was a close connection between eugenics and biometrics (Searle 1976; Farrall 1985; Jones 1986; Mazumdar 1992; Soloway 1995; Fennell 1996). In the United States, eugenics and genetics were intertwined (Pernick 1996; Dowbiggin 1997; Selden 1999; Kline 2001; Black 2003), just as they were in the Soviet Union, where eugenics occasionally served to legitimize genetics as a Bolshevist science (Adams 1990a). Yet no attempt will be made here to construct a uniquely German path. A similarly close connection of eugenics with medicine and psychiatry emerged in other countries as well, for instance in Japan (Frühstück 1997).

From country to country, eugenics occupied a different position on the scientific spectrum. Not even the transnational networking of eugenics movements succeeded in resolving these differences. This was certainly not for lack of effort. The Racist International movement (Kühl 1997), which took shape after the first International Eugenics Congress in London in 1912, was dedicated primarily to the establishment and international acceptance of eugenics as a scientific field of its own. Accordingly, at the first session of the Permanent International Eugenics Committee in Paris in 1913, the French eugenicist Lucien March developed an international classification for eugenics to clarify the field's relations to other sciences and to outline the methodological canon of eugenics—experimentation, observation, statistics, and genealogy. Yet none of these attempts at unification had any real success. The International Federation of Eugenic Organizations (IFEO), founded in 1925, attempted to standardize intelligence tests and methods for measuring characteristics of the human body, largely to no avail. The research questions and methods of eugenics ultimately depended on specific national scientific cultures (Kühl 1997, pp. 38–39, 76–77).

In Germany, the genetic foundation for eugenics was laid not in experimental genetics, but in anthropology and psychiatry. With research on blood groups and the dactyloscopy of races, anthropology attempted to build a bridge to human genetics. The Kaiser Wilhelm Institute for Anthropology, Human Heredity and Eugenics, founded in 1927 under the direction of Eugen Fischer, led the way (Lösch

1997; Schmuhl 2005). Even more important was psychiatric genetics, as practiced by Ernst Rüdin and his Department for Genealogy and Demography at the German Research Institute for Psychiatry founded in Munich in 1917. Beginning in the mid-1920s, Rüdin and his staff collected data about the population of certain regions of southern Bavaria with the express goal of developing criteria for the heritability of common mental illnesses like schizophrenia, manic depression, and epilepsy (Weber 1993; Roelcke 2003).

The Munich school set the tone in psychiatric genetics. Their methodological approach was characterized by a retreat into statistics. Because breeding experiments on humans were prohibited, Rüdin combined genealogical and statistical methods to produce an "empirical genetic prognosis." This raised a serious problem for eugenics. While in principle, an empirical genetic prognosis could be useful for epidemiologists, it did not provide racial hygienists with reliable diagnostic or prognostic information in any individual case. As such, it was hardly suitable as a tool of negative eugenics, which, by its very nature, called for practical applications. All over the world, eugenics thought of itself as an applied science, marked by its dual character as not only a science, but as a sociopolitical movement influencing a nation's population, health care, and social policies. In Germany racial hygiene's relationship to medical practice was especially pronounced, thanks to its close links to psychiatry, especially to the psychiatry practiced in sanatoriums and mental hospitals (see for example, Walter 1996; Kersting and Schmuhl 2004).

As a consequence, many German racial hygienists called for drastic eugenic measures, especially involuntary mass sterilizations, although they freely admitted to each other that the hereditability of most diseases and disabilities had not yet been sufficiently investigated by psychiatric genetics (see for example, Fischer 1933). The empirical genetic prognosis was considered just as certain as risk calculations for life insurance, and in their opinion, this would have to suffice as a basis for practical eugenic action.

In advancing such arguments, were scientists abandoning their scientific roots? This question misconstrues the character of an applied science. Scientific recommendations for policy are always the product of multiple factors: scientific knowledge, an assessment of a measure's practical utility, the expectation of its political enforceability and its cultural acceptance, and, finally, the appraisal of its ethical permissibility. Eugenics was by no means a *pseudo*-science that could be clearly demarcated from the methodologically precise field of psychiatric genetics. Indeed, the two fields complement each other. Eugenics constitutes what may be regarded as the hinge between psychiatric genetics, on the one hand, and politics, on the other. Thus, it always conforms to a dual logic: the logic of the sciences and the logic of politics.

In the late 1920s, the results of the empirical genetic prognosis seemed to offer sufficient proof that genes play a central role in the emergence of mental illnesses, even if in most cases the genetic path could not be traced back precisely. From the contemporary perspective, the thesis of the heritability of mental illnesses was a plausible assumption, yet extensive parts of it required additional empirical confirmation. The fact that this assumption was sometimes presented as a fact to the outside world—to the general public, but also to the state—is by no means unusual. Even today this procedure is part and parcel of everyday science (Roelcke 2003). Most eugenicists believed that the empirical proof for the apparently self-evident fact would be supplied retroactively, sooner or later (see for example, Weber 1993, pp. 220–223).

Because of these beliefs and assumptions, many racial hygienists and psychiatric geneticists advocated the legalization of sterilizations, even during the late Weimar Republic; some even argued that sterilization should be made compulsory. Most scientists were not deterred by the fact that a significant number of the people who did *not* suffer from a hereditary disease would be made infertile because the available diagnostics were imperfect. On the contrary, with a view to the inadequacies of the empirical genetic prognosis, these scientists argued that the circle of patients should be expanded to ensure that the entire target group would be sterilized. Their motto was that it was better to sterilize too many than too few (Bock 1986, pp. 238–240). As discussed above, ethical reservations did not generally play a role. However, most racial hygienists and psychiatric geneticists still refrained from publicly demanding compulsory sterilization for fear of strong social resistance, and most racial hygienists agreed that compulsory measures by the state, without broad public acceptance, would doom eugenics to failure in the long run. At the same time, they hoped for a swift political shift, for they sensed that they were under tremendous time pressure: The widely predominant idea of a rapidly degenerating gene pool of the German nation or of the white race demanded vigorous action (Weingart et al. 1988, pp. 291–306).

By referring to other countries, the German racial hygienists attempted to lend more urgency to their demands. Until 1933, racial hygienists viewed the United States as the "promised land" because of its pioneering role in eugenic sterilization. Indiana legalized involuntary sterilization on eugenic grounds for institutionalized "confirmed criminals, idiots, rapists and imbeciles" in 1907 ("Laws of State of Indiana," pp. 377–378). Sixteen states had sterilization laws on the books by the end of the First World War, and more states followed in the 1920s (Laughlin 1922; Reilly 1991; Trombley 1988). Sterilization laws were also passed in the Swiss canton of Waadt (1928), in Denmark (1929), in the Canadian provinces of Alberta (1928) and British Columbia (1933), and in the Mexican state of Vera Cruz (Huonker 2003; Ritter 2009; Hansen 1996; McLaren 1990; Dowbiggin 1997; Stepan 1991, pp. 130–133). Most of these laws also allowed involuntary sterilization. Developments abroad were monitored closely in Germany, especially the sterilization legislation in the United States, and used to legitimize the German program (see for example, von Hoffmann 1913; Blasbalg 1932; Steinwallner 1937/38).

The situation in Germany changed fundamentally in 1933 when eugenics became National Socialist state policy. One of the government's first official acts, on July 14, 1933, was to enact the Law for the Prevention of Offspring with Hereditary Diseases, commonly called The Sterilization Law. This law increased compulsory sterilizations to a level never achieved in any other country. By 1945 around 400,000 sterilizations and about 30,000 abortions were performed (Bock 1986, pp. 237–238). In this atmosphere, the Third Reich must have appeared to racial hygienists as the land of boundless opportunity. In the drafting of the law, racial hygienists advocated broad indications for sterilization, including the vague diagnosis of "congenital deficiency." By including this diagnosis, they hoped to include as many patients with inherited mental disability as possible, even if it meant sterilizing many patients with mental disabilities from other causes, such as antenatal or perinatal trauma (Bock 1986, pp. 239–241).

This concept was *not unscientific*. Rather, it was a train of thought in which scientific findings, supposedly necessary sociobiological circumstances, and political possibilities were weighed against each other. Racial hygienists and psychiatric geneticists also believed they were under tremendous time pressure to prevent irreversible damage to the German national gene pool. In addition, they had optimistic expectations that scientists would make rapid progress in psychiatric genetics and further refine their eugenic diagnostic instruments (see for example, Rüdin 1934).

While the eugenics movement in Germany advanced along more or less the same paths as in other countries up to 1933, the movement took a new path under National Socialism. Elsewhere, as the understanding of genetics advanced, eugenic movements came under pressure and tended to become less scholarly, evolving into scientifically informed political movements. In contrast, the interlocking of human genetics, psychiatry, and demographics in the Third Reich promised success and recognition for racial hygiene.

On the international level, this trend to the politicization of science and the "scientization" of politics was often sharply criticized. On the other hand, after the passage of its sterilization law, National Socialist Germany replaced the United States as the model state in the international eugenics world. Eugenicists all over the world, citing the German example, demanded legal provisions for eugenic sterilization. In fact, sterilization laws were passed in several European countries in the 1930s: Norway (1934), Finland (1935), Sweden (1935), Estonia (1936), Latvia (1937), and Iceland (1938) (Steinwallner 1937/38; Binswanger and Zurukzoglu 1938; Broberg and Roll-Hansen 1996; Turda and Weindling 2007; Felder and Weindling 2013). The number of American states with sterilization laws reached an apex of 32 in 1937, but, as of 1945, only about 40,000 sterilizations were performed in the United States, approximately 10 percent of the number performed in Nazi Germany (Bock 1986, p. 242).

With the collapse of the Third Reich, racial hygiene was discredited because of the mass sterilizations and even more because of the euthanasia program, which claimed around 300,000 victims (Faulstich 2000). Some racial hygienists succeeded in finding refuge in the apparently "value-neutral" field of human genetics, but eugenic ideas persisted beneath the surface. Well into the 1960s, German expert circles discussed a new sterilization law, and these same experts, instated as scientific authorities, rejected compensation for the victims of sterilization (see for example, Kaminsky 2005; Cottebrune 2012). In some countries like Sweden and Japan, official sterilizations continued into the recent past. Germany continued to unofficially sterilize mentally disabled persons, a practice that was finally directed into legally controlled channels through the new guardianship law passed in the 1980s (Schenk 2013).[1]

The rapid pace of development in the field of prenatal diagnosis now makes it possible to accurately predict some hereditary disabilities and diseases before birth, developments about which Nazi geneticists and racial hygienists could only dream. A "neo-eugenics" is emerging, which is fundamentally different from its predecessor in that it is no longer imposed on society "from above" by science or the state, but rather originates "from below," from society, which will compel governments and scientists to react (see for example, Fuchs 2008). In this sense, it would seem that eugenics is a chapter of our history that has yet to be closed.

References

Adams, M.B. 1990a. Eugenics in Russia 1900–1940. In *The wellborn science: Eugenics in Germany, France, Brazil, and Russia*, ed. M.B. Adams, 153–216. New York: Oxford University Press.
———. 1990b. Toward a comparative history of eugenics. In *The wellborn science: Eugenics in Germany, France, Brazil, and Russia*, ed. M.B. Adams, 217–231. New York: Oxford University Press.
Binswanger, L., and S. Zurukzoglu (eds.). 1938. *Verhütung erbkranken Nachwuchses: Eine kritische Betrachtung und Würdigung*. Basel: Schwabe.
Black, E. 2003. *War against the weak: Eugenics and America's campaign to create a master race*. New York: Four Walls Eight Windows.
Blasbalg, J. 1932. Ausländische und deutsche Gesetze und Gesetzentwürfe über Unfruchtbarmachung. *Zeitschrift für die gesamte Strafrechtswissenschaft* 52: 477–496.
Bock, G. 1986. *Zwangssterilisationen im Nationalsozialismus. Studien zur Rassenpolitik und Frauenpolitik. (Reprint: 2010)*. Opladen: Westdeutscher Verlag.
Broberg, G., and N. Roll-Hansen (eds.). 1996. *Eugenics and the welfare state: Sterilization policy in Denmark, Sweden, Norway, and Finland*. East Lansing: Michigan State University Press.
Cottebrune, A. 2012. Eugenische Konzepte in der westdeutschen Humangenetik, 1945–1980. *Journal of Modern European History* 10(4): 500–518.
Dowbiggin, I.R. 1997. *Keeping America sane: Psychiatry and eugenics in the United States and Canada*. Ithaca, NY: Cornell University Press.
Farrall, L.A. 1985. *The origins and growth of the English eugenics movement 1865–1925*. New York: Garland.
Faulstich, H. 2000. Die Zahl der "Euthanasie"-Opfer. In *"Euthanasie" und die aktuelle Sterbehilfe-Debatte. Die historischen Hintergründe medizinischer Ethik*, 8th ed, ed. A. Frewer and C. Eickhoff, 218–234. Frankfurt/Main: Campus.
Felder, B., and P. Weindling (eds.). 2013. *Bio-politics, race and nation in interwar Estonia, Latvia and Lithuania 1918–1940*. New York: Rodopi.

---

[1] The German Federal Ministry of Justice estimates that about 1,000 mentally handicapped women and girls were sterilized every year before the change of legislation. During a current research project on the living conditions of mentally handicapped people from 1945 to 1995 in the diaconic institution of Bethel (Bielefeld), we have interviewed "inmates" who said that sterilization was the common instrument of birth control until the 1970s.

Fennell, P. 1996. *Treatment without consent: Law, psychiatry and the treatment of mentally disordered people since 1845*. London: Routledge.

Fischer, E. 1933. Ist die menschliche Erblehre eine hinreichende Grundlage eugenischer Bevölkerungspolitik? *Archiv für Kriminologie* 93: 79–80.

Frühstück, S. 1997. *Die Politik der Sexualwissenschaft. Zur Produktion u. Popularisierung sexologischen Wissens in Japan 1908–1941*. Wien: Institut für Japanologie der Universität Wien.

Fuchs, R. 2008. *Life science. Eine Chronologie von den Anfängen der Eugenik bis zur Humangenetik der Gegenwart*. Berlin: LIT-Verlag.

Hansen, B.S. 1996. Something rotten in the state of Denmark: Eugenics and the ascent of the welfare state. In *Eugenics and the welfare state: Sterilization policy in Denmark, Sweden, Norway, and Finland*, ed. G. Broberg and N. Roll-Hansen, 9–76. East Lansing, MI: Michigan State University Press.

Huonker, T. 2003. *Diagnose: "moralisch defekt". Kastration, Sterilisation und Rassenhygiene im Dienste Schweizer Sozialpolitik und Psychiatrie 1890–1970*. Zürich: Orell Füssli.

Jones, G. 1986. *Social hygiene in twentieth century Britain*. London: Croon Helm.

Kaminsky, U. 2005. Zwischen Rassenhygiene und Biotechnologie. Die Fortsetzung der eugenischen Debatte in Diakonie und Kirche, 1945–1969. *Zeitschrift für Kirchengeschichte* 116: 204–241.

Kersting, F.-W., and H.-W. Schmuhl. 2004. *Quellen zur Geschichte der Anstaltspsychiatrie in Westfalen, Bd. 2: 1914–1955*. Paderborn: Ferdinand Schöningh.

Kline, W. 2001. *Building a better race: Gender, sexuality, and eugenics from the turn of the century to the baby boom*. Berkeley: University of California Press.

Kroll, J. 1983. *Zur Entstehung und Institutionalisierung einer naturwissenschaftlichen und sozialpolitischen Bewegung: Die Entwicklung der Eugenik/Rassenhygiene bis zum Jahre 1933*. Tübingen: Phil. Diss.

Kühl, S. 1997. *Die Internationale der Rassisten. Aufstieg und Niedergang der internationalen Bewegung für Eugenik und Rassenhygiene im 20. Jahrhundert*. Frankfurt: Campus.

Laughlin, H.H. 1922. *Eugenical sterilization in the United States*. Chicago: Psychopathic Laboratory of the Municipal Court of Chicago.

Laws of the State of Indiana. 1907. *Laws of the State of Indiana, Passed at the sixty-fifth regular session of the general assembly. Acts of 1907*, 377–378. Indianapolis, In: William B. Burford.

Lösch, N.C. 1997. *Rasse als Konstrukt: Leben und Werk Eugen Fischers*. Frankfurt: Lang.

Mazumdar, P.M.H. 1992. *Eugenics, human genetics and human failings: The Eugenics Society, its sources and its critics in Britain*. London: Routledge.

McLaren, A. 1990. *Our own master race: Eugenics in Canada, 1885–1945*. Toronto: McClelland & Stewart.

Pernick, M.S. 1996. *The black stork: Eugenics and the death of "defective" babies in American medicine and motion pictures since 1915*. New York: Oxford University Press.

Peukert, D.J.K. 1989. Weber contra Ploetz: Der historische Ort des Werturteilsstreits in der Vorgeschichte der Deutschen Barbarei. In *Max Webers Diagnose der Moderne*, ed. D.J.-K. Peukert, 92–102. Göttingen: Vandenhoeck & Ruprecht.

Ploetz, A. 1911. Ziele und Aufgaben der Rassenhygiene. *Deutsche Vierteljahrsschrift für öffentliche Gesundheitspflege* 43: 164–192.

Proctor, R.N. 1988. *Racial hygiene: Medicine under the Nazis*. Cambridge/Mass.: Havard University Press.

Reilly, P.R. 1991. *The surgical solution: A history of involuntary sterilization in the United States*. Baltimore: Johns Hopkins University Press.

Ritter, H.J. 2009. *Psychiatrie und Eugenik. Zur Ausprägung eugenischer Denk- und Handlungsmuster in der Schweizerischen Psychiatrie, 1850–1950*. Zürich: Chronos.

Roelcke, V. 1999. *Krankheit und Kulturkritik. Psychiatrische Gesellschaftsdeutungen im bürgerlichen Zeitalter (1790–1914)*. Frankfurt: Campus.

Roelcke, V. 2003. Programm und Praxis der psychiatrischen Genetik an der Deutschen Forschungsanstalt für Psychiatrie unter Ernst Rüdin. Zum Verhältnis von Wissenschaft, Politik und Rassebegriff vor und nach 1933. In *Rassenforschung an Kaiser-Wilhelm-Instituten vor und nach 1933*, ed. H.-W. Schmuhl, 38–67. Wallstein: Göttingen.

Rüdin, E. 1934. *Erblehre und Rassenhygiene im völkischen Staat*. Munich: Lehmann.

Schenk, B.-M. 2013. Behinderung, Genetik, Vorsorge. Sterilisationspraxis und humangenetische Beratung in der Bundesrepublik. Zeithistorische Forschungen 10: 433–454.

Schmuhl, H.-W. (2005). *Grenzüberschreitungen. Das Kaiser-Wilhelm-Institut für Anthropologie, menschliche Erblehre und Eugenik, 1927–1945*. Göttingen: Wallstein. English edition: Schmuhl, H.-W. (2008). *The Kaiser Wilhelm Institute for Anthropology, Human Heredity and Eugenics, 1927–1945. Crossing boundaries* (trans: Richter, S.). Dordrecht/Niederlande: Springer.

———. 2009. *Kulturrelativismus und Antirassismus. Der Kulturanthropologe Franz Boas (1858–1942)*. Bielefeld: Transcript.

Schmuhl, H.-W., and V. Roelcke (eds.). 2013. *"Heroische Therapien". Die deutsche Psychiatrie im internationalen Vergleich, 1918–1945*. Göttingen: Wallstein.

Searle, G.R. 1976. *Eugenics and politics in Britain 1900–1914*. Leyden: Noordhoff International Publications.

Selden, S. 1999. *Inheriting shame: The story of eugenics and racism in America*. New York: Teachers College Columbia University.

Simmel, G. (ed.). 1911. *Verhandlungen des ersten Deutschen Soziologentages vom 19.–22. Oktober 1910 in Frankfurt*. Tübingen: Mohr.

Soloway, R.A. 1995. *Demography and degeneration: Eugenics and the declining birthrate in twentieth-century Britain*. Chapel Hill: University of North Carolina Press.

Steinwallner, B. 1937/38. Rassenhygienische Gesetzgebung und Massnahmen im Ausland. *Fortschritte der Erbpathologie, Rassenhygiene und ihrer Grenzgebiete* 1: 193–260.

Stepan, N.L. 1991. *"The hour of eugenics": Race, gender and nation in Latin America*. Ithaca: Cornell University Press.

Trombley, S. 1988. *The right to reproduce: A history of coercive sterilization*. London: Weidenfeld & Nicolson.

Turda, M., and P. Weindling (eds.). 2007. *Blood and homeland: Eugenics and racial nationalism in central and southeast Europe, 1900–1940*. Budapest: Central European University Press.

von Gruber, M., and E. Rüdin (eds.). 1911. *Fortpflanzung, Vererbung, Rassenhygiene. Katalog der Gruppe Rassenhygiene auf der Internationalen Hygiene-Ausstellung 1911 in Dresden*. Munich: Lehmann.

von Hoffmann, G. 1913. *Die Rassenhygiene in den Vereinigten Staaten von Nordamerika*. Munich: Lehmann.

Walter, B. 1996. *Psychiatrie und Gesellschaft in der Moderne. Geisteskrankenfürsorge in der Provinz Westfalen zwischen Kaiserreich und NS-Regime*. Paderborn: Ferdinand Schöningh.

Weber, M.M. 1993. *Ernst Rüdin: Eine kritische Biographie*. Berlin: Springer.

Weindling, P. 1989. *Health, race, and German politics between national unification and Nazism, 1870–1945*. Cambridge: Cambridge University Press.

Weingart, P. 1999. Science and political culture: Eugenics in comparative perspective. *Scandinavian Journal of History* 24: 163–177.

Weingart, P., J. Kroll, and K. Bayertz. 1988. *Rasse, Blut und Gene. Geschichte der Eugenik und Rassenhygiene in Deutschland*. Suhrkamp: Frankfurt.

Weiss, S.F. 1987. *Race hygiene and the rational management of national efficiency: Wilhelm Schallmayer and the origins of German eugenics 1890–1920*. Berkeley: University of California Press.

Chapter 4
Medical Ethics and Medical Research on Human Beings in National Socialism

Florian Bruns

Introduction

The moral beliefs of Nazi doctors were of great significance for Nazi medicine and its brutal consequences. Research on the moral underpinnings of Nazi medicine has been neglected compared with the examination of the Nazi medical crimes themselves. Part of the reason for this academic void is the fact that focusing on Nazi mass murders rather than on Nazi norms and values makes it easier to regard National Socialism as an exceptional, unrepeatable event (Caplan 2010; Ofstad 1989, p. 16). Reflecting on medical ethics does not seem as alien to us as killing thousands of patients; however, the Nazis did both. Of course, one hesitates to speak of Nazi ethics, as ethics itself has a positive connotation, although the term alone does not have a normative meaning. In fact, National Socialism teaches us that even ethical reasoning can be corrupted (Haas 1988; Weikart 2004).

When we think about Nazi medical crimes in the concentration camps, there is one question that is posed time and again: Why did doctors and nurses, who are obligated to provide their patients with special care and protection, so flagrantly contravene the most elementary precepts of their professions? Why did doctors involved in human experimentation in concentration camps lack all empathy with their subjects? These, of course, are some of the most important but also most difficult questions in the context of Nazi medicine. Was it the perpetrators' abnormal, sadistic personality structures that predisposed them to commit crimes? We have learned that the answer was usually 'no'. Most of the doctors were quite normal, ordinary people (e.g., see Jensen and Szejnmann 2008; Browning 1992). Was it simply obedience to a superior? Yes, in some cases, but in general, coercion was not as important a factor as claimed by the defendants in the trials after war. Was it the golden opportunity for ambitious scientists to advance their field without regard for

F. Bruns, Dr. med., MA (✉)
Institute for the History of Medicine, Charité University Hospital, Thielallee 71, D-14195
Berlin, Germany
e-mail: florian.bruns@charite.de

S. Rubenfeld and S. Benedict (eds.), *Human Subjects Research after the Holocaust*,
DOI 10.1007/978-3-319-05702-6_4, © Springer International Publishing Switzerland 2014

consequences? In part, yes. Some doctors used the legal vacuum in the concentration camps as a license to sin. Was it the doctors' personal ambitions to get ahead, to advance their careers? Yes, to some extent (e.g., see Weindling 2004, p. 174).

Although all of these explanations have some merit, they are not sufficient. How is it possible that, after the war, almost none of the perpetrators showed any remorse (Gilbert 1947; see also McLaughlin 2012)? And how could it be that some of the defendants in the Nuremberg Doctors' Trial invoked, of all things, ethical arguments to justify their actions? Apparently, there is also an ethical dimension of the medical crimes—and we need to look at this dimension to gain more insight into the thinking of Nazi doctors (see Schmidt 2009). In this chapter, I will focus on some aspects of research ethics. I will first describe the relation between human and animal experimentation within the Nazi era. I will then discuss one characteristic feature of Nazi medical ethics: the unequal, one could also say the selective, application of ethical standards to patients and people depending on their moral status in the Nazi ideology. Last, I will demonstrate how Nazi medical ethics was taught to physicians and students at that time.

Human Experimentation and Animal Welfare

Nazi ethics defined itself as antimodernist and close to nature. Nature, biology, and Darwinism served as points of reference for this ethics and were seen as intrinsically good. As a consequence, nature and animal protection were promoted by National Socialists. Anti-vivisection leagues already existed in many German cities before the Nazis came to power in 1933. The movement against animal experiments had its own history in Germany as well as in other countries (e.g., see Rupke 1990; Paton 1993; Lisner 2009). The Nazis took advantage of the anti-vivisection movement by taking up some of its popular claims that were shared by many people. In the Nazi view, animal experimentation was considered an unethical exploitation of Mother Nature and was labeled as a "dreadful epitome of Jewish-materialistic academic medicine" (Germanicus 1933, p. 405).

Germany passed laws in 1933 and 1935 for the protection of animals and the preservation of nature. The animal protection law of 1933 restricted the use of animals for scientific research and stimulated a discourse on moral issues of scientific experimentation. The central point of this discussion was, *nota bene*, not the moral respect for human beings but for animals. The law for the protection of animals was one of the first laws passed during the Third Reich, and its ideological consequences should not be underestimated. Hoping to recruit the animal rights movement for National Socialism, Hermann Göring presented himself as the Reich's animal welfare activist. In 1933, a drawing appeared in a satirical magazine showing Göring marching in front of laboratory animals who raise their forelegs in a Hitler salute in gratitude for saving their lives (see Johnson 1933).

At first glance, from a moral standpoint, nothing can be said against an animal protection law. But the Nazis had political aims in mind. Supporting the cause of the animal rights activists helped to win these people over for National Socialism. Furthermore, animal protection served as a justification to prohibit kosher butchering, a Jewish ritual that has inflamed the passions for hundreds of years. Discrimination and anti-Semitic actions could be performed now under the guise of animal protection. The widespread disgust for animal cruelty was redirected toward Jewish people and their traditional butchering practice based on exsanguination without stunning the animal. Nazis together with many veterinarians condemned kosher slaughter as painful and cruel (Sax 2000, pp. 139–150).

In the medical field, putting animal experiments under the strict control of the Minister of the Interior complicated the use of animals for scientific experimentation for the next 12 years of Nazi regime (see Jütte 2002, p. 173). Punishment of even negligent violators of the law with a prison sentence seemed excessive, too. These restrictions, however, did not lead directly to the lethal experiments on humans, but the imbalance between human and animal rights was flagrant. Even in 1942, when human experimentation in several concentration camps was under way, a long article in the widely read medical journal *Deutsches Ärzteblatt* emphasized once more the immorality of torturing animals for scientific purposes. Almost ten years after enactment of the animal protection law, at the same time that hundreds of camp inmates were exposed to cruel and lethal experiments, the authors of that article highlighted and praised paragraphs of the law as outstanding signs of German culture (see Giese and Zschiesche 1942).

The striking overvaluation of animal life compared with human life blurred the moral distinction between the two species. In fact, animal life was deemed to have higher value than Jewish life. The Nazis passed a law regarding the humane transport of cattle but had no scruples about suffocatingly packing human beings in the same railroad cars for transport to death camps (see Sax 2000, p. 114–115). As *Life* magazine concisely described, prisoners, instead of animals, served as "human laboratory animals" in the experiments in concentration camps ("Human Laboratory Animals," 1947, February 24, p. 81). The main reason for preferring human subjects experimentation was to quickly obtain valuable and applicable scientific results for military medical or racial hygiene purposes. The fact that it was legal and easier for researchers to experiment on humans than animals highlights the Nazis' disregard for selected human life in their medical ethics.

The bizarre prioritization of animal rights above human rights was noted by Telford Taylor, chief prosecutor at the Nuremberg Doctors' Trial in his opening statement. Taylor recited the strict regulations of the 1933 law for protection of animals in detail, including the prohibition or severe limitation of painful or harmful operations or treatments, especially the use of cold, heat, or infection. Considering that National Socialism regarded it "as a sacred duty of German science to keep down the number of painful animal experiments to a minimum," Taylor concluded:

> If the principles announced in this law had been followed for human beings as well, this indictment would never have been filed. It is perhaps the deepest shame of the defendants that it probably never even occurred to them that human beings should be treated with at least equal humanity. ("Trials of War Criminals" 1950, p. 71)

The devaluation of human life compared to animal life also manifested itself on a semantic level. The Polish women who suffered in the notorious sulfonamide drug experiments in Ravensbrück concentration camp referred to themselves as *Kaninchen* ("rabbits"), a designation that blended self-irony with bitterness. Other camp inmates as well as the henchmen of the SS (*Schutzstaffel*) also spoke of the women as *Kaninchen* (see Klier 1994; Martin 1994; Walz 2001; Weindling 2004, p. 14), which is the German expression for guinea pigs. Even in the Nuremberg Doctors' Trial in 1946–1947, Herta Oberheuser, a physician who performed the experiments in Ravensbrück, continued to refer to her victims not as human beings but as guinea pigs.

Oberheuser, the only female defendant in the case, was involved in the sulfon-amide drug experiments initiated and supervised by the high-ranking SS surgeon and professor at Berlin University, Karl Gebhardt, in 1942 and 1943 (for Gebhardt, see Silver 2011). Gebhardt instructed Oberheuser to deliberately inflict wounds on the Polish women who were as young as she was. She rubbed wood, rusty nails, slivers of glass, dirt, or sawdust into the iatrogenic wounds to simulate the combat wounds of German soldiers. The goal was to investigate the efficacy of sulfon-amides, a new class of antibiotics, for the prevention of gas gangrene in combat wounds (see Ebbinghaus and Roth 2001; Roelcke 2009). Gebhardt's actual inten-tion was to prove that sulfonamides could *not* prevent gas gangrene, thereby asserting the superiority of surgery over antibiotics at the front lines and rebutting the charge that he improperly treated the badly wounded Reinhard Heydrich by withholding the new class of drugs. Heydrich, a leading organizer of the Holocaust and right-hand man of the head of the SS, Heinrich Himmler, supposedly died of septic shock a few days after a bomb attempt in Prague in 1942 (Gerwarth 2011).

The victims of the sulfonamide experiments suffered greatly, and more than 60 of them died from wound complications or lethal injections by Oberheuser, or they were murdered by SS troops to destroy evidence of the experiments. Some who survived served as prosecution witnesses at the Nuremberg Doctors' Trial and photographs of them baring their mutilated lower legs in court now belong to the iconography of the Nuremberg War Crime Trials.

The Shrinking Realm of Morality

Compared to the strictness of the animal protection law, the regulations concerning human subjects experimentation appear rather mild. The regulatory framework the Nazis inherited from the Weimar Republic consisted only of guidelines and did not stipulate any sanctions. In 1931, after fierce discussions in several German clinics and a lively debate in the press and parliament about unethical human subjects experimentation, the *Reichsgesundheitsrat* (Reich Health Council) issued detailed guidelines for new therapeutic and scientific human subjects research (for a trans-lated version, see Schmidt and Frewer 2007, pp. 333–335). These guidelines were developed primarily by the Jewish physician and Social Democrat Member of

Parliament, Julius Moses, a sharp critic of unethical human experimentation. Moses, a strong opponent of the Nazis, died ten years later in the ghetto of Theresienstadt (Eckart and Reuland 2006).

The 1931 guidelines clearly distinguished between therapeutic and nontherapeutic research and advocated standards that were unparalleled at the time, such as the obligation to obtain patient consent, the protection of children and socially disadvantaged people, and the prohibition of experiments on the dying. These directives were among the most comprehensive and most advanced research regulations at the time. Basic elements of the modern concept of informed consent can be found in these guidelines (see Eckart and Reuland 2006; von Engelhardt 2007; Tröhler 2007). The guidelines applied to everyone in Germany who was or could be a patient or subject of clinical research.

The innovative nature of regulations regarding human subjects and animal research was the only analogy between the two. For example, the animal protection law received great public attention when it was published, whereas the human subjects research guidelines received little public attention. Today they are seen as important ethical principles without great effect because they did not prevent the medical crimes in the concentration camps. However, this is an ahistorical ex post facto perspective: The guidelines were not designed for the prevention of atrocities of that scale. In 1931 nobody could anticipate the existence of concentration camps in Germany, not even Julius Moses, who clearly foresaw many other events.

After the Nazis came to power, the guidelines of 1931 were applied only to German and Aryan *Volksgenossen,* or people who belonged to the German *Volk* and, therefore, to the realm of moral obligation. The guidelines were explicitly invoked, for example, at the Berlin-based Robert Koch Institute in 1937 by Hans Reiter, the president of this important research facility and a very ardent Nazi, when he prohibited measles experiments on German children (see Hinz-Wessels 2008). Four years later, Reiter had no qualms about initiating human subjects experimentation in concentration camps. He designed typhus experiments that killed more than 200 people, mostly non-German prisoners, in the Buchenwald concentration camp (Wallace and Weisman 2003), a world apart from the Robert Koch Institute in the German capital. For SS doctors like Reiter, it was clear that the safeguards of the ethical research guidelines did not apply to Jews, Russian prisoners of war, and others at the lowest level in the Nazi hierarchy of human value.

Reiter's behavior accorded with Nazi medical ethics, which favored and protected only people in a defined realm and excluded others. According to this particular concept of morality, which was introduced and rigorously enforced within the medical profession, selected groups of human beings without a moral status—Jews, Russian prisoners of war, and mentally ill or disabled Germans—were no longer considered part of the protected community. This community became smaller and smaller, especially in the war years, when utilitarian considerations gained greater acceptance. Referring to a permanent national emergency in times of total war, the Nazi leadership sacrificed incurably sick or "useless" people for the benefit of others who met the requirements of the Nazi ideology for inclusion in the German *Volk*, such as healthy and productive citizens who could contribute to the war effort. At the Nuremberg Doctors' Trial, Karl Brandt, Hitler's personal physician and one of the leading figures in the Nazi medical administration, stated: "... [the] ethical obligation has to stop behind the totalitarian nature of war" (cited in Schmidt 2007, p. 373).

Teaching Nazi Medical Ethics

Considering the medical atrocities in Nazi Germany, one might expect that Nazi officials and physicians would spurn medical ethics. The Nazi regime and its medical functionaries, however, were intensely interested in a National Socialist version of medical ethics and in aligning a large number of German doctors and medical students with their health care policy by continued political training. Prospective medical doctors were a particularly important target group for indoctrination into Nazi thinking. Medical students were seen as predestined for future implementation of racial ideology and Nazi health care policy.

How were medical students and doctors indoctrinated with Nazi medical ethics? Since the Nazi *Weltanschauung,* or worldview, was not yet a part of the medical curriculum, the Nazis introduced the subject *Ärztliche Rechts- und Standeskunde* (Medical Law and Professional Studies) in April 1939, months before the start of World War II. They stated that, starting in the winter semester of 1939/40, Medical Law and Professional Studies was an obligatory course in the last semester of study. Nazi medical officials established a lecturer for the new subject at all 29 medical faculties in the German Reich. In the course of the war, the office of the *Reichsärzteführer* (Reich Physicians' Leader) gave lecturing posts for this subject either to doctors not affiliated with a university (most of them general practitioners) or to professors of forensic medicine. All of the external lecturers and almost all of the forensic pathologists were members of the Nazi Party (NSDAP) (Bruns 2009, pp. 102–112). Many had joined the NSDAP much earlier than 1933 and enthusiastically supported Nazi health care policy. For instance, the Tübingen lecturer Eugen Stähle, a neurologist who joined the NSDAP in 1927, was simultaneously the coordinator of the "euthanasia" campaign in Württemberg—one can easily imagine what he taught his students. Others served in the hereditary health courts, where decisions were made regarding forced sterilization.

The most prominent of these lecturers was Rudolf Ramm, who taught the new subject, Medical Law and Professional Studies, at the University of Berlin. He was a general practitioner who earned his stripes by expelling all Jewish doctors from Vienna in 1938. In the following years, he held influential positions in the Nazi health administration. Robert Proctor sees Ramm as "the leading Nazi medical ethicist" (1992, p. 17). In addition to teaching classes at the University of Berlin, Ramm was editor-in-chief of the previously mentioned journal *Deutsches Ärzteblatt*, in which he published an essay on the "Solution of the Jewish Question in Europe" in 1941 (see Ramm 1941).

In 1942 Ramm published a textbook on *Ärztliche Rechts- und Standeskunde* in which he outlined the National Socialist version of medical ethics and the mission of doctors in the Nazi state. The book was sold out within a year, received very positive reviews in almost all German medical journals, and a second extended edition was published in 1943. Ramm's textbook offers an approximate picture of what medical students learned in the *Ärztliche Rechts- und Standeskunde* course. He openly stated that Nazi medical ethics made a radical break from traditional forms of ethics, for example, by shifting the emphasis of ethical concern and medical care from the individual to the *volk*. Ramm emphasized that Nazism brought the "reinstatement of a high level of professional ethics" (1942, p. 69). He welcomed the fact that "the profession had been extensively cleansed of politically unreliable elements foreign to our race" (1942, p. 69). Ramm also addressed the "problem of euthanasia" and argued explicitly for the "mercy killing" of the disabled:

> These creatures merely vegetate and constitute a serious burden on the national community. They not only push down the living standard of the rest of their family members because of the cost of their care but also need a healthy person to take care of them throughout their lives. (1942, pp. 103–104)

Ramm referred only indirectly to human subjects research by noting that people had to make sacrifices for the sake of the nation. He denounced the bygone Weimar welfare state and its liberal democracy as weak and full of compassion for the wrong people, meaning the sick and needy. In Hitler's state, Ramm made clear, every person had a "moral duty to stay healthy" (1942, p. 148).

The Nazis also reached out to practicing physicians with an obligatory program of continuing medical education. In addition to the constant influence exerted by professional press publications aligned with the party's interests, the Nazis introduced new, obligatory, advanced training for doctors, which was intended to promote technical and ideological "education." In 1935, they opened in Alt-Rehse the *Führerschule der Deutschen Ärzteschaft,* a special school to promote Nazi medical ideology and health care policy to German doctors (Hansson et al. 2011, pp. 68–79). Similarly, the regime indoctrinated other health care professionals, such as midwives. Midwives were taught to serve as "guardians of the nation," and from 1939 on they were obliged to report any disabled newborn to medical or public authorities in order to determine their eligibility for the euthanasia program (Lisner 2006, pp. 267–278).

Meanwhile the medical students had to deal with another compulsory subject that found its way into medical curricula in 1939: the History of Medicine. The idea was to use history to legitimize racist Nazi medicine, to distinguish a certain "German character" of medicine, and to convey a new interpretation of traditional Hippocratic ethics. Starting shortly after their rise to power in 1933, Nazi medical functionaries took the first steps to ensure that medical history would be used for their political purposes. In turn, medical historians quickly discovered the benefits derived from the Nazis' interest in their field, and the fortunes of medical historiography in Germany did, indeed, experience an upswing. This change in fortune manifested itself not only by the opening of new institutes but by the inclusion of medical historiography as a mandatory subject in the medical school curriculum in 1939. Medical history was also integrated into the training courses for SS doctors at the Medical Academy in Graz, where the SS had built its own institute for the history of medicine (see Bruns 2009).

In a number of publications, Nazi medical historians tried to establish a direct historical line from the prior, great achievements of German medicine to Nazi medicine. For example, they characterized the German-Swiss Renaissance physician Paracelsus (1493–1541) as an early anti-Semite to provide both a historical foundation and a moral justification for their activities (see Gottlieb 1941).

Hitler or Hippocrates?

Although taking the Hippocratic Oath after graduation from medical school was not customary in Germany in the twentieth century, this ancient oath often has been looked upon as the most important document of medical ethics, regardless of the fact that its historical origin is unclear. The essence of the oath—not to harm patients, never to bring about their death even if the patient demands it, and always to maintain medical confidentiality—fits very well with Christian ethics, and medical professionals felt bound by its ethical maxims (e.g., see Smith 1996; Leven 1998; Timmermann 2002). Even the Nazis could not risk ignoring that tradition if they wanted the support of doctors and the population at large. One could ask if Hitler worked "with—or without—Hippocrates" (Rütten 1996; see also Hoedeman 1991). The answer depended upon the respondent's perspective.

While most Nazi medical ethicists avoided officially disputing the validity of the Hippocratic Oath, a significant number of doctors did not attribute any particular significance to the oath. Consider, for example, the concentration camp doctors who, when questioned after the war, claimed that their SS oath of allegiance to Hitler seemed much more real and binding than any vague rituals or ceremonies performed at medical school graduation (see Lifton 1986, p. 207). Beyond this, many of the Nazi doctors doubted that this ancient tradition had any meaning for the medicine of the modern age. Karl Brandt questioned the validity of a text that was 2,000 years old for contemporary medicine, and he was convinced that Hippocrates would formulate the oath differently today (see Schmidt 2007, p. 375). Others did not cast doubt on the basic validity of Hippocratic ethics, but they did dispute its applicability in the war—Brandt's codefendant in Nuremberg, SS hygienist Joachim Mrugowsky, was in this group. Like Brandt, he was responsible for lethal experiments on concentration camp prisoners. During the Nuremberg Doctors' Trial, Mrugowsky stated that these experiments were not carried out on sick persons, but on prisoners who, from his point of view, were healthy. Therefore, they "were not patients of the doctor in terms of medical ethics or in terms of the understanding of the relationship between a doctor and a patient. This is the reason why it would be possible to apply what we comprehend as medical ethics to this case only in a very limited sense" (cited in Bruns 2009, p. 163). Mrugowsky added that he had never seen the Hippocratic Oath during his studies and was not required to swear it.

It became evident in the Nuremberg Doctors' Trial that the Hippocratic Oath, given its ambiguous and somewhat antiquated formulations, was not a good witness for the prosecution (see Leven 1998). The court recognized that the numerous interpretations of the oath meant that it was not possible to read the oath as a document valid for all time. In addition, the Nuremberg judges acknowledged that the oath said nothing about human experimentation or autonomy of the individual. Thus, the Hippocratic Oath was simply sidelined by the Nazi defendants and "seemed not to be a deterrent of any significance for those who participated" in the concentration camp experiments (Pellegrino 2010, p. 14).

Conclusion

We should avoid the temptation to dismiss Nazi ethics as an oxymoron simply because Nazi doctors and scientists did things regarded today as deeply immoral. It is disturbing to acknowledge that, although its values and the priority of these values were totally different from ours, Nazi ideology was permeated with ethical concerns. Analysis of the Nazi medical crimes reveals neither a lack of commitment to ideals nor the absence of ethical discourse. Many doctors were quite convinced that their actions were necessary and, therefore, even right and good. Their concepts of morality provided the ethical underpinnings of Nazi medicine, and these concepts were powerful and convincing enough to transform doctors and nurses into murderers. Indeed, the prosecution at the Nuremberg Doctors' Trial had difficulty refuting some of these concepts, such as an extreme form of utilitarian thinking (see Caplan 2010, pp. 89–91; Schmidt 2007, p. 375; Weindling 2004, pp. 261–293).

Nazi medicine did have a system of ethical values. The well-being of the imagined community of the people's body (*Volkskörper*) outweighed the life of the individual. A pure race, a powerful and healthy nation, and, in times of war, gaining knowledge in military medicine outweighed the lives of humans who were deemed unworthy because of sickness or race. The Nazis did not abolish existing ethical guidelines for human subjects experimentation but applied them only to German citizens who belonged to the people's community (*Volksgemeinschaft*). Only people with value to society could claim a moral status protecting them against encroachments. The Nazis revised traditional morality by removing those outside the ethnic or racial community from the realm of moral obligation, thereby abolishing the idea of a universal medical ethics applicable to everyone (see Koonz 2003). The systematic devaluation of persons, based upon race, illness, or other subjective criteria, provided the ideological foundation for the murder of the sick, called *Euthanasie*, and for the lethal experiments in the concentration camps.

The Nazis focused mainly on animal rights in their discussion of the ethical boundaries of scientific and medical experimentation. Whereas the discussion on vivisection gained public attention and culminated in a rigorous law for the protection of animals, the debate on ethical human subjects experimentation, which had been very lively in the 1920s and early 1930s, faded from the spotlight. As a consequence, the Nazis gave far less consideration to the well-being of human research subjects than to the well-being of laboratory animals, especially during the war.

Beginning in 1939, compulsory courses in Medical History and in Medical Law and Professional Studies injected Nazi ideology and medical ethics into medical school curricula. It is a bitter irony that the Nazis were, therefore, the first to introduce obligatory courses in ethics in medical school curricula.

In the last few decades, we have learned that unethical human subjects research is by no means a specific phenomenon of National Socialism. Utilitarian thinking in medicine has not ceased, either. That is why the disturbing legacy of Nazi medical ethics should always remind us of the dangers that lurk even in our contemporary medical culture.

References

Browning, C. 1992. *Ordinary men. Reserve police battalion 101 and the final solution in Poland.* New York: HarperCollins.

Bruns, F. 2009. *Medizinethik im Nationalsozialismus. Entwicklungen und Protagonisten in Berlin (1939–1945).* Stuttgart: Steiner.

Caplan, A.L. 2010. The stain of silence: Nazi ethics and bioethics. In *Medicine after the Holocaust: From the Master Race to the human genome and beyond,* ed. S. Rubenfeld, 83–92. New York: Palgrave Macmillan.

Ebbinghaus, A., and K.H. Roth. 2001. Kriegswunden. Die kriegschirurgischen Experimente in den Konzentrationslagern und ihre Hintergründe. In *Vernichten und Heilen. Der Nürnberger Ärzteprozeß und seine Folgen,* ed. A. Ebbinghaus and K. Dörner, 177–218. Berlin: Aufbau.

Eckart, W.U., and A. Reuland. 2006. First principles: Julius Moses and medical experimentation in the late Weimar Republic. In *Man, medicine and the state. The human body as an object of government sponsored medical research in the 20th century,* ed. W. Eckart, 35–47. Stuttgart: Steiner.

Germanicus [pseudonym]. 1933. Dein Führer – Dein Vorbild. *Die Weiße Fahne* 14(6): 404–406.

Gerwarth, R. 2011. *Hitler's hangman: The life of Heydrich.* New Haven, CT: Yale University Press.

Giese, C., and A. Zschiesche. 1942. Wissenschaftliche Versuche an lebenden Tieren. Die §§ 5 bis 8 des Tierschutzgesetzes. *Deutsches Ärzteblatt* 72: 94–97.

Gilbert, G.M. 1947. *Nuremberg diary.* New York: Farrar, Straus & Company.

Gottlieb, S. 1941. Paracelsus als Kämpfer gegen das Judentum. *Deutsches Ärzteblatt* 71: 326–328.

Haas, P.J. 1988. *Morality after Auschwitz: The radical challenge of the Nazi ethic.* Philadelphia: Fortress.

Hansson, N., T. Maiboum, and P.M. Nilsson. 2011. The "Führerschule" in Alt Rehse: A character school for the physicians of National Socialist Germany. *Vesalius* 17: 68–79.

Hinz-Wessels, A. 2008. *Das Robert-Koch-Institut im Nationalsozialismus.* Berlin: Kadmos.

Hoedeman, P. 1991. *Hitler or Hippocrates: Medical experiments and euthanasia in the Third Reich.* Lewes, Sussex: Book Guild.

Human Laboratory Animals. 1947, February 24. *Life,* 22: 81–84.

Jensen, O., and C.-C.W. Szejnmann (eds.). 2008. *Ordinary people as mass murderers: Perpetrators in comparative perspectives*. Basingstoke: Palgrave Macmillan.

Johnson, A. 1933. Eine Kulturtat—Heil Göring! *Kladderadatsch* 86(36), n.p.

Jütte, D. 2002. Die Entstehung und Auswirkungen des nationalsozialistischen Reichstierschutzgesetzes von 1933. *Berichte des Institutes für Didaktik der Biologie der Westfälischen Wilhelms-Universität Münster*, Suppl. 2, 167–184.

Klier, F. 1994. *Die Kaninchen von Ravensbrück. Medizinische Versuche an Frauen in der NS-Zeit*. Munich: Knaur.

Koonz, C. 2003. *The Nazi conscience*. Cambridge: Harvard University Press.

Leven, K.-H. 1998. The invention of Hippocrates: Oath, letters and Hippocratic corpus. In *Ethics codes in medicine: Foundations and achievements of codification since 1947*, ed. U. Tröhler and S. Reiter-Theil, 3–23. Aldershot, UK: Ashgate.

Lifton, R.J. 1986. *The Nazi doctors: Medical killing and the psychology of genocide*. New York: Basic Books.

Lisner, W. 2006. *"Hüterinnen der Nation". Hebammen im Nationalsozialismus*. Frankfurt: Campus.

———. 2009. Experimente am lebendigen Leib: Zur Frage der Vivisektion in deutschen und britischen medizinischen Wochenschriften 1919–1939. *Medizinhistorisches Journal* 44: 179–218.

Martin, D. 1994. "Versuchskaninchen" – Opfer medizinischer Experimente. In *Frauen in Konzentrationslagern: Bergen-Belsen; Ravensbrück*, ed. C. Füllberg-Stolberg, M. Jung, R. Riebe, and M. Scheitenberger, 113–122. Bremen: Edition Temmen.

McLaughlin, D. 2012. I've never encountered a single Nazi who expressed remorse. Interview with Efraim Zuroff. *The Irish Times*. Retrieved from http://www.shoahlegacy.org/monitor/ive-never-encountered-single-nazi-who-expressed-remorse. Accessed 4 Dec 2013.

Ofstad, H. 1989. *Our contempt for weakness. Nazi norms and values – and our own*. Stockholm: Almqvist & Wiksell.

Paton, W.D.M. 1993. *Man and mouse: animals in medical research*. Oxford: Oxford University Press.

Pellegrino, E.D. 2010. When evil was good and good evil: Remembrances of Nuremberg. In *Medicine after the Holocaust: From the Master Race to the human genome and beyond*, ed. S. Rubenfeld, 11–16. New York: Palgrave Macmillan.

Proctor, R.N. 1992. Nazi doctors, racial medicine, and human experimentation. In *The Nazi doctors and the Nuremberg Code: Human rights in human experimentation*, ed. G.J. Annas and M.A. Grodin, 17–31. New York: Oxford University Press.

Ramm, R. 1941. Die Lösung der Judenfrage in Europa. *Deutsches Ärzteblatt* 71: 156.

———. 1942. *Ärztliche Rechts- und Standeskunde. Der Arzt als Gesundheitserzieher*. Berlin: Walter de Gruyter.

Roelcke, V. 2009. Die Sulfonamid-Experimente in nationalsozialistischen Konzentrationslagern: Eine kritische Neubewertung der epistemologischen und ethischen Dimension. *Medizinhistorisches Journal* 44: 42–60.

Rupke, N.A. (ed.). 1990. *Vivisection in historical perspective*. London: Routledge.

Rütten, T. 1996. Hitler with —or without—Hippocrates? The Hippocratic Oath during the Third Reich. *Koroth: The Israel Journal of the History of Medicine and Science* 12: 91–106.

Sax, B. 2000. *Animals in the Third Reich: Pets, scapegoats, and the Holocaust*. New York: Continuum International.

Schmidt, U. 2007. *Karl Brandt: The Nazi doctor. Medicine and power in the Third Reich*. London: Hambledon Continuum.

———. 2009. Medical ethics and nazism. In *The Cambridge world history of medical ethics*, ed. R.B. Baker and L.B. McCullough, 595–608. Cambridge: Cambridge University Press.

Schmidt, U., and A. Frewer (eds.). 2007. *History and theory of human experimentation: The Declaration of Helsinki and modern medical ethics*. Stuttgart: Steiner.

Silver, J.R. 2011. Karl Gebhardt (1897–1948): A lost man. *The Journal of the Royal College of Physicians of Edinburgh* 41: 366–371.

Smith, D.C. 1996. The Hippocratic Oath and modern medicine. *Journal of the History of Medicine* 51: 484–500.

Timmermann, C. 2002. A model for the new physician. Hippocrates in interwar Germany. In *Reinventing Hippocrates*, ed. D. Cantor, 302–324. Aldershot, UK: Ashgate.

Trials of War Criminals Before the Nuernberg Military Tribunals under Control Council Law No. 10. (1950). The Medical Case, vols 1–2. Washington: US Government Printing Office.

Tröhler, U. 2007. The long road of moral concern: Doctors' ethos and statute law relating to human research in Europe. In *History and theory of human experimentation. The Declaration of Helsinki and modern medical ethics*, ed. U. Schmidt and A. Frewer, 27–54. Stuttgart: Steiner.

von Engelhardt, D. 2007. The historical and philosophical background of ethics in clinical research. In *History and theory of human experimentation: The Declaration of Helsinki and modern medical ethics*, ed. U. Schmidt and A. Frewer, 55–70. Stuttgart: Steiner.

Wallace, D.J., and M.H. Weisman. 2003. The physician Hans Reiter as prisoner of war in Nuremberg: A contextual review of his interrogations (1945–1947). *Seminars in Arthritis and Rheumatism* 32: 208–230.

Walz, L. 2001. Gespräche mit Stanisława Bafia, Władysława Marczewska und Maria Plater über die medizinischen Versuche in Ravensbrück. In *Vernichten und Heilen. Der Nürnberger Ärzteprozeß und seine Folgen*, ed. A. Ebbinghaus and K. Dörner, 241–272. Berlin: Aufbau.

Weikart, R. 2004. *From Darwin to Hitler: Evolutionary ethics, eugenics, and racism in Germany*. New York: Palgrave Macmillan.

Weindling, P. 2004. *Nazi medicine and the Nuremberg trials. From medical war crimes to informed consent*. Basingstoke: Palgrave Macmillan.

Chapter 5
Sulfonamide Experiments on Prisoners in Nazi Concentration Camps: Coherent Scientific Rationality Combined with Complete Disregard of Humanity

Volker Roelcke

Since the end of World War II, human subjects medical research in Nazi concentration camps has been regarded not only as atrocious but also as scientifically obsolete, even pseudoscientific, or simply crime disguised as medicine (e.g., Cohen 1990; Nadav 2009, pp. 121–123).[1] According to this point of view, such evaluation of the scientific status of research would make any further deliberations on the ethics of this research unnecessary. If this perspective were true, then the historical cases could tell us something about the psychopathology of the perpetrators or about the pressure imposed upon medical scientists from outside political forces, but they could not teach us anything about the moral ambiguities of medical research more generally, or about the temptations for and moral frailties of good scientists.

In contrast to this long-standing point of view, recent historiography has documented many instances of medical research in Nazi concentration camps (as well as in psychiatric institutions and hospitals in the German occupied territories) that, if judged in terms of the scientific standards of the time, had scientific validity, or at least rationality, for the research aims or the applied methodology.[2]

This chapter is a synthesis of two previously published German-language articles by the author: Die Sulfonamidexperimente in nationalsozialistischen Konzentrationslagern. Eine kritische Neubewertung der epistemologischen und ethischen Dimension, *Medizinhistorisches Journal*, 44 (2009) and Fortschritt ohne Rücksicht. Menschen als Versuchskaninchen bei den Sulfonamid-Experimenten im Konzentrationslager Ravensbrück. In Eschebach, I., Ley, A. (Eds.). *Geschlecht und "Rasse" in der NS-Medizin* (pp. 101–114), 2012.

[1] For recent examples in German historical literature, see Henke (2008, p. 9); Süss (2011, p. 285).

[2] Compare, for instance, the research at the Dachau camp in the context of aviation medicine by Roth (2001). On the eugenically motivated research of Josef Mengele in the context of human genetics, see Massin (2003). On genetic research in the context of psychiatry, see Roelcke et al. (1994) and Roelcke (2000). For diverse examples and aspects of bacteriological research, see Weindling (2000) and various case histories in Eckart (2006).

V. Roelcke, Prof. Dr. med. (✉)
Institute of the History of Medicine, Giessen University, 35392 Giessen, Germany
e-mail: Volker.roelcke@histor.med.uni-giessen.de

S. Rubenfeld and S. Benedict (eds.), *Human Subjects Research after the Holocaust*,   51
DOI 10.1007/978-3-319-05702-6_5, © Springer International Publishing Switzerland 2014

These research activities cannot, therefore, be dismissed as having nothing to do with the professionalism and ethics of medical scientists. Rather, they point to profound issues about the ethics of human subjects research. At the core of these issues is the following question: Under what conditions are sane and rational medical scientists, trained in an internationally acclaimed system of academic medicine, prepared to prioritize their aim to produce new medical knowledge to such an extent that they completely disregard the subjectivity, suffering, and indeed humanity of their research "objects"?

The medical experiments investigating the efficiency of sulfonamides in war surgery are an interesting example for describing and analyzing the issues at stake here. Until very recently, they were classified as "outdated," in scientific terms and as marked by unnecessary brutality.[3] These experiments took place in the Dachau, Sachsenhausen, and Ravensbrück concentration camps beginning in June 1942. Because of the high mortality rate from extensive, gangrenous war wounds, the experiments investigated whether the traditional surgical therapy of immobilization, excising large amounts of tissue, and surgical dressing were still the optimum procedures, or whether the progression of the condition and survival rates might be improved significantly by two additional methods—local or systemic administration of sulfonamides (the first antibiotics, *avant la lettre*) or, alternatively, a homeopathic treatment favored by Reichsführer-SS Heinrich Himmler.

There is no doubt that these experiments—as described in detail below—were performed with extreme ruthlessness and brutality. It is also quite obvious that the core legal and ethical conditions for such experiments stipulated in the Reich directives for medical research on human subjects of 1930–1931 (which remained in force during the Nazi period), namely, informed consent and no exploitation of social hardship (Reichsministerium des Inneren, 1931), were completely ignored. For this reason alone, the experiments were inadmissible, no matter how they were carried out.

However, the usual images and interpretations of these experiments, and their dismissal as unacceptable in previous accounts, are based on an entirely different argument. For instance, the recent detailed standard depiction and analysis by Ebbinghaus and Roth (2001) repeatedly insists that the experiments were "completely superfluous when viewed from the scientific state of military surgery at the time" (p. 189) because the question of the "pharmacological and clinical effect" (p. 214) of an additional application of sulfonamides for treating infected wounds "had long been answered" (p. 215). This answer had been achieved through

---

[3] So reads the authors' final assessment in the most detailed reconstruction and analysis of these experiments available until recently (Ebbinghaus and Roth 2001, pp. 213 and 214). Similar assessments may be found in Silver (Silver 2011); Nadav (2009, p. 123); Cohen (1990); Strebel (2003); Hahn (2008, p. 459). At the Web site for an exhibition on human experiments in concentration camps, "Gewissenhaft – Gewissenlos" ["Conscientious – Unconscionable"] (the Institute for the History of Medicine at the University of Erlangen, n.d.), there is a more differentiated, but not explicitly critical assessment compared to that by Ebbinghaus and Roth. Paul Weindling (2004, pp. 11–15) not only described the context of the experiments but also sketched their practice; yet the scientific rationality for the experiments is not addressed directly.

"animal experiments and . . . clinical case reports of the consulting surgeons on the front" (p. 215). Therefore, the questions pursued in the experiments were "medically and scientifically obsolete" (p. 218), and the experiments senseless. Overall, the German surgeons were supposedly scientifically "backward" compared with their enemies' surgeons because the Allies had already recognized the efficacy of sulfonamides around 1940 and introduced them into clinical practice (Ebbinghaus and Roth 2001, p.183).

Ebbinghaus and Roth assumed that the clinical case reports from the front, in combination with the experiments on an animal model (the mouse), were sufficient to declare the question of systematic clinical research on sulfonamides in infected war wounds to be obsolete.[4] They dismissed a contemporary critique of the methods used in previous studies published by Martin Kirschner, professor of surgery in Heidelberg, in 1942 in the journal *Der Chirurg* [the Surgeon] as "polemics loaded with resentment" against the new therapeutic option (see also Ebbinghaus and Roth 2001, pp. 186–187).

Such vehement insistence on the obsolescence of the experiments is astonishing, and raises a whole range of questions: What kind of clinical or scientific evidence was available at the time when the experiments began in June 1942? Did the Allied military medical services really have the results of systematized clinical studies at their disposal? Would there really be a justification for these experiments if the scientific questions under investigation were up to date? Would, accordingly, this mere scientific justification of the experiments entail an ethical justification?

The interpretation of the experiments as pursuing an answer to an obsolete research question, and, therefore, as outdated and unnecessary research is usually supported by reference to Gerhard Domagk's demonstration of sulfonamides' effectiveness in 1932, and by noting that when the experiments began in several concentration camps in 1942, the Allies were routinely treating extensive war injuries with sulfonamides (Ebbinghaus and Roth 2001). Yet even a superficial review of these claims indicates that Domagk, in his first publication on the effect of sulfonamide in 1935, merely reported the results of experiments in *animals* (Domagk 1935). What is more, the Allies' use of sulfonamides does not necessarily mean that this praxis was based on evidence generated in accordance with the scientific standards valid during that period. In fact, in his recently published monograph on the history of sulfonamide therapy, John Lesch (2007) documents that, at least in the judgment of contemporary witnesses, the recommendation of May 1941 by the Subcommittee on Surgical Infections of the United States Surgeon

---

[4] They do this although the authors, in an introductory depiction of the broader context of military surgery, explicitly emphasize how the rapidly changing trench warfare, lack of essential equipment, etc., made fully inadequate working conditions prevalent throughout most of the front, necessitating continuous improvisation. At best, the courses of disease observed under such conditions could be partly documented, but the conditions precluded the performance of planned studies compliant with the methodological standards valid since the mid-1930s (see more detailed discussion below).

General regarding the use of sulfonamides was based not on newly available scientific evidence on the treatment of extensive war injuries, but primarily on the enthusiasm of civilian surgeons about this new therapeutic option.[5]

These counterarguments, therefore, raise the question whether the issue investigated in the concentration camp experiments—the effectiveness of sulfonamides for treating gas gangrene infected wounds—was really obsolete at the time the experiments were performed. To answer this question, I will now reconstruct the development of medical knowledge on the therapeutic effects of sulfonamides up to the beginning of the experiments in 1942, and analyze the methodological considerations on the basis of which the historical actors planned and implemented their investigations precisely the way they did. This historical reconstruction will pay particular attention to the form of scientific evidence that was used to come to conclusions about the efficiency of the new drug—the epistemology of the experiments.

Despite the focus on the scientific rationality of the experiments in the following reconstruction and analysis, neither the great brutality with which they were conducted, nor broader contexts such as the attempt on the life of Reinhard Heydrich, the *Reichsstatthalter* in Prague and deputy to Heinrich Himmler, will be ignored.[6] Yet previous historical portrayals and interpretations of the sulfonamide experiments lack a precise look at the specifically scientific rationality of the involved physicians. Reconstructing the sulfonamide experiments under this particular perspective leads to revision of the previous historical knowledge and to a correction of the long-standing point of view mentioned in the opening paragraph of this chapter. Also, the chronology of events as reconstructed here shows that Heydrich's death had, if at all, only a very late and minor influence on the development of the sulfonamide experiments.

---

[5] "[...] this advice was based not on new evidence, of which there was none, but 'solely upon the current trend of opinion of civilian surgeons, among whom a wave of enthusiasm for this procedure had begun to develop'" (Lesch 2007, p. 215). Lesch, however, does not broach the subject of sulfonamide experiments in German concentration camps.

[6] After the assassination attempt, Heydrich died in the beginning of June 1942 as a consequence of extensive infection to the wounds. He had not been treated surgically or with the newest available sulfonamide drugs. His death led to discussions about the appropriate therapy among the highest leadership of the Nazi state. These details have been known since the documentation of the Nuremberg Doctors' Trial by Alexander Mitscherlich and Fred Mielke (1949/1978). For more recent references, see Ebbinghaus and Roth (2001); Weindling (2004, pp. 11–15); and Hahn (2008, pp. 458–462).

The Methodology and Praxis of the Experiments

The following section will first summarize the praxis of the research efforts, in order to then reconstruct their rationality.[7]

Parallel series of tests were conducted in three concentration camps at essentially the same time. In Dachau the team led by Reichsarzt SS Ernst Grawitz, an internist, performed two series of tests on 27 prisoners with purulent wound infections and on 16 previously healthy prisoners who were infected by injections of pus to investigate homeopathic therapy. As the majority of the involuntary research subjects did not survive the experiments, adjunct homeopathic therapy appeared to be discredited as a treatment of patients with gas gangrene. However, doubts about the methods of investigation, particularly questions about an adequate control group, undermined the validity of the experimental results.

Therefore, the adjunct homeopathic therapy was then compared with adjunct sulfonamide therapy, and patients who had been treated only surgically were observed as "controls." The design of the study was defined precisely beforehand. The 60 prisoners involved were forced to participate without any previous information about the goals of the experiments, let alone consent. They were grouped into pairs on the basis of similar general parameters ("constitution") and then lots were drawn to divide them into either a homeopathic (B) or a sulfonamide treatment group (A). For 20 probands (10 A + 10 B), the adjunct therapy was begun on the fourth day after the artificially induced wound infections; for another 20 on the tenth day; and for another 20 on the first day. Eleven of the 60 prisoners died of the infection during the experiments. Overall, it became apparent that those who received sulfonamide earlier and at a sufficiently high dosage experienced a milder progression and fewer complications than did those who received the other therapy.

Due to the scarcity of sources, the praxis of the experiments in the Sachsenhausen concentration camp can be reconstructed only in fragments. With the participation of or perhaps under the direction of the Chief Camp Physician Emil Schmitz, simulated gangrenous wounds were created in 25 prisoners by making 15–20 cm incisions in the thigh and then applying pus, straw, dust from the barracks, and other dirt. The infected wounds remained open and were treated with an unknown antiseptic. It is no longer possible to determine whether there were comparison or control groups. On the basis of postwar historical research, it can be concluded that around 20 of the involuntary probands died as a consequence of the experiments.

The most extensive test series took place in the Ravensbrück camp under the direction of Karl Gebhardt, who was the chief surgeon of the Waffen-SS, personal friend of Himmler, and medical director of the nearby Hohenlychen SS military hospital (Hahn 2008). To the extent the data can be reconstructed, the probands

---

[7] If not otherwise noted, the reconstruction of the experimental practice follows Ebbinghaus and Roth (2001), especially pp. 196–211; it is quite detailed to enable an understanding of the underlying rationality.

were 20 men who were transferred to the concentration camp and 74 young, healthy, female, Polish resistance fighters who had just been captured. During the Nuremberg Doctors' Trial, the participating camp physician Hertha Oberheuser reported that the staff referred to the experimental subjects as "rabbits" (guinea pigs) (Poltawska 1968; Ebbinghaus 2001; Walz 2001).[8]

First of all, pretests were performed on three healthy prisoners to perfect the technique of producing a standardized research model of gas gangrene infections. For this purpose, one muscle group on the lower leg of each prisoner was laid open surgically, part of the muscle was contused, and a bloodless zone was produced by adrenaline injections. Various contaminants were inserted into this zone: gauze swabs with a mixture of streptococci and staphylococci, parablackleg pathogens, or, in the third case, gas gangrene pathogens and earth. However, it turned out that this method did not simulate war wounds infected with gas gangrene (Mitscherlich and Mielke 1949/1978).

Therefore, in a second and third series of tests the wounds of 15 prisoners were infected with a higher bacterial count of the same pathogens as well as additional coli bacteria and also some with wood shavings. A group of the experimental victims remained untreated as a control group, while the others received local treatment with two different kinds of sulfonamide powder.[9] Despite the massive tissue damage produced by both the infection and the subsequent surgical interventions, all experimental victims survived, although nearly all of them suffered permanent physical and mental consequences (Mitscherlich and Mielke 1949/1978, pp. 134–137).

However, the operating surgeon Fritz Fischer concluded that while it was possible to use this method to produce localized gas gangrene infections, these were not serious enough to correspond to full-blown war wounds. In another series of tests with nine prisoners, the experimental conditions were intensified, first by introducing wood shavings, then by preventing blood supply to the infected muscle, and finally by placing glass splinters in the wounds. Now the victims infected with the gas gangrene pathogens exhibited the full-blown clinical picture of war wounds, and even those infected with streptococci and staphylococci were at risk for sepsis. At this point, Gebhardt declared that the simulated war wounds were of sufficient quality that, "definitive conclusions [could be] reached about therapy using chemotherapeutic agents in combination with surgical interventions" (as quoted in Ebbinghaus and Roth, 2001, p. 206).

The researchers then systematically administered various therapies to the test groups assigned to the different types of infections. One experimental victim in each group remained untreated by drug therapy; the second received locally applied

---

[8] Paul Weindling (2004) has documented the role that the victims had in the emergence of the Nuremberg Medical Trial and the formation of the Nuremberg Code. He also referred to the medical consequences of the experiments and the problematic policy of recompensation of the German state toward the victims (pp. 11–20, 336–338, and passim).

[9] According to the postwar statement of Karl Gebhardt, the specific sulfa variants were Katoxyn and Marfanil: see Mitscherlich and Mielke (1949/1978).

sulfonamide powder; the third, the local application of sulfonamide plus sulfonamide tablets; and the fourth and final received supplementary sulfonamide injections in addition to the tablets and local application. Up to this point, the drug applications used were established sulfonamide preparations. Additional experiments presumably tested other newly developed variants of sulfonamide (Cibazol, Ultraseptyl). The courses of infection in the victims treated with drug therapy did not differ significantly from those of the probands treated only surgically.[10] Five of the experimental victims died; six others were executed after their last examinations (Ebbinghaus and Roth 2001, p. 206).

Available Knowledge on the Effects of Sulfonamide from 1932 to 1942

In 1932 Gerhard Domagk, a physician employed by the I.G. Farbenindustrie chemical company, along with the chemists Fritz Mietzsch and Josef Klarer, discovered the antibacterial effect of Prontosil, the first sulfonamide preparation, while performing experiments on animals. Apparently, in order to secure marketing opportunities, they did not publish their discovery until it was tested on humans in 1935 (Domagk 1935; Klee and Römer 1935; Schreus 1935). Domagk received the Nobel Prize in medicine in 1939 for this breakthrough.

The first publications on clinical experimentation with sulfonamides reported patients' good tolerance of the drugs as well as their positive effect on a number of different streptococcal infections in both internal medicine and dermatology. Staphylococcal infections also responded to sulfonamides, albeit not to the same extent. Exact numbers of subjects or criteria for inclusion in the clinical tests were not reported; similarly, no controlled experiments were performed with other treatment methods or placebos. Yet such standards for the methodology of clinical research had been formulated several years previously by Paul Martini, an internist in Bonn, and at least a number of elements of this methodology (for instance, the inclusion of numbers of cases, of criteria for including test subjects, formation of comparison groups) were also used in praxis by several, although certainly not all, contemporary clinical studies (Martini 1932).[11]

Other medical disciplines, such as otorhinolaryngology and gynecology, subsequently reported new therapeutic indications for sulfonamides. In 1937, Eleanor Bliss and Perrin Long published the first report of using sulfonamide successfully to treat mice infected with *clostridium welchii*, the presumptive pathogen of gas

---

[10] The historians Ebbinghaus and Roth (2001, p. 206) trace this back not only to the catastrophic general conditions in follow-up care but also to the fact that all probands were treated with dosages too low and for an insufficient period of time. However, they do not refer to the exact dosages applied, nor do they refer to "standard" dosages, and no sources are given for their interpretation.

[11] On Martini and his methodology of clinical trials, see Stoll et al. (2005).

gangrene. They also reported on animal experiments that might possibly explain sulfonamides' mode of action. In contrast to initial assumptions, these tests suggested that the specific effect of sulfonamides on the streptococcus consisted not in increased activity by the white blood cells and, therefore, from a strengthening of the immune system of the host organism, but more probably in the inhibition of bacterial growth (bacteriostasis). They were also expressly cautious regarding the generalization of these results to the treatment of other pathogens (Bliss and Long 1937, p. 1527). The first published report describing three successfully treated cases of gas gangrene in humans (although the clinical diagnosis was not completely certain) appeared in the same year (Bohlmann 1937).[12] A publication about the successful sulfonamide treatment of gas gangrene infections in two women with induced abortions or miscarriages appeared in July 1939 (Sadusk and Manahan 1939). Thus, when WWII began in 1939, there were only isolated case reports on the effectiveness of sulfonamides for gas gangrene infections in humans.

Even two years later, in 1941, while the effect of sulfonamides was firmly established in internal medicine, otorhinolaryngology, gynecology, and dermatology especially for infections with streptococci, pneumococci, meningococci, and staphylococci, no consensus can be detected in the existing reports of infected surgical wounds. An overview article on "chemotherapy" (meaning therapy with sulfonamides) in surgery published in April 1941 by Erich Schneider, director of the surgery department at the municipal hospital in Frankfurt/Oder, referred to the relatively meager number of important publications in surgery compared to other medical disciplines and to a lack of uniformity in the results. In early 1942, Werner Wachsmuth, consulting surgeon to the *Wehrmacht* (German Army) and chief field surgeon, made similarly hesitant and ambivalent statements about the surgical experiences from the German military (Wachsmuth 1942).

Precisely because the effectiveness of sulfonamides for infected wounds, especially for gas gangrene, had not been clarified at this point in time, the question was once again approached systematically by the Düsseldorf dermatologist Theodor Schreus and his working group through animal experimentation. Using a mouse model of gas gangrene infection created by the insertion of infected earth into opened muscle, Schreus was initially able to prove that suitable sulfanilic compounds, especially sulfanilic acid, *did* show a prophylactic effect when administered in high doses at the same time the wound was contaminated (Schreus and Peltzer 1941). In a second series of experiments designed to optimize the animal model of gas gangrene infection, the opened muscles were crushed to create ischemic and necrotic tissue that promoted the infection. Even under these conditions, the effect of administrating either previously investigated or newly developed sulfonamides were demonstrated as long as they were administered in a sufficiently high dose at the time of the wound infection or no more than four hours afterward (Schreus and Schümmer 1941).

---

[12] In 1938 the first in vitro experiments on *clostridium welchii*, the infectious agent for gas gangrene, were described (Spray 1938).

In another publication Schreus' working group reported the methodological difficulties that resulted from the different bioavailabilities of sulfonamide drugs: bioavailabilities varied widely for oral, intravenous, and subcutaneous administration in the animal experiments. For example, the effective bacteriostatic effect of the individual preparations on the gas gangrene pathogen could be compared to only a limited extent, even when administered to the animals in a uniform manner. Comparison of the direct effect of sulfonamides on the pathogen would, therefore, first have to be studied in vitro on a culture of the pathogen in an artificial medium such as liver bouillon and blood agar plates. In addition, the virulence of the pathogens in the in vitro experiments would have to be standardized to the greatest possible extent through multiple passages through animals (Schreus et al. 1941).

Schreus concluded in this December 1941 publication that the significance of the results obtained from the animal model was not sufficient to justify testing the new substances on humans. On the contrary, he initially distanced himself, methodologically speaking, even further from human trials by conducting another extensive analysis of the effectiveness of sulfonamides in vitro and not in either a human or animal.[13] This new analysis yielded very different results: Some completely new substances that were not yet on the market appeared to be very effective, while others like the Mesudin/Marfanil, which IG Farben and Domagk's group had just brought to market, appeared to show no practical effect. However, the results of these in vitro experiments by no means implied direct transferability to humans.

Prompted by data like these, Domagk stated in a review article in early 1942 that while the therapeutic application of sulfonamide preparations may well have been considered established in many clinical disciplines and for many types of pathogens in a number of applications areas, especially for infected wounds, "the opinions about the value of the sulfonamide compounds remain divergent" (p. 257).[14] Domagk also reviewed relevant publications from the international literature and emphasized that American scientists had already, at least, elucidated the important details of the sulfa drugs' mode of action.[15]

A few weeks later, Martin Kirschner, a Heidelberg professor of surgery, echoed Domagk's ambivalence about sulfonamide therapy for surgical infections. His discussion noted that the previous "success reports" about the treatment of infected wounds and especially gas gangrene infections, which he had judged critically,

---

[13] The order of the experiments can be reconstructed from the order of the three publications (titled „Mitteilungen"II-IV). In addition, the exact time of the laboratory work is directly addressed in the texts (Schreus and Peltzer 1941, p. 531; Schreus and Schümmer 1941, p. 705; Schreus et al. 1941, p. 1233). Ebbinghaus and Roth claim that the last experimental series was the first (2001, p. 185).

[14] Domagk summarized mostly the positive case reports, with reference to the concrete clinical conditions; he did not recount to the same extent the existing skeptical or critical reports.

[15] Domagk himself undertook further investigations on the effects of sulfa drugs in gas gangrene infections in the animal model in 1942. Together with the animal experiments of Schreus, this is an indicator that a number of questions on the exact indications, application forms, dosages, and side effects of the various sulfa preparations were apparently not completely clarified (see Domagk 1943, as quoted in Lesch 2007, pp. 96–97).

were based primarily on animal experiments and on unsystematic clinical observations. Kirschner also pointed out that many of the previous reports disregarded the quite common phenomenon of infected wounds healing spontaneously. He then asked what conditions must be fulfilled to prevent a false judgment about the value of new therapies *before* their clinical application became routine and formulated an entire list of demands (Kirschner 1942, pp. 445–446): advance planning of clinical studies with the formation of comparable treatment and control groups "under the same external and internal conditions," with the control group differing from the treatment group "only by the treatment factor to be tested"; an alternating random allocation of probands to the *verum* or control group; large numbers of cases allowing a statistical evaluation; and the participation of "as many as possible" investigators to exclude too subjective an evaluation by only one (potentially biased) investigator. The initial conditions for all probands were also to be as uniform as possible with regard to the size and form of the wound and the pathogen.[16] Kirschner (1942, pp. 444, 451) pointed out that the positive reports about the treatment of infected wounds with sulfonamides rarely satisfied even one of these methodological requirements.

Domagk's review of the contemporary positive reports confirmed Kirschner's critique in full (see list of references in Domagk, 1942, p. 287). That the methodological standards listed by Kirschner did indeed correspond to contemporary state of the art in the methodology of clinical research was documented in the book published in 1932 by Paul Martini and cited above, *Methodenlehre der therapeutischen Untersuchung*, which portrayed all of the requirements mentioned (and more) in great detail and discussed their value.

Kirschner himself expanded his critique of previous progress reports by adding his own prospective and comparative study with defined criteria of inclusion. In this study (1942, pp. 447–450), he investigated the potential of sulfonamide to minimize postoperative infectious complications in patients undergoing aseptic operations and found no significant differences between the sulfonamide group and the control group. He reached the following conclusion: The existing reports about positive experiences with sulfonamides on *humans* are fully inadequate methodologically and thus not sufficient on their own to make binding therapy recommendations, for instance in military surgery. The previously submitted results from *animal tests* with pure pathogen cultures may have been conclusive on their own, but did not correspond to "natural conditions" in humans with polybacterial infections, "substantial contamination and other damage from accidental wounds" (Kirschner 1942, p. 450).

These conclusions mean that in early 1942 a leading representative of German academic surgery evaluated the available knowledge on the efficacy of sulfonamide therapy in patients with infected war wounds as insufficient on methodological grounds, and therefore, he argued for more research before routinely treating war

---

[16] With this, Kirschner urged for a prospective study design with defined criteria of inclusion for the research subjects, the constitution of control groups, randomization, as well as statistically significant sizes of proband groups and methods of statistical analysis.

wounds with sulfonamides. A similar debate on the efficacy of the new sulfa drugs was going on in Britain. The British Medical Research Council published a systematic review in 1940 in which the authors urged further laboratory and clinical investigation to establish the effects of the new sulfa variants. In 1941, microbiologist Frank Hawking of the National Institute of Medical Research in London published the results of his experimental investigations on the efficacy of sulfa powder for prevention of gas gangrene infection in guinea pigs. His methodology was less elaborated than that of Schreus' group in Düsseldorf, but Hawking also concluded that more in vitro laboratory research was needed on the bioavailability and other properties of the various sulfa preparations. In August 1942, the *Lancet* published an article about the treatment of infected wounds by local application of sulfathiazole in 15 human research subjects. The authors reported mainly positive results, but they also posed questions about the comparability of the individual cases because the wound features, such as extent, depth, and pathogen mix, differed from patient to patient (Green and Parkin 1942). This apparent methodological shortcoming restricted the value of the results. In the September 1942 issue of the *Bulletin of War Medicine*, a journal published by the British Medical Research Council summarizing medical research relevant to the war effort, all 11 abstracts in the section on pharmacology were devoted to the efficacy of sulfa drugs in war wounds, with quite diverse results.

Thus, when the sulfonamide experiments in the concentration camps began in the summer of 1942, knowledge about the efficacy of the new drugs was by no means complete, and the questions pursued in the camp experiments were not obsolete. Rather, there were divergent assessments of the efficacy of sulfonamides in treating infected wounds, and the positive conclusions drawn from successful case histories at the front were widely contested. Because the medical challenges of infected war wounds were perceived as pressing by both the German and the Allied military, two options were at hand: either introduce sulfonamides into routine practice despite incomplete evidence, or initiate additional research. In this situation, the questions examined and the methodology applied in the experiments on the effect of sulfonamide in Dachau and Ravensbrück[17] picked up precisely where the publications by Schreus, Domagk, and Kirschner from the years 1941 and 1942 had left off. The experiments attempted, before broad clinical application of these drugs, to translate Schreus' methodologically convincing experiments from the animal model to a human model by controlling and standardizing as many parameters as possible and to apply the *methodological* standards demanded by Kirschner.[18] At the same time, the *ethical* standards that applied to humans as opposed to animals were brutally shunted aside, leading the medical researchers in

---

[17] This refers in fact only to the issues related to the efficiency of sulfa drugs in addition to conventional surgical therapy in Dachau and Ravensbrück, not for the experiments using homeopathic treatment in Dachau. For the sulfonamide experiments in Sachsenhausen, the scarcity of available sources does not allow an adequate historical reconstruction of the exact methods applied and the results.

[18] The rationality of the medical researchers is already hinted at in Weindling (2004, p. 12).

the concentration camps to follow a predetermined plan to duplicate the animal experiments in human subjects. In practical terms, this disregard of medical ethics meant creating defined wounds in comparable treatment and control groups of humans, infecting the wounds with pathogens of standardized virulence, further contaminating and irritating the wounds by inserting wood shavings and glass splinters, and then, at defined points in time after the infection/contamination, administering the various sulfonamide preparations.

It is clear, therefore, that the questions and the experimental *design*, to the extent that they can be reconstructed, corresponded to the contemporary state of scientific knowledge and to the methodological standards at the time the experiments were started in summer 1942. It is also clear from the brutal *praxis* of the experiments— replicating animal experiments in humans directly without any modification—that the investigators fully ignored the animal–human boundary.

As documented above, in early 1942 both German and British physicians perceived an urgent need for further laboratory and clinical research on the efficiency of sulfa drugs in the treatment of infected wounds. Initiating sulfonamide experiments on concentration camp prisoners in June 1942 may be understood as a rational response to the pressing medical problems at the front and the availability of unlimited human research subjects in Nazi concentration camps. The death of Reinhard Heydrich from a widespread wound infection in the beginning of June 1942, a few days before the start of the experiments, contributed little if anything to the conditions that led to the experiments. When Gebhardt introduced Heydrich's death at the Nuremberg Doctors' Trial as an explanation for the pressure he felt as Heydrich's treating physician to settle questions about treatment of war wounds, he may have been attempting to divert responsibility for the experiments to outside political forces and to obscure a fuller view of the ambivalent rationality of Nazi human subjects research.

Epistemological and Ethical Aspects of the Animal–Human Boundary: Methodological Care Versus Moral Concern

The sulfonamide experiments in the Nazi concentration camps, especially in Ravensbrück, cannot be dismissed as scientifically senseless and obsolete. They can be understood in their *epistemological* dimension as a coherent methodological step *after* the conclusive proof of effectiveness in the animal model and *before* the availability of sufficient evidence for routine and obligatory use in the German military. Upon conclusion of the established and successful methodology of the animal model with control of as many parameters as possible, a corresponding "human model" of gangrenous war wounds was implemented in the concentration camps in accordance with the scientific principles of bacteriology and experimental pharmacology.

This "human model" presumed not only the validity of the same epistemological conditions for animals and humans but also the validity of the same ethical conditions for animals and humans: The humans in the concentration camp experiments were not treated any differently than animals. It is apparent that these human test "objects" had a dual status that was, from the perspective of the scientists, particularly suitable for closing the knowledge gap between animal and human. Biologically and for the scientific purposes of the experiments, the scientists regarded the test subjects as human, but for social and ethical purposes, the scientists denied the subjects' status as human. The language of the medical perpetrators is consistent with this denial, as indicated by the designation of their human subjects as rabbits (*Kaninchen*). In this way the validity of the knowledge obtained from animals could be tested on human bodies without having to take into account the ethical standards that apply to humans. In abstract *epistemological* terms, in the search for knowledge, the intermediate methodological step of purposely infecting standardized, simulated war wounds in healthy humans with pathogens, just like in animals, was completely rational and coherent; but this step was possible only if the scientist ignored the difference in moral status between animals and humans.

In contrast to the methodically logical and coherent Nazi approach, the American call for the routine and obligatory administration of sulfonamides for war wounds in 1941 (Domagk 1942, p. 287; Lesch 2007, p. 215) *without* this intermediate step, viewed exclusively from the *epistemological* perspective, must be understood as backsliding well behind existing methodological standards. The behavior of the researchers in the concentration camps could be pointedly described as methodologically accurate without consideration for humans, and that of the American physicians and military medicine authorities as consideration for humans without sufficient methodological accuracy.

Thus, the sulfonamide experiments illustrate again that human subjects research during the Nazi period was frequently methodologically logical, but beyond moral boundaries (Eckart 2006; Massin 2003; Roelcke et al. 1994; Roelcke 2000; Roth 2001; Weindling 2000). Indeed, the approach pursued by the German investigators was only possible in the concentration camps of the Nazi state, a realm where existing legal and ethical standards could be ignored. The experiments could not possibly have been done on "normal" adult German citizens. The choice of location for the experiments is proof that the scientists were conscious of the legal and ethical inadmissibility of the experiments. In fact, the results of the experiments were made public only in a military medicine forum with highly restricted access and not in any of the standard journals (Gebhardt and Fischer 1943). Aside from the war context—confidentiality of "strategically vital knowledge" and severe paper shortages—it is highly probable that one of the reasons for the restricted publication of the data was the awareness of the brutality of the experiments, which would have been quite unacceptable under normal circumstances.[19]

---

[19] The results were reported by Gebhardt and his medical assistant Fritz Fischer at *Arbeitstagung Ost der Beratenden Fachärzte* [Working Conference East of the Consulting Physicians to the *Wehrmacht*] (Gebhardt and Fischer 1943); see also the statement by Fischer during the Nuremberg Medical Trial, documented in Mitscherlich and Mielke (1949/1978); Ebbinghaus and Roth (2001, note 127).

In conclusion, the sulfonamide experiments in Dachau and Ravensbrück may not be condemned because they pursued scientifically obsolete questions. If obsolescence was an adequate explanation, these experiments would represent scientifically "bad" medical research, which from today's perspective has nothing to do with modern, scientifically "good" medical research. The portrayal of Nazi medical experiments as scientifically "bad" research or pseudoscience indicates, however, a conscious or unconscious strategy to evade the real issues at stake. For example, this portrayal and subsequent evasion of the real issues is broadly accepted by bioethicists: While the brutality of medical research in National Socialism is, of course, wrong, and the participation of important physicians in such experiments is regrettable, what happened in this historical chapter of medical research supposedly has nothing to do with the ethical problems encountered in research today.

Yet, in fact, the historical case reconstructed here shows that, as far as the research question and methodology were concerned, the research was by no means obsolete, but rather, according to the standards of the day, up-to-the-minute. The Nazi sulfonamide experiments demonstrate that the topicality of research questions, the high quality of the scientific methodology, and the probable utility of the experimental results are insufficient reasons to declare human subjects research ethically acceptable.

The sulfonamide experiments also raise questions as to whether a concept of scientific method or epistemology can ever lead to an adequate historical reconstruction of biomedical research when it considers only theoretical and methodological principles and rationalities of knowledge production, and, at the same time, ignores ethical and anthropological premises and concrete social and political contexts in the *praxis* of knowledge production. A potential implication would be, for instance, that scientific rationality must always be viewed and analyzed in connection with the anthropological premises and the hierarchies of values connected with the scientific activity.

Finally, the sulfonamide experiments also demonstrate that historical reconstructions that concentrate primarily on the social, political, and moral dimension of medical activity and, at the same time, ignore the scientific rationality of the historical actors, can result in a portrayal that takes no account of similarities or parallels to medicine today. This type of historical reconstruction misses important opportunities for ethical reflections proceeding from the historical example.

References

Bliss, E., and P. Long. 1937. Observations on the mode of action of sulfanilamide. *Journal of the American Medical Association* 109: 1524–1528.

Bohlmann, H.R. 1937. Gas gangrene treated with sulfanilamide. Report of three cases. *Journal of the American Medical Association* 109: 254.

British Medical Research Council. 1940. *Notes on the diagnosis and treatment of gas gangrene.* MRC War Memorandum No. 2. London.

———. 1942. *Bulletin of War Medicine.* London.

Cohen, N. 1990. Medical experiments. In *Encyclopedia of the Holocaust*, ed. I. Gutman, 957–966. New York: Macmillan.

Domagk, G. 1935. Ein Beitrag zur Chemotherapie der bakteriellen Infektionen. *Deutsche Medizinische Wochenschrift* 61: 250–253.

———. 1942. Die Sulfonamidpräparate und ihre therapeutische Auswertung. *Die Medizinische Welt* 16(257–262 and 283–287).

———. 1943. Jahresbericht [annual report]. Betriebsberichte: Abteilung für Experimentelle Pathologie und Bakteriologie. (103/11, 1-20). Bayer Firmenarchiv, Leverkusen, Germany.

Ebbinghaus, A. 2001. Zwei Welten. Die Opfer und die Täter der kriegschirurgischen Experimente. In *Vernichten und Heilen. Der Nürnberger Ärzteprozess und seine Folgen*, ed. A. Ebbinghaus and K. Dörner, 219–240. Berlin: Aufbau.

Ebbinghaus, A., and K.H. Roth. 2001. Kriegswunden. Die kriegschirurgischen Experimente in den Konzentrationslagern und ihre Hintergründe. In *Vernichten und Heilen. Der Nürnberger Ärzteprozess und seine Folgen*, ed. A. Ebbinghaus and K. Dörner, 177–218. Berlin: Aufbau.

Eckart, W.U. (ed.). 2006. *Man, medicine, and the state. The human body as an object of government sponsored medical research in the 20th Century*. Stuttgart: Steiner.

Gebhardt, K., and F. Fischer. 1943. [Report]. 3. *Arbeitstagung Ost der Beratenden Fachärzte* [Working Conference East of the Consulting Physicians to the *Wehrmacht*]. May 24–26, 1943. Militärärztliche Akademie, Berlin, Germany.

Green, H.N., and T. Parkin. 1942. Local treatment of infected wounds with sulphathiazol. *Lancet* 240: 205–210.

Hahn, J. 2008. *Grawitz, Genzken, Gebhardt. Drei Karrieren im Sanitätsdienst der SS*. Klemm & Oelschläger: Münster.

Hawking, F. 1941. Prevention of gas-gangrene infections in experimental wounds by local application of sulphonamide compounds. *British Medical Journal* 1(4181): 263–268.

Henke, K.-D. 2008. Einleitung. In *Tödliche Medizin im Nationalsozialismus. Von der Rassenhygiene zum Massenmord*, ed. K.-D. Henke, 9–29. Böhlau: Cologne.

Institute for the History of Medicine at the University of Erlangen. n.d. "Gewissenhaft – gewissenlos" ["Conscientious – Unconscionable"]. Exhibition on human experiments in concentration camps. [website] Retrieved from http://www.gesch.med.unierlangen.de/gewissen/index1.htm. Accessed 12 Oct 2013.

Kirschner, M. 1942. Die Chemotherapie chirurgischer Infektionskrankheiten. *Der Chirurg* 13: 443–457.

Klee, Ph., and H. Römer. 1935. Prontosil bei Streptokokkenerkrankungen. *Deutsche Medizinische Wochenschrift* 61: 253–255.

Lesch, J.E. 2007. *The first miracle drugs: How the sulfa drugs transformed medicine*. Oxford: Oxford University Press.

Martini, P. 1932. *Methodenlehre der therapeutischen Untersuchung*. Berlin: Springer.

Massin, B. 2003. Mengele, die Zwillingsforschung und die 'Auschwitz-Dahlem Connection'. In *Die Verbindung nach Auschwitz. Biowissenschaften und Menschenversuche an Kaiser-Wilhelm-Instituten*, ed. C. Sachse, 201–254. Göttingen: Wallstein.

Mitscherlich, A., and F. Mielke (eds.). 1978. *Medizin ohne Menschlichkeit. Dokumente des Nürnberger Ärzteprozesses*. Frankfurt: S. Fischer. (Original work published in 1949 as *Wissenschaft ohne Menschlichkeit*. Heidelberg: Lambert Schneider). English edition: Mitscherlich, A., and F. Mielke (eds.). 1949. *Doctors of infamy: The story of the Nazi medical crimes* (trans: Norden, H.). New York: Schuman.

Nadav, D.N. 2009. *Medicine and Nazism*. Jerusalem: Magnes Press.

Poltawska, W. 1968. *I boje sie snów*. Warsaw: Czytelnik. German version: Poltawska, W. 1994. *Und ich fürchte meine Träume*. Abensberg: Kral.

Reichsministerium des Inneren. 1931. Richtlinien für neuartige Heilbehandlung und für die Vornahme Wissenschaftlicher Versuche am Menschen. *Reichsgesundheitsblatt* 6: 174–175.

Roelcke, V. 2000. Psychiatrische Wissenschaft im Kontext nationalsozialistischer Politik und 'Euthanasie'. Zur Rolle von Ernst Rüdin und der Deutschen Forschungsanstalt für Psychiatrie.

In *Die Kaiser-Wilhelm-Gesellschaft im Nationalsozialismus*, ed. D. Kaufmann, 112–150. Göttingen: Wallstein.

———. 2009. Die Sulfonamidexperimente in nationalsozialistischen Konzentrationslagern. Eine kritische Neubewertung der epistemologischen und ethischen Dimension. *Medizinhistorisches Journal* 44: 42–60.

———. 2012. Fortschritt ohne Rücksicht. Menschen als Versuchskaninchen bei den Sulfonamid-Experimenten im Konzentrationslager Ravensbrück. In *Geschlecht und "rasse" in der NS-Medizin*, ed. I. Eschebach and A. Ley, 101–114. Berlin: Metropol-Verlag.

Roelcke, V., G. Hohendorf, and M. Rotzoll. 1994. Psychiatric research and "euthanasia": The case of the psychiatric department at the University of Heidelberg, 1941–1945. *History of Psychiatry* 5: 517–532.

Roth, K.H. 2001. Tödliche Höhen. Die Unterdruckkammer-Experimente im Konzentrationslager Dachau und ihre Bedeutung für die luftfahrtmedizinische Forschung des "Dritten Reichs". In *Vernichten und Heilen. Der Nürnberger Ärzteprozess und seine Folgen*, ed. A. Ebbinghaus and K. Dörner, 110–151. Berlin: Aufbau.

Sadusk, J.F., and C.P. Manahan. 1939. Sulfanilamide for puerperal infections due to Clostridium welchii. *Journal of the American Medical Association* 113: 14–17.

Schneider, E. 1941. Ergebnisse der modernen Chemotherapie in der Chirurgie und ihre topographischen Bedingungen. *Medizinische Klinik* 37: 385–387.

Schreus, H.Th. 1935. Chemotherapie des Erysipels und anderer Infektionen mit Prontosil. *Deutsche Medizinische Wochenschrift* 61: 255–256.

Schreus, Th., and E. Peltzer. 1941. Chemoprophylaxe des Gasbrandes, II. Mitteilung. *Klinische Wochenschrift* 20: 530–535.

Schreus, Th., and H. Schümmer. 1941. Chemoprophylaxe des Gasbrandes. III. Mitteilung. *Klinische Wochenschrift* 20: 705–708.

Schreus, Th., A. Brauns, and H. Schümmer. 1941. Vergleich der Wirkung verschiedener Sulfonamidverbindungen auf die Gasbrandinfektion durch Kulturerreger, IV. Mitteilung. *Klinische Wochenschrift* 20: 1233–1236.

Silver, J.R. 2011. Karl Gebhardt (1897–1948): A lost man. *Journal of the Royal College of Physicians Edinburgh* 41(4): 366–371.

Spray, R.S. 1938. Bacteriostatic action of Prontosil soluble, sulphanilamide, and disulfanilamide on the sporulating anerobes commonly causally associated with gas gangrene. *Journal of Laboratory and Clinical Medicine* 23: 609–614.

Stoll, S., V. Roelcke, and H. Raspe. 2005. Gibt es eine deutsche Vorgeschichte der Evidenz-basierten Medizin? Methodische Standards therapeutischer Forschung im beginnenden 20. Jahrhundert. *Deutsche Medizinische Wochenschrift* 130: 1781–1784.

Strebel, B. 2003. *Das Konzentrationslager Ravensbrück*. Paderborn: Schönigh.

Süss, W. 2011. Versuche der Wiedergutmachung. In *Medizin und Nationalsozialisums*, ed. R. Jütte, 283–294. Göttingen: Wallstein.

Wachsmuth, W. 1942. Erfahrungsberichte aus dem chirurgischen Sonderlazarett des Oberkommandos der Wehrmacht. *Der Deutsche Militärarzt* 7: 65–69.

Walz, L. 2001. Gespräche mit Stanislawa Bafia, Wladyslawa Marczewska und Maria Plater über die medizinischen Versuche in Ravensbrück. In *Vernichten und Heilen. Der Nürnberger Ärzteprozess und seine Folgen*, ed. A. Ebbinghaus and K. Dörner, 241–274. Berlin: Aufbau.

Weindling, P.J. 2000. *Epidemics and genocide in eastern Europe*. Oxford: Oxford University Press.

———. 2004. *Nazi medicine and the Nuremberg trials. From medical war crimes to informed consent*. Houndmills/Basingstoke: Palgrave Macmillan.

Chapter 6
Stages of Transgression: Anatomical Research in National Socialism

Sabine Hildebrandt

> *The same suspension of empathy that was so necessary a part of a physician's task was also, in other contexts, the root of all monstrosity.*
>
> (The Eye in the Door, Pat Barker 1995)

Introduction

German anatomists used bodies of victims of the National Socialist (NS) regime for teaching and research purposes in all German anatomical departments from 1933 to 1945 and for some years thereafter. This history has been the focus of several studies over the last few years, which confirmed a ubiquitous anatomical practice that was professionally accepted at the time but is considered unethical in modern anatomy (Hildebrandt 2009a, b, c; 2013a). The gradual process of transformation of traditional anatomical procedures and attitudes to inhumane acts committed by anatomists and facilitated by a criminal government has not been explored in detail. However, the development of atrocities in anatomy can be traced through four stages of ethical transgression: (1) the initial transgression of the invasion of the intact dead human body for the purposes of science; (2) new transgressions inherent in the changes in body procurement specific to the NS period; (3) the escalation of these transgressions in anatomical research; and (4) the ultimate transgression—anatomical research on still living but "future dead"[1] persons.

---

[1] "Future dead" will henceforth refer to living people who were used for research purposes by anatomists with knowledge of their research subjects' certain death because of NS policies and practices.

S. Hildebrandt, MD (✉)
Instructor in Pediatrics; Lecturer on Global Health and Social Medicine; Boston Children's Hospital, Harvard Medical School; General Pediatrics, Department of Medicine, 333 Longwood Avenue, LO234, Boston, MA 02115, USA
e-mail: sabine.hildebrandt@childrens.harvard.edu

S. Rubenfeld and S. Benedict (eds.), *Human Subjects Research after the Holocaust*,   67
DOI 10.1007/978-3-319-05702-6_6, © Springer International Publishing Switzerland 2014

The Anatomical Method: The Initial Transgression of a Taboo

Anatomy as a medical discipline is based on the premise that knowledge of the structure of the human body is essential for effective medical practice. Traditionally, it postulates that anatomical knowledge must be gained by the dissection of human bodies. In most human societies, such a "cutting up" of human bodies would be considered an unbearable violation under any other than these scientific circumstances. Thus, anatomists need to be granted permission by society to commit the transgression of a taboo under the special circumstances of the advancement of medical knowledge (Winkelmann 2007). Jones stated: "Dissection and autopsies exist on the edge of the cultures that allow them, walking a fine line between acceptance and repulsion" (1998, p.101) and pointed out that there "are limits on what a culture will accept" (p. 102). Historically, not all members of society found the anatomical method acceptable, as demonstrated by protests against situations in which body procurement was perceived as coercive (Sappol 2002; Hildebrandt 2008) or the scientific reductionist and mechanistic view of human life was not shared (Bergmann 2008). Only within the last decades has the problem of coercive body procurement been resolved in many countries after the development of voluntary body donation programs (Garment et al. 2007).

The work of an anatomist thus begins with the breaking of a taboo, of which the beginner in the field and the sensitive dissector are still aware each time he or she puts a scalpel on the skin of a human body to be dissected. The anatomist learns to overcome this initial reluctance by the development of a clinical detachment, which is one of the core elements in the socialization of a future physician, and without which objective observation and effective medical practice are impossible (Warner and Rizzolo 2006; Böckers et al. 2010). Indeed, the phenomenon is so distinct that William Hunter, the great eighteenth century anatomist, called this change of thinking and emotions "a certain inhumanity" (Richardson 1987), a strong term that implies the potential for further transgressions unless the societal boundaries of the special permission given to anatomists are clearly defined. Grodin described the psychological development of clinical detachment, which can lead to compartmentalization and dehumanization of the patient (or body donor in anatomy), and stated: "Medicine as a profession contains the rudiments of evil, and some of the most humane acts of medicine are only small steps away from real evil" (Grodin 2010, p. 58).

The "inhumanity" of the anatomist stems from two sources: first, from the destructive practical method of "disassembling" a human body against all taboos, and second, from the scientific reductionist view of human life. This "inhumanity" or clinical detachment is sometimes interpreted as a distinct loss of empathy. However, in the ideal case, the anatomist and physician will learn to counteract a potentially overbearing clinical detachment by self-reflection and a balance with natural empathy in order to become a fully effective and humane practitioner (Swick 2006; Warner 2009; Böckers et al. 2010; Hildebrandt 2010). In the worst-case scenario, societal permission under a criminal regime like National Socialism can enhance this "inhumanity" and lead to medical atrocities.

New Transgressions Specific to the NS Period

Body Procurement Before 1933 and Changes After 1933

For many centuries, traditional sources for anatomical body procurement were regulated by specific laws and, in many countries, included so-called unclaimed bodies. These were bodies of persons whose relatives did not claim them for burial. They were delivered to the anatomical departments from communal hospitals, poor houses, psychiatric institutions and prisons, and included persons who committed suicide and those who were executed (for the history of anatomical body procurement see Richardson 1987; Sappol 2002; Stukenbrock 2001, 2003; Halperin 2007; Hildebrandt 2008). In cases where the body of an executed person was not directly delivered to an anatomical department, it was established practice in Germany in the early twentieth century that anatomists were present at executions, either in a special room close to the execution chamber or within the same, to remove tissues from the bodies of the executed as quickly as possible after death (Hildebrandt 2013d).

All of the traditional sources were still available after 1933; however, many of them changed considerably in character due to new laws implemented by the NS government (Hildebrandt 2013a). Among the deceased psychiatric patients were now patients who had been killed within the framework of the so-called "euthanasia"[2] program (Schönhagen 1992; Grundmann and Aumüller 1996; Aumüller and Grundmann 2002; Redies et al. 2005; Redies et al. 2012; Eckart 2012). In one unprecedented secret action, Curt Elze, chairman of the anatomical department at the University of Würzburg, accepted the bodies of 80 euthanasia victims offered and provided by Werner Heyde, one of the leaders of the euthanasia program (Blessing et al. 2012; Klee 2012). Among the prisoners were those who died in prison, persons killed by the Gestapo (*Geheime Staatspolizei*, Secret Service) elsewhere, inmates of the various new kinds of detention camps for forced laborers, and prisoners of war and of concentration camps. All 14 anatomical departments for which detailed documentation exists accepted bodies from these sources (Hildebrandt 2013a). Special NS laws denied families of Polish and Jewish prisoners the right to claim the bodies of their loved ones, and their bodies were often directly transported to the anatomical departments (Hildebrandt 2013a). Among the

---

[2] Whereas "euthanasia" usually refers to a voluntary death with dignity requested by patients or their surrogates at the end of life, in the context of National Socialism it meant the systematic selection and killing by the government of approximately 250,000 to 300,000 children and adult whose lives were considered unworthy of living.

suicides were increasing numbers of Jewish citizens, who sought this desperate solution to their persecution (Goeschel 2009; Kwiet and Eschwege 1984).

The most significant change for many anatomists came with the exponentially rising numbers of bodies of executed persons following new NS legislation concerning crimes punishable by death. While 184 men were executed during the Weimar Republic—women were not executed at all during this period—a minimum of 15,000 civilian and at least 15,000 military executions were carried out during the NS period, and these bodies were all potentially available to anatomical departments (Hildebrandt 2013b). Documentation now exists for the use of a minimum of 3,749 bodies of executed persons for anatomical purposes out of an estimated total number of 40,000–45,000 available bodies for anatomy from all sources between 1933 and 1945; the percentage of NS victims within this number is unclear at this point (Hildebrandt 2013a).

Reaction of Anatomists to "New Sources"

The history of anatomy is full of complaints by anatomists to government authorities concerning the lack of body supply and petitions for new legislation (Hildebrandt 2008). It was no different in the Third Reich, as demonstrated by the universities of Halle, Göttingen, and Cologne reminding the new government of the need for bodies for their anatomical departments in 1933, and by the petitions that followed from other anatomists (Noack and Heyll 2006). Anatomists were especially interested in the bodies of the executed because they were often young and healthy persons and, due to the planned time of death, tissues could be retrieved in a fresh state. This latter point was considered essential, as postmortem changes were common in certain delicate tissues, and the "freshness" of the "material" had become a gold standard for the quality of histological research as published in German-language anatomical journals (Hildebrandt 2013b). This histological quality standard had been established in Germany long before 1933 and existed even in times of scarcity of tissues from the executed during the Weimar Republic. After 1933, many execution sites allowed anatomists access to the bodies directly after execution in the execution facilities in cases where the families later claimed the bodies (Noack 2007). During the 1930s, the execution rates rose steadily and, during the war years, nearly exponentially as the government applied capital punishment to black marketers, looters, thieves, conscientious objectors, so-called "defeatists," and many others. Until then, only murderers, rapists, and traitors were subject to the death penalty (Hildebrandt 2013c).

What were the anatomists' reactions? Did they know whose bodies they were dissecting? Apart from the fact that, at least in Berlin, Hermann Stieve, chairman of anatomy at the Humboldt-University, was officially informed of the reasons for the death verdict, there are indications that many anatomists and medical students knew very well that they were dealing with the bodies of NS victims (Hildebrandt 2009b, 2013a, c). The anatomist and medical historian Robert Herrlinger remarked after the war that he had never questioned the righteousness of the NS legislation and believed he had been dealing with the bodies of common criminals. He also professed a distinct lack of curiosity as to the history of the person whose body he was dissecting; many of his colleagues shared this feeling with him (Hildebrandt 2013d). Anatomists who were also committed National Socialists had no reason to doubt the NS government and its legal decisions in the first place (Hildebrandt 2009a). But even if, for example, Ferdinand Wagenseil, head of the anatomical department at the University of Giessen and no friend of the regime, recognized and lamented the dismal fate of camp prisoners found dead by the side of a rail track, he still used the bodies of victims to fulfill his teaching obligations in the anatomical dissection course (Oehler-Klein et al. 2007). For many others, the increase in bodies from executed men, and now also women, opened up unforeseen new professional opportunities.

Anatomical Research with Bodies of the Executed: Escalation of Transgressions

A comparative study of German- and English-language anatomy journals revealed that between 1924 and 1933 only 1 percent of all publications in German-language anatomical journals referenced material from bodies of executed persons. However, this percentage rose to 2 percent from 1933 to 1938, to 7 percent from 1939 to 1945, and remained at 3 percent as late as 1951, six years after the end of the war. There was no similar development in English-language journals (Hildebrandt 2013b). Anatomists used the increased availability of the valuable "material" after 1933 for studies on tissues that were particularly affected by fast postmortem changes: Philipp Stöhr Jr., chairman of anatomy at the University of Bonn, studied autonomic ganglia and the adrenal gland; Wolfgang Bargmann investigated various tissues including the pituitary gland; Heinrich von Hayek explored the lungs; and Robert Herrlinger examined the spleen (Hildebrandt 2013b). These NS period papers often mentioned the source of their "material" explicitly as tissues from the executed, just as many anatomists, including Bargmann, had done so before 1933 (Bargmann 1931). However, small verbal additions seem to indicate a gradual transgression into new and ethically dubious territory. As early as 1935, Stöhr wrote that until then the known difficulties of conserving a greater number of intact adrenal glands had caused a lack of data concerning ganglion cells in this organ, and that he was now able to present results on 12 such organs from the bodies of executed persons (Stöhr 1935). Clearly, he seemed to imply, times had changed, and for the better. Hayek reported in his 1940 treatise on lung structure: "... of course, best suitable are lungs from younger executed persons, of which I had several at my disposal" (p. 405). He used this "ideal material" and data gained from it for many years to come (Hayek 1950; Hildebrandt 2013e). And, while in 1931 Bargmann had to collect tissue samples from six different departments all over

Germany to produce results on 14 executed persons, he was able to work with 10 pituitary glands from the executed in 1942 and 18 samples of intestinal tissues in 1944, all procured locally from the execution site in Königsberg (Bargmann 1931, 1942; Bargmann and Scheffler 1944). Indeed, the supply from bodies of the executed to the anatomical department at the University of Königsberg was so plentiful that at some point all storage spaces were filled to capacity, and only the bodies of women and young persons were accepted any longer (Maier 1973, p. 178). The situation was very similar at other anatomical departments, among them the University of Vienna (Mühlberger 1998).

Finally, there was a publication by Robert Herrlinger in 1947, which precipitated an unprecedented controversy about ethics in anatomy at the University of Würzburg ten years later. Herrlinger had written this paper in 1944 while still assistant to Hermann Voss, chairman of anatomy at the University of Posen/ Poznan. The German university at Posen was founded as a *Reichsuniversität* (university of the German Reich), intending it to be a "bulwark" against "foreign cultural influences" (Piotrowski 1984; Wróblewska 2000, p. 89; Baechler et al. 2005). Herrlinger's study itself was probably performed in 1943 and was only published much later because of the upheaval surrounding the end of the war. In the "Methods" section of his paper, Herrlinger included a precise description of the circumstances of recovery of blood and splenic tissues from eight executed men: "The investigations were carried out on eight male bodies, which were available for blood sampling and laparotomy within 40 to 80 seconds after death" (1947, p. 228). He continued by describing the contamination of the blood pulsing from the open carotids at the cut neck by foodstuffs from the likewise severed esophagus and described a simultaneous "careful" laparotomy. While in this postwar publication, the term *executed* was not used any longer, the details of the description left no doubt about the source of the "material" and Herrlinger's presence at the execution site. While Herrlinger's behavior and the ethics of anatomical body procurement during the Third Reich were extensively discussed in the Herrlinger controversy from 1957 to 1959, this early debate on the subject remained intramural at Würzburg University and was essentially kept secret (Hildebrandt 2013d).

In many ways, Stöhr, Bargmann, Hayek, and Herrlinger were following standard anatomical procedure with a motivation for the highest quality of work. However, as the examples show, a new quality, which might be called "increased clinical detachment," had crept into their work. Hermann Stieve was an expert of such "increased detachment" and the reductionist-mechanistic view of human life. Already in the 1920s, he had realized that his animal model experiment on the influences of chronic and acute stress on reproductive organs was perfectly reflected in the setting of human beings on death row. Chronic stress, in his interpretation, was represented by the long time of imprisonment, while acute stress was elicited through the announcement of the time of death. His work on the reproductive organs of executed men during the 1920s confirmed his results from the animal experiments (Winkelmann and Schagen 2009). However, women were not exe-cuted during the Weimar Republic, so Stieve seized the opportunities offered by the

NS regime, which had no qualms about the execution of women, not even pregnant ones. Stieve himself described his body supply as, "Material, the likes of which no other institute in the world possesses" (Schagen 2005, p. 45).[3] Charlotte Jünemann, a pregnant 24-year-old who was executed for manslaughter at the Berlin execution site in Plötzensee on August 26, 1935, was probably the first woman executed in the Third Reich. She became number one on a list of 174 women and 8 men whose names Stieve had written down after the war as persons whose bodies he had used for his research. Among these executed women were German, Polish, French, and Czech dissidents, forced laborers, Jehovah's witnesses, petty thieves, looters, and many others (Hildebrandt 2013c). There is clear evidence that Stieve was well aware of the so-called "crimes" of the persons he was dissecting including their identities, even though he would deny much of this after the war (Stieve 1952). He felt entirely justified in his work, as he declared in 1946:

> "The anatomist…only tries to retrieve results from those incidents [executions] that belong to the saddest experiences known in the history of Mankind. I have no reason to be ashamed of the fact that I was able to elucidate new data from the bodies of the executed; facts that were unknown before and are now recognized by the whole world." (as quoted in Schagen 2005, p. 49)

Interestingly enough, Stieve's case is one of only a few in which anatomists were asked to justify their work directly after the war. The only other known cases are Rudolf Spanner, Johann Paul Kremer, and August Hirt.

Spanner was chairman of anatomy at the University of Danzig and belonged to the anatomists who lobbied actively for the delivery of bodies of the executed to their institute, in his case by suggesting the building of an execution site in Danzig (Bussche 1989). He left Danzig precipitously in January 1945 before its liberation by the Allied Forces (Papers of Alfred Benninghoff, 1920–1950, letter from Eduard von Bargen to Alfred Benninghoff). On inspection by the new Polish authorities in March 1945, there were not only human bodies and body parts remaining at the institute but also some soap made from human fat (Neander 2006). This last discovery gave rise to a rumor of the manufacturing of soap from Jewish bodies in concentration camps, and entered the "collective memory of the Polish people" (Neander 2006, p. 65). During several postwar interrogations, Spanner conceded the production of small amounts of soap for the treatment of anatomical joint preparations, but he was not prosecuted. While Spanner's methods may have been within the realm of traditional anatomy, the general public in Poland interpreted his use of soap made from the fat of victims as an ethical, if not a criminal, failing when viewed in the context of the newly discovered NS atrocities.

---

[3] Unless otherwise noted, all translations are the author's.

Anatomical Research on the "Future Dead": Final Stages of Transgression

The traditional paradigm of anatomical work is that anatomists worked with dead human bodies, with the exception of studies of surface and functional anatomy with living human voluntary models. Experimentation was strictly postmortem. In most cases anatomists removed unaltered tissues, sometimes injected organs before removal, and at most, posed the joints or the whole body for questions of body mechanics (e.g., Hayek 1935; Hewel 1941). Anatomists were rarely interested in the history of the persons whose bodies they were investigating (Hildebrandt 2013d). Should they want any premortem information, they received it from relatives or medical records of the dead person. There is no documentation that even Stieve, who needed information on the medical and social history of the persons he dissected for his research, visited the prisoners before death. Instead, he used information garnered from prison records and prison personnel (Winkelmann and Schagen 2009).

Max Clara, chairman of anatomy at the University of Leipzig, and his group of colleagues there were among the most prolific users of bodies of the executed for research (Winkelmann and Noack 2010; Hildebrandt 2013b). They published 38 papers on "material" from unprecedented numbers of executed persons; in one case, livers from 38 persons were studied (Schiller 1942). These investigations still followed the traditional anatomical method of working with human bodies after death, however else they stretched ethical boundaries in terms of the sheer horror of working, for example, with organs from 38 executed persons at once. In 1942, Clara crossed this essential boundary of anatomical work by publishing for the first time the results of a medical experiment performed by him on a still living prisoner before execution (Clara 1942; Winkelmann and Noack 2010). Clara had been interested in the distribution of vitamin C throughout the human body and described in this paper the vitamin C content of nervous tissue in 15 "apparently healthy adult individuals of various ages, who without exception died suddenly after varying times of imprisonment; one 33¾-year-old male individual had received one tablet of Cebion (Merck) [vitamin C] four times daily for the last five days before his death" (1942, p. 362). Clara, who often contacted the authorities with his professional needs (Noack 2012), had apparently gained access to the prisoners on death row and received permission to have vitamin C tablets administered to this prisoner. While the tablet in itself was harmless, the coercive nature of this experiment on a death row prisoner has to be emphasized. Clara's experiment signifies the decisive transgression from anatomists' work with the dead to their work with the still living but future dead. The prisoner's death was part of Clara's anatomical research plan and, as such, resembles the medical experiments performed by the psychiatrist Carl Schneider on children within the euthanasia program in Heidelberg (Roelcke et al. 2001).

It seems that the war years promoted a radicalization among some anatomists in terms of their readiness to cross the traditional boundary between the land of the living and the land of the dead, as the example of Johann Paul Kremer demonstrates. Kremer was a senior anatomist and SS officer (*Sturmstaffel*, NS elite corps) at the University of Münster from 1927 to 1945, who had worked on the effects of hunger in amphibians (Kremer 1930, 1932, 1933, 1937, 1938a, b, 1939, 1941/42; 1942/43). A 1941 manuscript purportedly describing the inheritance of traumatic injuries (Kremer, 1942) brought him not only in conflict with his peers in anatomy, who already had doubts about his scientific work (Papers of Alfred Benninghoff, 31 January 1935, letter from Benninghoff to Geheimrat Schwoerer), but also with the NS party and the Gestapo because news of the possible heritability of war injuries was highly inopportune for the government (*Landgericht Münster* 1960).

In his function as SS officer, Kremer had been detailed to the SS Main Sanitation Office in Berlin and the Waffen-SS front-line unit hospitals in Prague and Dachau before he was ordered to work in Auschwitz starting in August 1942 (Strzelecka 2011). In Dachau Kremer had only been responsible for German personnel and had no contact with prisoners, but he may have been aware of medical experiments performed on them by other SS physicians (Mitscherlich and Mielke 1960/1997). In Auschwitz he had direct access to prisoners through the selection procedures of sick prisoners on the hospital wards, which were part of his duty (*Landgericht Münster* 1960). According to Kremer's recollections in his diary, after his arrival in the camp, he seized the opportunity to follow-up on his animal experiments on hunger with studies on human beings (Höss et al. 1984). He presented this idea to his chief at Auschwitz, Dr. Eduard Wirths, who, in Kremer's words, "told me that I could collect fresh living material (*lebensfrisches Material*) from prisoners who were put to death by phenol injection" (quoted in Strzelecka 2011, p. 161). He then set about to observe prisoners "and when one of them interested me because of a highly advanced state of starvation, I commanded the orderly to reserve that patient for me, and inform me of when that patient would be killed with the help of an injection" (p. 161). The prisoners were photographed before their murder, at which Kremer was often present. He then collected tissue samples of spleen, liver, and pancreas; put the tissues in preservative fluids; and took the photographs and samples back to Münster at the end of his Auschwitz assignment in November 1942 (Strzelecka 2011, pp. 161, 162; see also Höss et al. 1984). It is not known whether he performed any research on these tissues, as no publication of results exists. After the war, during his trial in Poland, Kremer reported that he did not know where the photographs and tissue samples had ended up (Strzelecka 2011, p. 162).

In 1947 Kremer was sentenced to death for his part in the selections at the train ramp in Auschwitz and on the hospital wards. He had sent at least 10,717 men, women, and children directly to their death (Rawicz 1972). In 1948 the verdict was converted into a life sentence, and in 1958 Kremer was extradited from Poland to Germany. There he was again put on trial, received a verdict of ten years in prison for aiding and abetting murder, but was set free in recognition of his prison time in Poland (*Landgericht Münster* 1960). He died in January 1965 in Münster (Klee 2003).

While Kremer's activities in Auschwitz were recognized as war crimes, his research there also signifies a distinct shift in the anatomical paradigm that had its equal only in the offenses against humanity committed by August Hirt (Lang 2013) and to a certain extent in Max Clara's experiment. Kremer was able to advance his studies on the morphological effects of hunger from the animal model to the human "system" due to the availability of starving prisoners at Auschwitz. There is no documentation that he had volunteered for or somehow influenced his deployment there; however, he realized the "scientific potential" of his new workplace soon after his arrival. As his diaries reveal, he had no ethical compunction about choosing and using victims of the NS system for his work (Höss et al. 1984). The inhumanity of Auschwitz was reflected and multiplied in the inhumanity of this anatomist and his coercive medical experiments. The scientifically mediocre anatomist Kremer was easily able to follow the scientific method within the settings of the criminal NS justice system; and, by disregarding any ethical considerations, he transgressed the traditional boundaries of anatomical work to follow the new and monstrous concept of working with the future dead.

However offensive Kremer's activities were, there is no evidence that he had a concrete research plan when he arrived at Auschwitz. The true master and theoretician of anatomical work with the future dead was the academically successful and politically active August Hirt, chairman of anatomy at the University of Strasbourg and an SS officer. Early on he realized the potential new opportunities for research offered by the murderous reality of the NS regime (Lang 2013). In the mid- to late 1930s, Hirt had changed the direction of his scientific investigations from questions of basic anatomy and microscopic technology to several new fields, among them the exploration of the effect of warfare agents on the human body and questions of "racial" anthropology. The *SS-Ahnenerbe*, Heinrich Himmler's organization for research on Germanic heritage and supremacy, supported his activities financially as well as administratively (Pringle 2006).

In 1942 Hirt performed coercive medical experiments with the warfare agent mustard gas on prisoners in the concentration camp Struthof/Natzweiler, which led to the death of many prisoners (Mitscherlich and Mielke 1960/1997, p. 219). This research was the continuation of work on animal models Hirt had performed in 1941 (Kasten 1991, p. 190). While the mustard gas experiments were outside the realm of traditional anatomical work, his plans for a so-called "racial" skeleton collection belonged in the anatomical category of physical anthropology. Hirt saw such a collection as an extension of the famous craniological collection created by Gustav Schwalbe, which was housed at the anatomical department in Strasbourg (Lang 2007).

In June 1943, at Hirt's behest, anthropologist Bruno Beger, a member of Himmler's personal staff, selected 86 Jewish prisoners, from Auschwitz. The prisoners were transported by train to Struthof/Natzweiler and murdered there in a gas chamber in August 1943. Their bodies were finally transferred to the anatomical department at the University of Strasbourg (Pringle 2006; Lang 2007, 2013). Hirt himself had handed the Cyclon B gas capsules for the murder of these prisoners to Josef Kramer, the director of the concentration camp (Lachman 1977). However, Hirt clearly never produced the skeleton collection because the bodies and body parts of these victims were still in storage at the anatomical department when the

Allied Forces liberated Strasbourg in November 1944. August Hirt was the only anatomist whose work was referred to at the Nuremberg Doctors' Trial in 1947 (Mitscherlich and Mielke 1960/1997). The prosecutor did not know at the time that Hirt had committed suicide in June 1945 (Kasten 1991, p. 195). A court in Metz, also apparently unaware of Hirt's suicide, sentenced him to death in absentia in 1953 (Lang 2007, p. 212).

While Hirt's non-anatomical medical experiments alone qualified him as a war criminal, his ideas for the future of anatomy and his first experimental steps in this direction mark him as a truly innovative and most dangerous thinker. Initial proposals for the "racial" skeleton or skull collection went back to November 1941 (Kasten 1991, p. 183), and in early 1942, he formulated his revolutionary idea of creating a skeleton collection on the basis of the selection of still living but "future-dead" persons. These plans, which became known as "Auftrag Beger" (Commission Beger), included a "special deputy, commissioned with the collection of the material" and the instructions that, "Following the subsequently induced death of the Jew whose head must not be damaged, he [the deputy] will separate the head from the torso" (Hirt to Himmler, February 9, 1942, quoted in Kasten 1991, p. 184). And in 1942, while Hirt was still waiting for the implementation of this plan, he had a greater vision for all of German anatomy in his sight, which was apparently discussed at a meeting of German anatomists in Tübingen from November 5 to 7, 1942. In a report about the gathering to the *SS-Ahnenerbe*, Hirt wrote: "There [at the meeting] a proposal was made that anatomists should collect and process materials as already specified in 'Auftrag Beger.' Others are gradually becoming aware that something could happen here. . ." He added in handwriting: "I have been commissioned to compose guidelines for the collection of materials for all German anatomists" (quoted in Lang 2013, p. 374). The formal meetings of the Anatomical Society (*Anatomische Gesellschaft*), the official body of anatomists in German-speaking countries with an international membership, were suspended during the war, so Wetzel's meeting was organized outside the normal curriculum of the society. Indeed, very little is known about this gathering in Tübingen, which was hosted by the active National Socialist Robert Wetzel, chairman of anatomy at the University of Tübingen, in his role as leader of the Reich Lecturers' Association. According to Wetzel, 43 anatomists accepted his invitation and discussed both the new opportunities presented for anatomical material and the founding of an exclusively German anatomical association as opposed to the traditional international Anatomical Society (Hildebrandt 2013f). It has to be assumed that the anatomists who assembled there were, in general, accepting of the NS regime and certainly of the new professional opportunities offered by it because Wetzel and all the persons whose names he mentioned in correspondence concerning this event were part of an NS friendly group within the Anatomical Society (Hildebrandt 2013f). Therefore, it seems that at least a certain subsection of German anatomists did, indeed, discuss Hirt's new ideas of body procurement of the future dead in an open conversation and even commissioned him to create guidelines for such research with the future dead. The future dead became a potential research "opportunity" to be exploited by any anatomist willing to commit this final transgression. Hans-Joachim Lang was the first to point out this decisive paradigm shift in German anatomy brought about by Hirt's ideas from 1942 (Lang 2013, p. 373). It is probably no coincidence that Clara, Kremer, and Hirt committed this final transgression in anatomy in the middle of the war, when German society had become imbued with inhumanity and brutality ranging from NS governmental violence to the carnage of a multi-front war and the devastation of bombed cities.

It is possible that the turbulences of the later stages of war prevented the further use of "extraordinary opportunities for cadaver delivery" (Lang 2013, p. 373) after 1943. Also, there is documentation that at least two anatomists prevented the killing of victims on site in their institutes. In the first case at the University of Giessen in 1942, an unsuccessfully executed Polish forced laborer was still alive at delivery by the Gestapo. The anatomist Ernst von Herrath insisted on having the prisoner transported to a hospital, but the officers refused to do so. They removed the prisoner from the institute and proceeded to kill him shortly thereafter in a different location (Oehler-Klein et al. 2012). In 1944 a similar incident occurred at the anatomical department of the University of Erlangen. The anatomist Johannes Hett had previously complained about delayed transports of bodies of executed persons from the execution site to the institute and had petitioned for faster transfer times. In March 1944, he received a delivery of four bodies of executed victims and a still living Polish prisoner, whom his guards proposed to kill inside the anatomical facilities to ensure freshness of the "material." Hett is said to have indignantly denied such a proceeding. The victim was, therefore, taken away from the anatomical department, killed in a nearby wood, and his body then returned to the institute (Wendehorst 1993, pp. 237–238). Herrath's and Hett's examples demonstrate a certain reticence of these anatomists against crossing from the land of the dead to the land of the living. While they readily worked with bodies of the executed, they were not willing to let the killing spread from execution sites to anatomical departments. Hett's and Herrath's behavior may be interpreted as a refusal to have dealings with the future dead within the realm of anatomy. Finally, it can be said that by the end of the war there were some German anatomists who were ready to transgress the borders of the anatomical paradigm, while others remained within the traditional boundaries of anatomy.

Conclusion: Stages of Transgression

Acknowledging the initial transgression in anatomy, which is the societally accepted violation of the dead human body for scientific reasons, the following gradually worsening stages of transgression specific to anatomists in National Socialist Germany can be identified:

- Victims of the NS regime became part of all traditional sources of body procurement in anatomy and were used for purposes of anatomical education and research by all German anatomical departments.

- Many anatomists used the "opportunities" given by increased numbers of executed victims for research and presented them as a positive development.
- Just as with his research on the reproductive organs of men on death row in the 1920s, Stieve recognized in the situation of women sentenced to death in NS Germany a reflection of his previous strictly planned research designs in animal experiments. He exploited the plight of great numbers of NS victims for his scientific purposes. However, there is no documentation that he—like, for example, his colleague Clara—ever approached a still living prisoner before death or caused any prisoner's death to further his own research interests like Kremer or Hirth.
- Clara used a still living prisoner, whose execution date was already determined and who was thus a future-dead person, for a planned experiment.
- Kremer selected living prisoners for his experiments without prior planning for it and created future-dead persons by this selection because their death, even in Auschwitz, might not have been inevitable.
- Hirt had living prisoners selected according to a strict research plan, thus creating future dead by causing their death.

None of this research was "pseudoscientific," an epithet often used in the past by those physicians (and anatomists) who would have liked to distance themselves from NS medical crimes by claiming that these offenses had been committed by only a few criminal dilettantes, and thus, they themselves were not responsible (Schleiermacher and Schagen 2012). The research by the anatomists presented in this chapter, even in its various stages of transgression, was still based on the traditional anatomical scientific method and cannot be dismissed as scientifically irrelevant, certainly not at the time of its publication. The histological publications by Bargmann, Hayek, and Clara were typical for their field; Stieve was integrating innovative ideas of functional morphology in his work. Even Hirt's plan for a "racial" skull collection still fit within the realm of physical anthropology in the 1930s. However, much of this anatomical research had become an "experimental medicine in epistemologically conclusive form ... [which] had lost its moral boundaries" (Roelcke 2009, p. 46). While, for example, the transition from animal to human experiment in Stieve's anatomical research was scientifically justified, he never considered possible ethical shortcomings in terms of his responsibility for the prisoners in his research design. In that respect, he was very much a scientist of his time, even before 1933 in his work with tissues from executed men. Roelcke called this phenomenon a "methodological carefulness" that was "care-less" (*entsorgt*)[4] (2009, p. 47) of any ethical implications and neglected "moral care" (2009, p. 43), thereby pointing out the scientists' utmost concern for correct scientific methodology in combination with a complete lack of care for the human "objects" of their

---

[4] The verb *entsorgen* has a double meaning as used by Roelcke: The usual translation is "to dispose of." However, in the literal sense, it means *ent-sorgen*, "to rid oneself of care for someone or something." The translation as "care-less" is meant to emphasize the attitude of the anatomists that excluded care for the victims.

studies. The same "care-less" attitude towards their victims can be observed in Kremer's and Hirt's transition from animal experiments to coercive human experiments.

The phenomenon itself was not specific to either the NS period or anatomists alone. However, the murderous nature of the NS regime and the specific suscepti-bility of anatomists complemented each other. Anatomists, in need of societal permission for their work through legislative authority, were officially and legally given the opportunity to work not only with bodies procured through traditional sources but also with bodies of all those persons who were eliminated from society according to the biologistic and racist ideology of the regime. In addition, they were given official permission by authorities like Himmler to work with the future dead, and some anatomists did so without any empathy. In this situation, the anatomists became morally complicit with the NS regime in their use of material gained from wrongfully executed victims because, according to Jones, there exists an "indissol-uble ethical link" between the "origin of the material" and "what later can be done with it" (Jones 2007, p. 339; also on the subject of complicity, Miller 2012). At the same time, anatomists were familiar with the process of clinical detachment through their work so that, for many of them, a "suspension of empathy" towards the persons whose bodies they dissected, may not have been hard to come by, as Herrlinger's example showed. Indeed, anatomists reported after the war that they had not cared that some of the persons, whose bodies they dissected, were executed victims (Aharinejad and Carmichael 2013). This example of "care-lessness" illus-trates again the phenomenon of "methodological care versus moral care" (Roelcke 2009, p. 43).

Anatomy's development during the National Socialist period may be interpreted as an extreme manifestation of the "experimentalization of biology and medicine in the nineteenth and twentieth century," when a "categorical separation and discon-nected perception of science, ethics and politics in the modern biosciences devel-oped" (Roelcke 2009, p. 16, p. 47). The specific conditions of the NS regime unleashed the "aggressive and destructive potential [of a] medical science that is primarily focused on the gain of scientific knowledge and blanks out the humanity . . . of the research subject" (Roelcke 2012, p. 101). In 1947, the psychi-atrist Alice von Platen-Hallermund, an observer at the Nuremberg Doctors' Trial, noticed this phenomenon and remarked that atrocities are possible when scientists lose their focus on the well-being of individual patients and instead allow the scientific question to dominate their thinking (Schleiermacher and Schagen 2012, p. 174).

While National Socialism led to specific forms of transgressions within anat-omy, the underlying forces behind these transgressions are not specific to the Third Reich but are still active in modern anatomy and medicine. While the scientific method has to remain central to new medical insights, medicine's focus, and that includes anatomy, has to reintegrate scientific inquiry with the ethical and political dimensions of human nature. Modern anatomy can be a professional model for medical education as the future physician learns to acknowledge the development of clinical detachment and to balance it with empathy (Hildebrandt 2010). Thus, to paraphrase Pat Barker, a "suspension of empathy," if kept in balance, may not have to lead to "monstrosity" any longer (Barker 1995, p. 165).

Acknowledgement The author would like to thank William Seidelman for a critical discussion of the manuscript.

References

Aharinejad, S.H., and S.W. Carmichael. 2013. First hand accounts of events in the laboratory of Prof. Eduard Pernkopf. *Clinical Anatomy* 26(3): 297–303.

Aumüller, G., and K. Grundmann. 2002. Anatomy during the Third Reich. The Institute of Anatomy at the University of Marburg, as an example. *Annals of Anatomy* 184: 295–303.

Baechler, C., F. Igersheim, and R. Racine. 2005. *Les Reichsuniversitaeten des Strasbourg et de Poznan et les Resistances Universitaires 1941–1944.* Strasbourg: Presse Universitaires de Strasbourg.

Bargmann, W. 1931. Über Struktur und Speicherungsvermögen des Nierenglomerulus. *Zeitschrift für Zellforschung und Mikroskopische Anatomie* 14: 73–137.

———. 1942. Über Kernsekretion in der Neurohypophyse des Menschen. *Zeitschrift für Zellforschung und Mikroskopische Anatomie* 32(3): 394–400.

Bargmann, W., and A. Scheffler. 1944. Über den Saum des menschlichen Darmepithels. *Zeitschrift für Zellforschung und Mikroskopische Anatomie* 33(1–2): 5–13.

Barker, P. 1995. *The eye in the door.* London: Plume-Penguin.

Bergmann, A. 2008. Taboo transgressions in transplantation medicine. *Journal of American Physicians and Surgeons* 13(2): 52–55.

Blessing, T., A. Wegener, H. Koepsell, and M. Stolberg. 2012. The Würzburg anatomical institute and its supply of corpses (1933–1945). *Annals of Anatomy* 194: 281–285.

Böckers, A., L. Jerg-Bretzke, C. Lamp, A. Brinkmann, H.C. Traue, and T.M. Böckers. 2010. The gross anatomy course: An analysis of its importance. *Anatomical Sciences Education* 3: 3–11.

Clara, M. 1942. Beiträge zur Histotopochemie des Vitamin C im Nervensystem des Menschen. *Zeitschrift für Mikroskopisch-Anatomische Forschung* 52: 359–392.

Eckart, W.U. 2012. *Medizin in der NS-Diktatur. Ideologie, Praxis, Folgen.* Cologne: Böhlau.

Garment, A., S. Lederer, N. Rogers, and L. Boult. 2007. Let the dead teach the living: The rise of body bequeathal in 20th-century America. *Academic Medicine* 82: 1000–1005.

Goeschel, C. 2009. *Suicide in Nazi Germany.* Oxford: Oxford University Press.

Grodin, M.A. 2010. Mad, bad, or evil: How physician healers turn to torture and murder. In *Medicine after the Holocaust. From the master race to the human genome and beyond,* ed. S. Rubenfeld, 49–65. New York: Palgrave MacMillan.

Grundmann, K., and G. Aumüller. 1996. Anatomen in der NS-Zeit: Parteigenossen oder Karteigenossen? Das Marburger anatomische Institut im Dritten Reich. *Medizinhistorisches Journal* 31(3–4): 322–357.

Halperin, E.C. 2007. The poor, the Black, and the marginalized as the source of cadavers in the United States anatomical education. *Clinical Anatomy* 20: 489–495.

Hayek, H. 1935. Das Verhalten der Arterien bei Beugung der Gelenke. *Zeitschrift für Anatomie und Entwicklungsgeschichte* 105(1): 25–36.

———. 1940. Die Läppchen und Septa interlobaria der menschlichen Lunge. *Zeitschrift für Anatomie und Entwicklungsgeschichte* 110(3): 405–411.

———. 1950. Die Muskulatur im Lungenparenchym des Menschen. *Zeitschrift für Anatomie und Entwicklungsgeschichte* 115(1): 88–94.

Herrlinger, R. 1947. Das Blut in der Milzvene des Menschen. *Anatomischer Anzeiger* 96: 226–235.

Hewel, J. 1941. Über die Beweglichkeit der menschlichen Leber. *Anatomischer Anzeiger* 90 (22/24): 273–296.

Hildebrandt, S. 2008. Capital punishment and anatomy: History and ethics of an ongoing association. *Clinical Anatomy* 21: 5–14.

———. 2009a. Anatomy in the Third Reich: An outline, part 1. National Socialist politics, anatomical institutions, and anatomists. *Clinical Anatomy* 22: 883–893.

———. 2009b. Anatomy in the Third Reich: An outline, part 2. Bodies for anatomy and related medical disciplines. *Clinical Anatomy* 22: 894–905.

———. 2009c. Anatomy in the Third Reich: An outline, part 3. The science and ethics of anatomy in National Socialist Germany and postwar consequences. *Clinical Anatomy* 22: 906–915.

———. 2010. Developing empathy and clinical detachment during the dissection course in gross anatomy. *Anatomical Sciences Education* 3: 216.

———. 2013a. Current status of identification of victims of the National Socialist regime whose bodies were used for anatomical purposes. *Clinical Anatomy*, published online September 2013. doi: 10.1002/ca.22305

———. 2013b. Research on bodies of the executed in German anatomy: An accepted method that changed during the Third Reich. Study of anatomical journals from 1924 to 1951. *Clinical Anatomy* 26(3): 304–326.

———. 2013c. The women on Stieve's list: Victims of National Socialism whose bodies were used for anatomical research. *Clinical Anatomy* 26: 3–21.

———. 2013d. The case of Robert Herrlinger: A unique postwar controversy on the ethics of the anatomical use of bodies of the executed during National Socialism. *Annals of Anatomy* 195: 11–24.

———. 2013e. Wolfgang Bargmann (1906–1978) and Heinrich von Hayek (1900–1969): Careers in anatomy continuing through German National Socialism to postwar leadership, *Annals of Anatomy*, published online 4/23/13, http://dx.doi.org/doi:10.1016/j.aanat.2013.04.003.

———. 2013f. Anatomische Gesellschaft from 1933 to 1950: A professional society under political strain – The Benninghoff papers. *Annals of Anatomy*, published online June 5, 2013. http://dx.doi.org/doi:10.1016/j.aanat.2013.05.001.

Höss, R., P. Broad, and J.P. Kremer. 1984. *KL Auschwitz seen by the SS. (Selection, elaboration, and notes by Bezwinska, J. & Czech, D.).* New York: Howard Fertig.

Jones, D.G. 1998. Anatomy and ethics: an exploration of some ethical dimensions of contemporary anatomy. *Clinical Anatomy* 11: 100–105.

———. 2007. Anatomical investigations and their ethical dilemmas. *Clinical Anatomy* 20: 338–343.

Kasten, F.H. 1991. Unethical Nazi medicine in annexed Alsace-Lorraine: The strange case of Nazi anatomist Professor Dr. August Hirt. Historians and archivists. In *Essays in modern German history and archival policy*, ed. G.O. Kent, 173–208. Fairfax, VA: George Mason University Press.

Klee, E. 2003. *Das Personenlexikon zum Dritten Reich. Wer war was vor und nach 1945.* Frankfurt: S. Fischer Verlag GmbH.

———. 2012. *Was sie taten- was sie wurden. Ärzte, Juristen und andere Beteiligte am Kranken- oder Judenmord.* Frankfurt: Fischer Taschenbuch Verlag.

Kremer, J. 1930. Die histologischen Veränderungen der quergestreiften Muskulatur der Amphibien im Hungerzustande. *Zeitschrift für mikroskopisch-anatomische Forschung* 21: 184–349.

———. 1932. Die fortlaufendenVeränderungen der Amphibienleber im Hungerzustande. *Zeitschrift für mikroskopisch-anatomische Forschung* 28: 81–156.

———. 1933. Die morphologische Gestaltung der Gallensekretion. *Zeitschrift für mikroskopisch-anatomische Forschung* 33: 486–524.

———. 1937. Zur Frage der Lokalisation des Hungerpigmentes in der Kaltblüterleber. *Anatomischer Anzeiger* 83: 316–330.

———. 1938a. Das Problem der Pigmentablagerung in der Leber und Milz der Kaltblüter und seine Beziehungen zur Frage des Blutabbaues und Eisenstoffwechsels. *Zeitschrift für mikroskopisch-anatomische Forschung* 44: 234–323.

———. 1938b. Über das Wesen und die Beduetung der pugmentierten Zellen in der Leber hungernder Kaltblüter. *Anatomischer Anzeiger* 85: 310–312.

———. 1939. Die zentrale Bedeutung der Leber im Pigment- und Eisenstoffwechsel der Reptilien. *Anatomischer Anzeiger* 88: 119–129.

———. 1941/42. Neue Fundamente der Zellen- und Gewebeforschung. Mikrokosmos, 51–52.

———. 1942. Ein bemerkenswerter Beitrag zur Frage der Vererbung traumatischer Verstümmelungen. *Zeitschrift für menschliche Vererbungs- und Konstitutionslehre* 25: 535–570.

———. 1942/43. Das Wesen und die Herkunft der mit der Zerstörung roter Blutkörperchen in Verbindung gebrachten eisenpigmenthaltigen. *Zellen der Milz. Mikrokosmos-Jahrbuch*, 36 (6/7): 77–80

Kwiet, K., and H. Eschwege. 1984. *Selbstbehauptung und Widerstand. Deutsche Juden im Kampf um Existenz und Menschenwürde 1933–1945*. Hamburg: Christians.

Lachman, E. 1977. Anatomist of infamy: August Hirt. *Bulletin of the History of Medicine* 51: 594–602.

Landgericht Münster. 1960. Das Urteil gegen Dr. Johann Paul Kremer. In *Justiz und NS-Verbrechen*, Band XVII. http://www.expostfacto.nl/junsvpdf/JuNSV500.pdf. Accessed 3 Dec 2013.

Lang, H.J. 2007. *Die Namen der Nummern. Wie es gelang, die 86 Opfer eines NS-Verbrechens zu identifizieren. Überarbeitete Ausgabe*. Frankfurt: S. Fischer Verlag.

———. 2013. August Hirt and "extraordinary opportunities for cadaver delivery" to anatomical institutes in National Socialism: A murderous change in paradigm. *Annals of Anatomy* 195: 373–380.

Maier, A. 1973. Erlebnisse eines Gefängnisseelsorgers. Königsberg 1939–1945. In *Unser Ermlandbuch 1973*, ed. Bischof-Maximilian-Kaller-Stiftung, 169–187. Osnabrück: A Fromm KG.

Miller, F.G. 2012. Research and complicity: the case of Julius Hallervorden. *Journal of Medical Ethics* 38: 53–56.

Mitscherlich, A., and F. Mielke. 1997. *Medizin ohne Menschlichkeit. Dokumente des Nürnberger Ärzteprozesses (First edition 1960)*. Frankfurt: Fischer Taschenbuch Verlag.

Mühlberger, K. 1998. II. Die Belieferung des anatomischen Instituts der Universität Wien mit Studienleichen in der Zeit von 1938–1946. In *Senatsprojekt der Universität Wien: Untersuchungen zur anatomischen Wissenschaft in Wien 1938–1945*, ed. Akademischer Senat der Universität Wien, 29–66. Unpublished report.

Neander, J. 2006. The Danzig soap case: Facts and legends around "Professor Spanner" and the Danzig Anatomic Institute 1944–1945. *German Studies Review* 29(1): 63–86.

Noack, T. 2007. Begehrte Leichen. Der Berliner Anatom Hermann Stieve (1886-1952) und die medizinische Verwertung Hingerichteter im Natinoalsozialismus. *Medizin, Gesellschaft und Geschichte: Jahrbuch des Robert-Bosch-Institutes* 26: 9–35.

———. 2012. Anatomical departments in Bavaria and the corpes of executed victims of National Socialism. *Annals of Anatomy* 194: 286–292.

Noack, T., and U. Heyll. 2006. Der Streit der Fakultäten. Die medizinische Verwertung der Leichen Hingerichteter im Nationalsozialismus. In *Geschichte der Medizin-Geschichte in der Medizin*, ed. J. Vögele, H. Fangerau, and T. Noack, 133–142. Hamburg: Literatur Verlag.

Oehler-Klein, S. (ed.). 2007. *Die Medizinische Fakultät der Universität Giessen im Nationalsozialismus und in der Nachkriegszeit: Personen und Institutionen, Umbrüche und Kontinuitäten*. Stuttgart: Franz Steiner Verlag.

Oehler-Klein, S., D. Preuss, and V. Roelcke. 2012. The use of executed Nazi victims in anatomy: Findings from the Institute of Anatomy at Giessen University, pre- and post-1945. *Annals of Anatomy* 194: 293–297.

Papers of Alfred Benninghoff. 1920–1950. Estate of Alfred Benninghoff. Universitätsarchiv Marburg, Philipps-Universität. Marburg, Germany.

Piotrowski, B. 1984. Die Rolle der "Reichsuniversitäten" in der Politik und Wissenschaft des hitlerfaschistischen Deutschlands. Materials of the International Symposium on Universities during World War II held at the Jagiellonian University on the 40th anniversary of "Sonderaktion Krakau," Crakow, Poland, October 22–24, 1979. In *Zeszytynaukowe Uniwersytetu Jagiellonskiego. Prace historyczne . Universitas Jagellonica Cracoviensis Acta scientiarum litterarumque. Schedae historicae DCXLIII (72)*, 66–486.

Pringle, H. 2006. *The master plan. Hitler's scholars and the Holocaust*. New York: Hyperion.

Rawicz, J. 1972. Foreword. In *KL Auschwitz seen by the SS*, ed. K. Smolén, J. Bezwińska, D. Czech, T. Iwaszko, and I. Polska, 5–32. Oświęcimiu: Państwowe Muzeum w Oświęcimiu.

Redies, C., R. Fröber, M. Viebig, and S. Zimmermann. 2012. Dead bodies for the anatomical institute in the Third Reich: An investigation at the University of Jena. *Annals of Anatomy* 194 (3): 298–303.

Redies, C., M. Viebig, S. Zimmermann, and R. Fröber. 2005. Origin of the corpses received by the anatomical institute at the University of Jena during the Nazi regime. *The Anatomical Record (Part B: The New Anatomist)* 285(1): 6–10.

Richardson, R. 1987. *Death, dissection and the destitute*, 2nd ed. Chicago: The University of Chicago Press.

Roelcke, V. 2009. Tiermodell und Menschenbild. Konfigurationen der epistemiologischen und ethischen Mensch-Tier-Grenzziehung in der Humanmedizin zwischen 1880 und 1945. In *Kulturgeschichte des Menschenversuchs im 20. Jahrhundert*, ed. B. Griesecke, M. Krause, N. Pethes, and K. Sabisch, 16–47. Frankfurt: Suhrkamp.

———. 2012. Fortschritt ohne Rücksicht. Menschen als Versuchkaninchen bei den Sulfonamid-Experimenten im Konzentrationslager Ravensbrück. In *Geschlecht und "Rasse" in der NS-Medizin*, ed. I. Eschebach and A. Ley, 101–114. Berlin: Metropol Verlag.

Roelcke, V., G. Hohendorf, and M. Rotzoll. 2001. Psychiatric research and "euthanasia"; The case of the psychiatric department at the University of Heidelberg, 1941–1945. *Psychoanalytic Review* 88(2): 275–294.

Sappol, M. 2002. *A traffic of dead bodies. Anatomy and embodied social identity in nineteenth–century America*, 1st ed. Princeton, NJ: Princeton University Press.

Schagen, U. 2005. Die Forschung an menschlichen Organen nach "plötzlichem Tod" und der Anatom Hermann Stieve (1886–1952). In *Die Berliner Universität in der NS-Zeit. Band 2: Fachbereiche und Fakultäten*, ed. R. von Bruch and R. Schaarschmidt, 35–54. Stuttgart: Franz Steiner Verlag Wiesbaden GmbH.

Schiller, E. 1942. Über den Fettgehalt der Leber beim gesunden Menschen. *Zeitschrift für mikroskopisch-anatomische Forschung* 51: 309–321.

Schleiermacher, S., and U. Schagen. 2012. Semantik als Strategie. Die Klassifizierung der Medizinverbrechen in Konzentrationslagern als Pseudowissenschaft. In *Geschlecht und "Rasse" in der NS-Medizin*, ed. I. Eschebach and A. Ley, 157–177. Berlin: Metropol Verlag.

Schönhagen, B. 1992. Das Gräberfeld X auf dem Tübinger Stadtfriedhof. Die verdrängte 'Normalität" nationalsozialistischer Vernichtungspolitik. In *Menschenverachtung und Opportunismus. Tübingen: Zur Medizin im Dritten Reich*, ed. J. Peiffer, 69–92. Tübingen: Attempto.

Stieve, H. 1952. *Der Einfluss des Nervensystems auf Bau und Tätigkeit der Geschlechtsorgane des Menschen*. Stuttgart: Thieme.

Stöhr, P. 1935. Zur Innervation der menschlichen Nebenniere. *Zeitschrift für Anatomie und Entwicklungsgeschichte* 104(5): 475–490.

Strzelecka, I. 2011. *Medical crimes. Medical experiments in Auschwitz*. Oświęcim: International Center for Education about Auschwitz and the Holocaust.

Stukenbrock, K. 2001. *Der zerstückte Coerper: Zur Sozialgeschichteder anatomischen Sektionen in der frühen Neuzeit (1650–1800)*. Stuttgart: Franz Steiner Verlag.

———. 2003. Unter dem Primat der Ökonomie? Soziale und wirtschaftliche Randbedingungen der Leichenbeschaffung für die Anatomie. In *Anatomie: Sektionen einer medizinischen Wissenschaft im 18. Jahrhundert*, ed. J. Helm and K. Stukenbrock, 227–239. Stuttgart: Franz Steiner Verlag.

Swick, H.M. 2006. Medical professionalism and the clinical anatomist. *Clinical Anatomy* 19: 393–402.

van den Bussche, H. 1989. *Im Dienste der "Volksgemeinschaft": Studienreform im Nationalsozialismus am Beispiel der ärztlichen Ausbildung*. Berlin: Dietrich Reimer Verlag.

Warner, J.H. 2009. Witnessing dissection: Photography, medicine and American culture. In *Dissection: Photographs of a rite of passage in American medicine 1880–1930*, ed. J.H. Warner and J.M. Edmonson, 7–29. New York: Blast Books.

Warner, J.H., and L.J. Rizzolo. 2006. Anatomical instruction and training for professionalism from the 19th to the 21st centuries. *Clinical Anatomy* 19: 403–414.

Wendehorst, A. 1993. *Geschichte der Friedrich-Alexander-Universität Erlangen-Nürnberg 1743-1993*. Munich: Verlag C.H. Beck.

Winkelmann, A. 2007. Die menschliche Leiche in der heutigen Anatomie. In *Grenzen des Lebens. Beiträge aus dem Institut Mensch, Ethik und Wissenschaft*, ed. S. Graumann and K. Grüber, 61–74. Berlin: LIT Verlag Dr. W. Hopf.

Winkelmann, A., and U. Schagen. 2009. Hermann Stieve's clinical-anatomical research on executed women during the "Third Reich". *Clinical Anatomy* 22: 163–171.

Winkelmann, A., and T. Noack. 2010. The Clara cell: A "Third Reich eponym"? *European Respiratory Journal* 36: 722–727. Supplementary online material accessible from www.erj.ersjournals.com

Wróblewska, T. 2000. *Die Reichsuniversitäten Posen, Prag und Strassburg als Modell Nationalsozialistischer Hochschulen in den von Deutschland besetzten Gebieten*. Torun: Marszalek.

Chapter 7
Nurses and Human Subjects Research during the Third Reich and Now

Susan Benedict and Cathy Rozmus

Soon after the end of World War II, the world became aware of the egregiously unethical, violent, and often capricious human subjects research on concentration camp prisoners in Germany and Poland that was documented in the Nuremberg Doctors' Trial ("Trials of War Criminals" [Nuremberg] 1946–1949). During this trial, the participation of physicians and scientists in these experiments was exposed to the world. It was, and still is, difficult for humankind to contemplate how graduates of some of the world's best universities became instruments of the state to further its racial and military agendas. While the Nuremberg Doctors' Trial highlighted the role of physicians in these experiments, the role of nurses remains largely unexplored.

Nurses who were prisoners in some of the concentration camps, as well as nurses employed by the *Schutzstaffel* (SS) and assigned to the camps, were involved in many of the experiments, particularly those at Auschwitz and Ravensbrück. In general, the experiments were of two categories: those of military importance and those related to racial hygiene policies. Military experiments included investigations of typhus, malaria, "phlegmon" (widespread inflammation due to infection), simulated war wounds, and exposure to high altitude and freezing water (Nuremberg 1946–1949, p. v; Weinberger 2009, p. 86). Experiments related to racial hygiene included investigations of methods to either sterilize large numbers of individuals or to change physical characteristics, such as eye color, to desirable

S. Benedict, CRNA, PhD, FAAN (✉)
Assistant Dean and Department Chair: Acute and Continuing Care; Professor and Director of Global Health; Co-director Campus-wide Program in Interprofessional Ethics; The University of Texas Health Science Center at Houston School of Nursing, 6901 Bertner Ave., Houston, TX 77030, USA
e-mail: Susan.C.Benedict@uth.tmc.edu

C. Rozmus, RN, PhD
PARTNERS Professor; Associate Dean for Academic Affairs; Assistant Vice President for Institutional Assessment and Enhancement; The University of Texas Health Science Center at Houston School of Nursing, 6901 Bertner Ave., Houston, TX 77030, USA
e-mail: Cathy.L.Rozmus@uth.tmc.edu

S. Rubenfeld and S. Benedict (eds.), *Human Subjects Research after the Holocaust*,
DOI 10.1007/978-3-319-05702-6_7, © Springer International Publishing Switzerland 2014

Aryan characteristics (Weinberger 2009, p. 87). Nurses participated in both types of experiments.

Auschwitz II

Two types of sterilization experiments were conducted at Auschwitz II (also known as Birkenau). The purpose of these experiments was to find a quick, clandestine method to sterilize both males and females, enabling them to function as slave-laborers in the rebuilding of Germany after the war while simultaneously preventing these "undesirables," such as Jews and Gypsies, from reproducing (Weinberger 2009, p. 89).

Victor Brack was an economist who had worked as Heinrich Himmler's chauffeur before becoming a member of the Reich Chancellory. On June 23, 1942, he declared in a letter to Himmler, *Reichsführer-SS* and Chief of the German Police including the Gestapo or secret state police:

> Among 10 million Jews in Europe there are, I figure, at least 2–3 million men and women who are fit enough to work. Considering the extraordinary difficulties the labor problem presents us with, I hold the view that those 2–3 millions should be specially selected and preserved. This can, however, only be done if at the same time they are rendered incapable to propagate. (Nuremberg 1946–1949, p. 721)

Two physicians, Drs. Horst Schumann and Carl Clauberg, were the primary physicians involved in these experiments. Dr. Schumann sought to sterilize men and women with radiation, whereas Dr. Clauberg sought to perfect the technique of sterilizing women by injecting caustic substances into their Fallopian tubes (Weinberger 2009, pp. 136, 151).

Dr. Horst Schumann's Experiments

Dr. Schumann, *SS-Sturmbannführer* (Michael and Doerr 2002, p. 391) and Lieutenant in the *Luftwaffe* (Air Force), began his experiments to determine the optimal dose of radiation to render prisoners both sterile yet able to work: Too little produced only temporary sterility, whereas too much often produced life-threatening burns. Brack proposed that 3,000 to 4,000 prisoners per day could be sterilized by radiation (Nuremberg 1946–1949, p. 736). Horst Schumann was given access to the prisoners, both male and female, of Auschwitz to carry out these radiation experiments.

Schumann arrived in Auschwitz on November 2, 1942, to begin his experiments (Czech 1989, p. 263). Initially, Schumann's sterilization experiments were conducted in Block 30 of Auschwitz II but later moved to Block 10 in Auschwitz I. The plan was to conduct the sterilizations without the prisoners' awareness.

A counter was set up that concealed the X-ray equipment. The prisoner was to stand at the counter and complete a questionnaire, which had been designed to take the amount of time required for the radiation. This plan, however, was soon replaced by more obvious methods. Female prisoners were irradiated while standing in front of the X-ray machine with plates placed on their abdomens and backs. Men had to place their genitals on a ceramic plate and were directly irradiated (Minney 1966, p. 115). Prisoners were not asked for their consent to be sterilized (Fride 1956).

Sonja Fritz, who was imprisoned in Auschwitz II solely because she was half-Jewish and married to a Jew, was appointed as Schumann's nursing assistant. She was initially assigned to difficult outdoor labor, but contact through prisoner friends enabled her to be reassigned as Schumann's main nursing assistant (S. Fritz, personal communication, August 27, 1999). Sonja was not told the nature of the assignment but was very excited to be assigned to indoor work, often the difference between life and death.

Every morning a group of 100 prisoners was delivered to Block 30 (Czech 1989, p. 263), the experimental block in Auschwitz II. Sonja's job was to record their prisoner numbers and nationalities and to keep the prisoners in order as they awaited an unknown fate. Each prisoner would be taken individually into the X-ray room to receive varying amounts of radiation to the ovaries or testicles. As Sonja described, the returning prisoners soon developed open wounds that became infected and would not heal. Some prisoners were irradiated so long in each session that they vomited on the way back to their barracks. When Sonja asked Schumann for permission to give the prisoner-subjects drinks of water, he would not allow it (personal communication, August 27, 1999).

A former prisoner, Gabriele Wertheimer, described her experience:

> We had to wait in front of the sick-block and were called up one by one. Nakedly we had to stand very still between two lamps, maybe for about 2–4 minutes. After all of us stood between these lamps, we were taken back to our block. We became very sick, had to vomit, had fever, and the skin turned red where the lamps had hit—lower abdomen and in the back.
>
> After a few days, I had to stand between those lamps again and a few days later for the third time. Only at that time was I told I had been x-rayed when standing between those lamps. I also was told the name of the person who had treated me. It was Dr. Schumann. (Wertheimer 1967, file TR, 10/2584)

Although most of Schumann's sterilization experiments were on young Jewish males, a former prisoner, Ima Spanjaard, described the experiences of some young Greek female prisoners:

> Four of these Greek girls died due to radiation burns. These burns looked horrible. The skin revealed blisters in all sizes, even small holes. Death occurred due to perforated intestines. Three of the Greek girls [who] had surgery died as a direct result of the surgery. A Dr. Dering performed surgery on 10 girls during one afternoon—all procedures were done with the same scalpel blade. Of those 10, 3 had died and the remaining 7 went to Birkenau within 6 months. (Spanjaard 1956, file 631A 556 R 990)

As Sonja realized what was happening to the prisoners, she began to sabotage the experiments by interrupting the electrical supply, a technique she learned from the Polish prisoners who worked with the X-ray machines. On the days when Sonja was able to interrupt the electrical supply, the experiments were cancelled because it took time to locate the repairmen and to make the necessary "repairs" (S. Fritz, personal communication, August 27, 1999).

Sonja described Schumann as always acting very proper and professional around her. He would, on occasion, leave bread or cigarettes prominently in sight in the trash can as a "gift." Sonja always refused these, leaving them on the X-ray table (S. Fritz, personal communication, August 27, 1999). By December 31, 1942, Schumann had performed, with Sonja's assistance, approximately 200 sterilization experiments on young Jewish men (Czech 1989, p. 294).

Dr. Carl Clauberg's Experiments

Dr. Carl Clauberg's goal for his experiments was the same as Schumann's: to sterilize women without their awareness and without recovery time. Clauberg was ordered to report to Himmler how long it would take to sterilize a thousand Jewish women by his method (Nuremberg 1946–1949, p. 700), which was to scar and occlude the Fallopian tubes by injecting caustic substances vaginally. In a letter dated June 7, 1943, Clauberg reported to Himmler that if he (Clauberg) had the necessary facilities and assistance, he could sterilize several hundred or possibly a thousand women per day (Nuremberg 1946–1949, p. 700).

Like Schumann, Clauberg began his experiments in Block 30 of Auschwitz II but moved to Block 10 in Auschwitz I, a two-story brick building sharing a courtyard with the punishment bunker of Block 11. It was in the courtyard between these two buildings that many executions were carried out. The windows on the side of Block 10 that overlooked the courtyard were boarded over to prevent the women of Block 10 from seeing the executions—a measure that was not always effective, as described by prisoner Hinda Tennenbaum née Elsztajn:

> At the side of Block 10 there was the Block for punishment—Block 11—also called the Bunker. That is where political prisoners, men and women, were shot. Every execution was done accompanied with music. The prisoner had to place himself against a black wall and he got a bullet into his neck. That block—just like ours—was totally barricaded against the outside. However, we managed to make small slits and openings between the boards to be able to look out. (Tennenbaum, n.d., file 02/1095)

Block 10's second floor consisted of two large rooms, called the *Revier* (short for *Krankenrevier,* or sickbay), which held the prisoner-subjects who were all Jewish and from 14 nations (Brewda 1946). Initially, only about 26 prisoner-subjects were in Block 10, but that number eventually increased to 300–500 (Valenska 1957). On the first floor, Clauberg's rooms were to the right and left of the middle hallway. The treatment room had two gynecological chairs, one examination table, and medical instruments; the X-ray room was on the left side (Münch 1956). Additionally, there was a room for caregivers, a room for female SS workers, and a bath and toilets (Brewda 1946).

Sylvia Friedmann was assigned to be Dr. Carl Clauberg's primary assistant. A young Jewish woman from Slovakia, she had attended nursing school in Presow (S. Friedmann, personal communication, August 28, 1999). Like Sonja Fritz, Sylvia acquired the assignment through fellow prisoners and was indeed happy about it. She had been working outdoors so the new assignment meant, "there would be no roll call and there would be a clean uniform and underwear." In this interview, Sylvia stated that she was told beforehand that she would be assisting with experiments on Jewish women but was not told the nature of the experiments (S. Friedmann, personal communication, August 28, 1999).

According to prisoner Anneliese Borinsky, many were selected upon arrival on the ramp and were asked:

> ... whether they would rather undergo an operation or have some injections. Those who chose the operations had part of their uterus removed for further examination in the lab, supposedly for studies in cancer prevention. These women were returned to Birkenau where they soon were gassed or where they soon died from other causes. For this reason, most of the women preferred the injections. (Borinsky 1945, file 02/313)

Because Clauberg's appearance was well known and evoked such horror and fear, caregivers and SS personnel, instead of Clauberg himself, actually made the selections of many of the women for Block 10 (Landstofova 1957).

There are conflicting statements about whether the women "consented" to be experimental subjects. One former Block 10 subject, Louise Pleskoff stated:

> Of course, regarding Clauberg or any other German physician, the inmates were never asked for their permission prior to being experimented on. We never received any kind of explanation. We were hardly treated any better than a piece of slaughtered meat. (Pleskoff 1956, file 631A 556 R 990)

However, Elisabeth de Jong, another Clauberg subject, reported that she and her sister-in-law were given a choice upon arrival in Block 10: agree to be experimented upon in an unspecified way or get on the waiting lorry for the gas chamber at Auschwitz II (E. de Jong, personal communication, April 2007). Another former subject, Irene Düring, drew a picture of the "consent form" that she had to sign. The form requested name, age, nationality, and number of gold teeth but did not specify nor describe the type of experiment (Düring 1991, p. 108).

The experiments were so painful that the nurses would sit on the victims' arms (de Leon 1956). The pain after each injection was described as being excruciating and lasting about 14 days, usually accompanied by high fever (Fride 1956).

Sylvia Friedmann's duties included caring for the women undergoing the experiments and recording their pain, fever, and menstrual cycles. She was known as the "death announcer" (Loewendorff-De Haff, n.d.) because she was the one who summoned the women for the experiments.

> In the beginning, we would hide whenever Sylvia would call out our names so that Clauberg could inject us. It was senseless and later on we gave up. Sylvia had told us, "If you do not get injected, you will be put away!" It was clear to us that this "away" stood for Birkenau and the gas. (Tonn, n.d., file 631A 556 R 990)

Another prisoner said Sylvia administered the intrauterine injections (de Leon 1956). Others described her as "demanding of absolute respect" (Münch 1956, file 631A 556 R 990), and brutal (Tennenbaum, n.d.), as well as beating the prisoners (Melzer 1956), being very mean to the women (Milner 1956), and hitting the women if they screamed during the injections. In a 1999 interview in which she implied that she saved prisoners' lives, Sylvia acknowledged hitting the women if they screamed during the injections because, "if the women had screamed, they would have been sent to the Birkenau and the gas" (S. Friedmann, personal communication, August 28, 1999).

There were two different categories of women who returned from Block 10 to Auschwitz II. Eva Landstofova stated:

> One category was those women who had refused to undergo the experiments. They were few in numbers. They were assigned to the *Sonderkommando*. They were separated from other women by heavy barbwire and they were under constant observation. They had to perform heavy-duty work. They were severely beaten and their food rations were cut. I know of 2 or 3 cases like this. Then I was still working with the commando to remove corpses and the foreman of the commando pointed out to me that these were the corpses who had been sent back from the "Clauberg" block because they had refused to take part in Clauberg's experiments. The other category was women who returned from Block 10 after the experiments. (Landstofova 1957, file 02/720)

Sylvia did ingratiate herself with Clauberg. Elisabeth de Jong reported that, at one point, Sylvia had Clauberg's victims knit sweaters for him while they were recovering from the experiments (E. de Jong, personal communication, April 2007). Sylvia was so privileged that she was able to have her mother be a special prisoner, exempt from the experiments, in Block 10 (Mouchova 1957). According to her mother, whenever she asked Sylvia about her work, Sylvia replied that she had to do it. If she did not do it, they all would be killed. Furthermore, if she would not do it, someone else would (Mouchova 1957). It is highly likely that had Sylvia not taken part in the experiments, others would have had the skills to assist Clauberg.

As Sylvia stated in her second interview with one of the authors (Benedict) in 2006, "Many of us did things that we should not have done. But, you have to understand, I was young and I wanted to live" (S. Friedmann, personal communication, August 24, 2006).

One former prisoner estimated that Clauberg experimented on around 1,000 women who were either sent to the gas chamber afterward or who died after a few days (Valenska 1957). According to Karl Brandt at the Nuremberg Doctors' Trial, "several thousand women were sterilized by Clauberg in Auschwitz" (Nuremberg 1946–1949, p. 701).

From a Subject's Perspective: Elisabeth de Jong

Elisabeth de Jong, along with her sister-in-law Lillian, arrived at Auschwitz on a transport from Amsterdam in September 1943 after being hidden by a family since 1941. They were selected immediately upon arrival and taken to Block 10. According to de Jong, upon arrival at Block 10, they were told to either sign a consent to be subjects in the experiments or get on the waiting truck to go to Auschwitz II (E. de Jong, personal communication, April 2007).

Both Elisabeth and Lillian signed the consent, which did not describe the experiments. The next morning, after seeing the suffering of the victims of the experiments, including the pain and open draining wounds, de Jong and her sister-in-law went to the block supervisor and told her that they had changed their minds. They would rather go to the gas chamber, but they were told that option was no longer available to them (E. de Jong, personal communication, April 2007). Elisabeth stated:

> The experiments started two or three days later, but they never let us know what it was all about. We had to go downstairs into a special room, you had to stretch out on a table, they strapped you down, and they started first with 24 injections in many parts of our body. Those injections left open sores on our bodies. Another day they injected some substance into the womb and ovaries, we didn't know what, that burned like hell and our pains were beyond endurance. Of course, all of this was done always without anesthetic. They also took biopsies from the womb. Our resistance was so low that our wounds and sores never healed.
>
> There was a woman doctor, Dr. Kleinova, in Block 10, a prisoner from Poland or Russia, who was forced to work with the Nazis. She tried to help us as much as possible. At night she would come to the room where we slept. She would gently wash and dress our wounds and try to console us. Another time she was supposed to inject blood into our veins but not our own blood type. When it was my turn, she deliberately let the needle fall out. The Nazis got angry but she insisted that the needle accidentally fell out. She was very, very good to me and perhaps to others too. I know without her kind help, I would not have survived. Our misery was indescribable. (E. de Jong, personal communication, April 2007)

As the Russian army approached Auschwitz, Drs. Schumann and Clauberg moved their sterilization experiments further west to Ravensbrück, where they used 120–140 Gypsy girls as young as eight years of age as subjects in the sterilization experiments (Tillion 1975).

Eva Justin: Nurse and Principal Investigator

Eva Justin was a rare example of a nurse who was the principal investigator of her own research. She was educated as a nurse and later became an associate of Dr. Robert Ritter of the research unit for Racial Hygiene and Population Biology at Berlin-Dahlem. She initially assisted with data collection and measuring the physical characteristics of Romani (Gypsy) children (Friedlander 1995, p. 250), but eventually undertook her own study of half-Romani children who were taken from their parents and raised in orphanages and foster homes without any contact with the Romani culture. Justin wanted to know if certain behavioral characteristics of Romani children could disappear or be attenuated by the absence of their parents. After the completion of Justin's studies, 39 children were deported to the "Gypsy Camp" at Auschwitz II and registered on May 12, 1944 (Czech 1989, p. 677). Dr. Josef Mengele was the doctor in charge of the Gypsy family camp (Czech 1989, p. 408). Under his orders, 2,897 men, women, and children were gassed, including most of the children from Justin's study, when the Gypsy camp was "liquidated" on August 2, 1944 (Czech 1989, p. 408).

Eva Justin's dissertation was approved by the German ethnologist Richard Thurnwald, and she was awarded a PhD in anthropology in 1944 from the University of Berlin (Friedlander 1995, pp. 250, 294). After the war, Eva Justin worked in Frankfurt as a child psychologist. A criminal investigation was opened against her in 1958, charging her with mass murder. However, in 1960 the investigation was closed without going to trial. Justin died of cancer in 1966 (Margalit 2002, pp. 133, 163).

Ravensbrück Concentration Camp

Ravensbrück differed from Auschwitz II in that SS-employed nurses did work in the prisoners' *Reviers*, whereas in Auschwitz II the care of sick prisoners was left to other prisoners. Thus, in Ravensbrück both SS-employed nurses and prisoner-nurses interacted with the prisoners.

The first medical experiments were done at Ravensbrück under the direction of Dr. Karl Gebhardt. The subjects were 75 Polish women who had been in the resistance; many were university students. The women, ages 18–30 years, had experimental surgeries to simulate war injuries. Long incisions were intentionally infected with glass, wood particles, and bacteria such as gas gangrene and tetanus. Some received no postoperative treatment, whereas others received various sulfanilamides (Nuremberg 1946–1949, pp. 356–357). Nursing care of these subjects, such as it was, was provided only by German nurses to maintain secrecy. One former prisoner stated:

> I was able to ascertain that the German nurses had never actually had any professional activity; they looked on the misery of the sick with the greatest indifference and even sarcasm and never did they make any effort whatever to help them even though it was possible for them to do so. It was a perfectly moral thing for the *Schwester* to beat the sick, indeed I have seen Schwester Lisa beating sick women without any reason at all. (Ravensbrück Trial 1947–1948, file WO 309/692/81961)

In other experiments, amputation techniques were devised for producing "spare parts" for wounded German soldiers. Gerda Quernheim, a prisoner-nurse, assisted with some of these experimental amputations. After the war, she served seven years in prison for her actions. She was released from prison in 1955, at the age of 47 (Tillion 1975).

Several of the women who survived the Ravensbrück experiments testified against the physicians at the Doctors' Trial at Nuremberg. (See Chaps. 4 and 5 in this book for more on the experiments conducted at Ravensbrück.)

Nurses assisted with the medical experiments not only at Ravensbrück and Auschwitz II but also at other camps throughout the Third Reich. The SS employed some of these nurses, whereas others were prisoner-nurses. Clearly the motivation of the two groups was vastly different, but two common elements prevailed within both groups: obedience to one's "superiors" and an abdication of the long-held nursing principle of providing care, not harm. It is easy to understand the prisoner-nurses' powerlessness to disobey. To have overtly done so would have cost their lives; yet at some risk to self, Sonja Fritz did what she could to intermittently interrupt the experiments by intentionally damaging the equipment. She also attempted to comfort the subjects by asking if she could, at least, give them drinks of water. The activities of Sylvia Friedmann are more vexing—she claimed that she was too afraid to sabotage Clauberg's experiments in any way (S. Friedmann, personal communication, August 28, 1999). It is interesting that after 60 years of reflection, she still believed that she saved women from the gas chamber by slapping them into silence. Not having been there, one cannot know.

Research Nurses Today

Today research nurses have, broadly speaking, some of the same roles as the nurses who assisted with the Nazi experiments. They may be principal investigators of their own studies, data collectors, and/or persons who obtain informed consent. Many monitor subjects' adherence to research protocols. In light of the unethical behavior of nurses under National Socialism, it is useful to examine each of these roles in today's human subjects research.

A common role for nurses in human subjects research is the provision of normal daily care to subjects or the administration of experimental drugs or treatments. In a double-blind study, the nurse will not know what drug the subject is receiving. The staff nurse, while not having any direct responsibility for the conduct of the research, may have ethical concerns about the study, such as a lack of subjects' understanding about their informed consent. Untoward events that are ignored or unrecognized by the principal investigator may also pose ethical problems for the staff nurse, and it may be difficult for the staff nurse to decide to whom these ethical issues should be addressed. While the chain of command regarding ethical issues is often clear in clinical or therapeutic care, there may be no clear chain of command in human subjects research. If, for example, the principal investigator is both the person to notify about an ethical problem with the study and the source of the problem, to whom does the staff nurse speak?

Another current role is as research nurse or research coordinator. The principal investigator often interviews and hires the research nurse, but the sponsor—federal government, philanthropic foundation, or pharmaceutical company, for example—

pays the nurse's salary. The nurse's responsibilities often include subject accrual, obtaining informed consent, collecting data, and administering or supervising experimental treatments. Because the nurse's employment often depends on the accrual of the requisite number of subjects, there is a potential conflict of interest between what might be best for the potential subject and best for the nurse's employment. If continued employment depends on the number of subjects accrued, will the nurse ensure that all potential subjects are fully informed and understand what they are consenting to do? Is there pressure to enroll subjects who do not meet the inclusion criteria or who should otherwise be excluded?

Another potential ethical issue for a research nurse is integrity of data. The research nurse may observe discrepancies between the data collected and the data reported. Again, the employment of the nurse may be in jeopardy if the integrity of the data is questioned, especially if the principal investigator is the object of the nurse's questions. In addition, the chain of command may be unclear if the nurse wishes to report possible unethical behavior by the principal investigator.

In some cases, the research nurse behaves unethically. In one such case, the nurse falsified and fabricated data in a cancer drug trial. If this misconduct had not been revealed, an ineffective drug could have been marketed and used on patients (Office of Research Integrity, n.d.). In this particular case, the nurse was found guilty of fraud in criminal court.

Habermann et al. reported in 2010 that research coordinators, mostly nurses, had observed "protocol violations, consent violations, fabrication, falsification, and financial conflict" (p. 51). The coordinators reported their observations in 70 percent of the cases, and, in some instances, either nothing was done or the coordinators lost their jobs after their report. In other cases, the coordinators did not report the misconduct because the perpetrators were their supervisors (Habermann et al. 2010).

Finally, a nurse may serve as principal investigator, responsible for all aspects of the study. As principal investigator, the nurse may feel external pressure from the funding agency either to increase accrual of subjects or to reach conclusions that benefit the funding agency. The nurse also may feel internal pressure to reach conclusions that will advance his or her research career. In fact, the nurse as principal investigator is subject to all of the enticements for ethical lapses that other principal investigators confront (Davis et al. 2007; Wells 2008). An example of a nurse principal investigator's involvement in research misconduct is that of the nurse faculty member who plagiarized grant applications and publications as well as falsified data. As a result of the National Institutes of Health Office of Research Integrity's findings, he lost his job and was excluded for three years from contracting with the US government and from serving on any advisory panel (Dahlberg 2011).

Nurses today have, overall, the same roles in research as nurses in Nazi Germany with the exception that there are many more nurse principal investigators. Whereas today's nurses face economic consequences for both reporting and perpetrating ethical lapses, some nurses in Nazi Germany, particularly prisoner-nurses, faced the ultimate consequence of losing their lives if they did not collaborate in egregious research.

References

Borinsky, A. 1945. *Statement of Anneliese Borinsky (file 02/313)*. Jerusalem, Israel: Yad Vashem.

Brewda, A. 1946. *Statement of Dr. Alina Brewda, Warsaw (file 19721/26)*. Vienna, Austria: Dokumentationsarchiv des österreichischen Widerstandes.

Czech, D. 1989. *Auschwitz chronicle*. New York: Henry Holt and Company.

Dahlberg, J. 2011. Office of Research Integrity. Case Summary: Weber, Scott. Retrieved from http://ori.hhs.gov/content/case-summary-weber-scott. Accessed 12 Dec 2013.

Davis, M.S., M. Riske-Morris, and S.R. Diaz. 2007. Causal factors implicated in research misconduct: Evidence from ORI case files. *Science and Engineering Ethics* 13: 395–414. doi:10.1007/s11948-007-9045-2.

de Leon, R. 1956. *Statement of Rosalinde de Leon, Arrondissements-Reichtbank to Almelo (file 631A 556 R 990)*. Wiesbaden, Germany: Hessisches hauptstaatsarchiv.

Düring, R. 1991. *Criminal experiments on human beings in Auschwitz and war research laboratories: Twenty women prisoners' accounts*. San Francisco: Mellen Research University Press.

Fride, Z. 1956. *Statement of Zijsa Fride in the matters of the District Ministry of Kiel against Professor Clauberg (file 631A 556 R 990)*. Wiesbaden, Germany: Hessisches hauptstaatsarchiv.

Friedlander, H. 1995. *The origins of Nazi genocide*. Chapel Hill: University of North Carolina Press.

Habermann, B., M. Broome, E.R. Pryor, and K.W. Ziner. 2010. Research coordinators' experiences with scientific misconduct and research integrity. *Nursing Research* 59(1): 51–57.

Landstofova, E. 1957. *Statement of Eva Landstofova in the criminal proceedings against the late Professor Carl Clauberg (file 02/720)*. Jerusalem, Israel: Yad Vashem.

Loewendorff-De Haff, C. n.d. *Statement of Cornelia Loewendorff-De Haff (file 631A 556 R 990)*. Wiesbaden, Germany: Hessisches hauptstaatsarchiv.

Margalit, G. 2002. *Germany and its Gypsies: A post-Auschwitz ordeal*. Madison: University of Wisconsin Press.

Melzer, S. 1956. *Statement of Schewa Melzer (file 631A 556 R 990)*. Wiesbaden, Germany: Hessisches hauptstaatsarchiv.

Michael, R., and K. Doerr. 2002. *Nazi-Deutsch Nazi German*. Westport, CT: Greenwood.

Milner, E. 1956. *Statement of Elfriede Milner, New York, NY (file 631A 556 R 990)*. Wiesbaden, Germany: Hessisches hauptstaatsarchiv.

Minney, R. 1966. *I shall fear no evil: The story of Dr. Alina Brewda*. London: William Kimber.

Mouchova, P. 1957. *Statement of Prizka Mouchova, testimony in the criminal proceedings against the late Professor Carl Clauberg, Slovakia (file 02/725)*. Jerusalem, Israel: Yad Vashem.

Münch, H. 1956. *Statement of Hans Münch (file 631A 556 R 990)*. Wiesbaden, Germany: Hessisches hauptstaatsarchiv.

Office of Research Integrity. n.d. U.S. Department of Health and Human Services. Findings and consequences of research conduct. Retrieved from http://ori.hhs.gov/education/products/RIandImages/misconduct_cases/findings_of_misconduct.pdf. Accessed 16 Dec 2013.

Pleskoff, L. 1956. *Statement of Louise Pleskoff (file 631A 556 R 990)*. Wiesbaden, Germany: Hessisches hauptstaatsarchiv.

Ravensbrück Trial IV. 1947–1948. *Public Record Office (file WO 309/692/81961)*. London: The National Archives.

Spanjaard, I.S. 1956. *Statement of Ima Schalom Spanjaard, Kleve, Germany (file 631A 556 R 990)*. Wiesbaden, Germany: Hessisches hauptstaatsarchiv.

Tennenbaum, H. n.d. *Statement of Hinda Tennenbaum née Elsztajn in the criminal procedure against Carl Clauberg (file 02/1095)*. Jerusalem, Israel: Yad Vashem.

Tillion, G. 1975. *Ravensbrück*, Trans. G. Satterwhite. Garden City, NY: Anchor Books.

Tonn, M. (n.d.). *Statement of Margot Tonn (file 631A 556 R 990)*. Wiesbaden, Germany: Hessisches hauptstaatsarchiv.

Trials of War Criminals before the Nuernberg Military Tribunals under control Council Law No. 10 (Vol. 1), (hereafter Nuremberg). 1946–1949. *The Medical Case. United States vs. Karl Brandt* et al. Washington, DC: Government Printing Office.

Valenska, K. 1957. *Statement of Katarina Valenska in the criminal procedure against Carl Clauberg. Prague (file 02/1095).* Jerusalem, Israel: Yad Vashem.

Weinberger, R. 2009. *Fertility experiments in Auschwitz-Birkenau: The perpetrators and their victims.* Saarbrücken, Germany: Südwestdeutscher Verlag für hochschulschriften Aktiengesellschaft & Co. KG.

Wells, J.J. 2008. Final report: Observing and reporting suspected misconduct in biomedical research. Office of Research Integrity. U.S. Department of Health and Human Services. Retrieved from ori.hhs.gov/sites/default/files/gallup_finalreport.pdf. Accessed 12 Dec 2013.

Wertheimer, G. 1967. *Statement of Gabriele Wertheimer in Landesgericht Frankfurt am Main, Germany (file TR 10/2584).* Jerusalem, Israel: Yad Vashem.

Chapter 8
Involuntary Abortion and Coercive Research on Pregnant Forced Laborers in National Socialism

Gabriele Czarnowski

At a meeting of the Medical Society of Styria on March 12, 1944, Karl Ehrhardt (1895–1993), head of the Graz University Women's Hospital, spoke to this scientific, professional audience about "Termination of pregnancy beyond the fourth to fifth month by intra-amnial injection of active agents" (p. 507). The report was published in the *Medizinische Klinik* (*Medical Clinic*) and the *Münchener Medizinische Wochenschrift* (*Munich Medical Weekly*), well-known medical journals. "Every so often," the meeting minutes began, "the physician is faced with the necessity of terminating a somewhat more advanced pregnancy (beyond the fourth to fifth month)" (Medizinische Gesellschaft Steiermark 1944, p. 507). Just a few lines down, the minutes noted that Ehrhardt had injected more than 50 women with "active agents" (p. 507), which raises the question of his definition of "every so often" and underlines the fact that he was, in fact, speaking of a large number of procedures. This intervention, the report went on to say, "could prove difficult, especially in the case of primiparas, and entailed significant stress for the pregnant women. The search for a simple and gentler method of abortion for more advanced pregnancies was thus justified" (p. 507). After this introduction, which suggested a caring physician responding appropriately to unnamed but urgent necessities, the rapporteur got down to the details:

Ehrhardt tested a procedure whose technique was based on the method of abdominal amniopunction ... with subsequent intra-amnial injection ... of colloidal thorium, which he presented at the 1937 German Gynecological Congress. Here, instead of thorium, he injects an agent that induces intrauterine fetal death. The procedure may be performed under local anesthesia and takes only two to three minutes. The following agents were tested: sulfonamides, opiates, scopolamine, evipan, stilbene, novocaine-adrenaline, gynergene, camphor, and dolantin, among others, in some cases with rapid success. In most cases, however, ... formalin was injected... which in fifty cases without exception led to intrauterine fetal death. The ejection of the dead fetus often occurred after just a few

G. Czarnowski, Dr.phil. (✉)
Institut of Social Medicine and Epidemiology, Medical University of Graz, 8010 Graz, Austria
e-mail: gabriele.czarnowski@medunigraz.at

S. Rubenfeld and S. Benedict (eds.), *Human Subjects Research after the Holocaust*,
DOI 10.1007/978-3-319-05702-6_8, © Springer International Publishing Switzerland 2014

days, but sometimes only after several weeks, for which reason, in cases of delayed ejection, labor was induced . . . . (Medizinische Gesellschaft Steiermark 1944, p. 507)

The account of the meeting documented Ehrhardt's extensive experiments on young women, apparently in direct connection with the terminations for which they were actually admitted to the hospital. But it also referred in passing to one of his central research areas, fetography.

In fact, I shall argue here, Ehrhardt was not searching for a gentle abortion method, as he claimed. The abortion procedure using the injection of lethal poisons into the uterus that he presented to his colleagues must be viewed in the context of his long-standing research on animals and pregnant women with the aim of developing radiological methods for viewing the placenta and the fetus or its internal organs. Ehrhardt had been experimenting with intrauterine imaging techniques since the early 1930s, first in the University Women's Hospital in Frankfurt am Main, Germany, and then as head of the University Women's Hospital in Graz, Austria. In 1939, SS influence ensured his transfer from Frankfurt am Main to the vacant chair of gynecology and obstetrics in Graz previously occupied by Hans Zacherl, who was dismissed shortly after the *Anschluss* (occupation and annexation) of Austria (Czarnowski 2012, pp.139–140).

What Is Fetography?

Fetography is a little-studied technology in the history of depictions of the unborn child, beginning with Samuel Thomas Soemmering's first anatomical representations in 1799 and ending with the invention of ultrasound imaging of the pregnant womb in the 1950s (Soemmering 1799; see also Duden 1993, p. 33; Oakley 1993, pp. 188–197). The emergence and development of fetography are closely associated with the discovery of X-rays and their promising uses in medical diagnostics, including the exploration of the uterus. Leaving aside the dangers, which were addressed for the first time as early as 1905, X-rays of pregnant women did not prove very productive. The fetus in utero was only visible on X-rays after the development of the skeleton during the fifth month, and the images showed little aside from bone structures. The details seen in X-ray images changed with the introduction of radiological contrast media, which, for example, rendered visible the contours of the fetus. All over the world, scientists and research groups experimented on various animal species and pregnant women in the race to find the most suitable and least harmful contrast media. In their choice of these contrast media, gynecologists followed their colleagues in other medical specialties such as internal medicine, pediatrics, and surgery, who were conducting experiments around the same time. Their aim was the radiological depiction of the inner organs and blood vessels for diagnostic purposes. Ehrhardt's experiments were inspired by the interdisciplinary radiological debate conducted, for example, in the journal *Fortschritte auf dem Gebiete der Röntgenstrahlen (Progress in the Field of*

X-Rays), in which he also published a paper (Ehrhardt 1933). His experiments must also be viewed in the context of the competition to find the most suitable and least harmful contrast media.

What interested Ehrhardt about fetography was not so much its significance as a diagnostic tool, although he did address the future practical uses of the method in his publications. His chief ambition was to fathom "the secrets of intrauterine life," and he claimed to have been the first to demonstrate radiologically that the fetus drinks and breathes in the womb. In conducting his experiment, he delayed by several days the abortions of patients who had come to the hospital to terminate their pregnancies in order to perform an amniopunction. Using a syringe, he removed a certain amount of amniotic fluid through the abdominal wall and injected the same amount of the X-ray contrast medium Thorotrast into the womb. Over the following hours and days (up to three days, or more), he used X-rays to monitor whether and to what extent the fetus's gastrointestinal tract and lungs became shaded and drew his conclusions from the results. At the end of the process, he surgically removed the fetus from the womb. He selected those methods that allowed him to extract the fetus intact and, if possible, still alive in its embryonic membrane. For the women involved, however, the *sectio parva* (hys-terotomy or Cesarean section) was a far more invasive procedure than a simple removal.

Ehrhardt borrowed the technique of amniopunction from the Dutch gynecologist de Snoo, head of the University Women's Hospital in Utrecht. De Snoo's experiments, as described in "The Drinking Child in the Uterus," did not involve X-rays or abortions. He had studied the excretion of methylene blue injected into the amniotic sac by testing the pregnant woman's urine (de Snoo 1937). Ehrhardt's method, in contrast, necessitated terminating the pregnancy at the end of the experiment, but not because of concern about potential harmful effects on the patient from Thorotrast's radioactivity. After the abortion, the final stage in his experiment, Erhardt could X-ray the fetus outside the womb, "observe" its "agonal movements and reflexes," or "study" the nature of its respiration "until it expires completely" (Ehrhardt 1939, p. 916). The use of the contrast media Thorotrast and Umbrathor to test radiologically the fetal drinking and respiratory function was, in Ehrhardt's own words, "a lucky experimental hit," at which he had arrived via the "detour of intravenous placentography" (1941, p. 117). Ehrhardt had studied the effect of Thorotrast on the placenta mainly in animal experiments and discovered that, as far as he could discern from X-rays, this substance did not penetrate the placenta, either from mother to fetus or from fetus to mother (1933, p. 412).

Thorotrast and Umbrathor had been on the market since 1929. Intravenously administered thorium is not excreted, but stored permanently in the body, which is constantly exposed to radiation because of thorium's long half-life of 14 billion years. In the United States, the Food and Drug Administration (FDA) had already warned against its use in 1933, and it was also controversial in Europe (Becker et al. 2006). In 1933, Ehrhardt discussed the issues of germ cell damage and the effect of Thorotrast "on the gravid organism" (1933, p. 6). He took no position on the possible long-term dangers to women from radiation exposure. When it came to

Thorotrast, Ehrhardt was no more reckless or irresponsible than many of his colleagues. But in Germany, he was the first to try it on pregnant women. Despite warnings, this contrast medium was used on thousands of patients in many countries, primarily for visualizing arteries, before it was banned in 1949–1950 because of possible lethal carcinogenesis suggested by the first epidemiological studies on Thorotrast (Becker et al. 2006, p. 6).[1] Although I do not know to what extent the long-term effects of thorium fetography have been studied, according to my private correspondence in 2013 with Gerhard van Kaick, the coordinator of the German Thorotrast study, there were no harmful effects on the pregnant women if Thorotrast was not given intravenously, but the procedure was dangerous because of the multiple X-rays.

The German Research Foundation (DFG) funded Ehrhardt's fetographic research. Between 1937 and 1945, he presented his findings to three scientific societies and published four papers on the subject. His publications appeared in respected medical journals.

Who Were the Subjects of Ehrhardt's Research?

In the early 1930s, before the Nazi's seizure of power, Ehrhardt's human research subjects were six patients referred to the Frankfurt University Women's Hospital for therapeutic abortions. After the beginning of the National Socialist forced sterilizations in the middle of 1934 and the eugenic abortions connected with it in 1935, he also misused pregnant women scheduled for sterilization for his scientific interests. In his first publication on fetography in 1937, under the heading "Experimental Design," he describes his treatment of an unmarried 21-year-old woman in her sixth month of pregnancy "who was to undergo sterilization and termination for eugenic reasons (because of congenital feeblemindedness)" (Ehrhardt 1937, pp. 1,699–1,700). This young woman and her aborted fetus were clearly victims of the Nazi sterilization law. Such patients considerably broadened the pool of individuals for his experiments (e.g., Bock 1986; Spring 2009).[2] The first study was published hundreds of young women from Eastern Europe who were admitted for abortions for political reasons increased the number of potential and actual research subjects many times over. These abortions on "racial" grounds were performed on forced laborers from Poland and from the countries of the Soviet Union. In contrast to therapeutic abortions on "hereditarily healthy" Germans, which were very restricted and thus occurred only rarely, terminations of the pregnancies of forced

---

[1] The first study was published in Denmark, followed by Portugal, Japan, the United States and in 1968 in Germany, with the last follow-up study done here in 2004.

[2] In Nazi Germany (including the occupied countries), about 400,000 men, women, youth, and children were forced to undergo sterilization according to the Law for the Prevention of Genetically Diseased Offspring.

laborers occurred on a massive scale. The German Medical Association monitored the procedures in both instances (Czarnowski 1997, pp. 130–131; 1999, p. 242). Most of these "racial" abortions took place in labor camps, and we never will know the exact number. They also were performed at various hospitals and even at university women's hospitals. In Germany thus far we know about abortions on forced laborers at the Freiburg University Women's Hospital (Link 1999, p. 395), the Würzburg University Women's Hospital (Dietl 2005, pp. 96–97), the Erlangen University Women's Hospital (Frobenius 2004), and the Cologne University Women's Hospital (Franken 2012). In the four university women's hospitals in occupied Austria, the number of abortions in each hospital were[3]: First Vienna University Women's Hospital, 8; Second Vienna University Women's Hospital, 3[4]; Innsbruck University Women's Hospital, no abortion on forced laborers could be found in the patient books[5]; and in Graz University Women's Hospital, between April 1943 and May 1945, pregnancy terminations on more than 500 women and girls were performed. They and their unborn children fell victim not just to politically coerced abortions, but also to abusive surgical interventions and clinical research of a kind not yet documented in any other university hospital in the Greater German Reich (Czarnowski 2008, pp. 56–59).

Ehrhardt, indeed, continued his fetographic experiments on pregnant forced laborers. That is clear from a personal notebook entry written by Dr. Hoff, the senior physician, in January 1944: ". . . the use of X-ray film is enormous (radiograms of the Russian fetuses!)" (Landgericht [LG] für Strafsachen, 19 Vr 20/1947). We also have evidence in the form of a protocol that survived the war in the desk drawer of the clinic director, who had long since escaped to Bavaria (LG für Strafsachen,19 Vr 20/1947). The experiments documented in the director's notes go well beyond anything previously recorded, beginning with the number of research subjects involved. The protocol covers the period from January to June 1944 and contains notes on experiments on 85 women and girls from the Ukraine, Russia, and Poland. Sixty of them were between the ages of 18 and 23 years old, and 63 were pregnant for the first time. Nearly two-thirds were past their fourth month, and four were already eight months pregnant. The protocol shows that Ehrhardt not only injected the women with various substances to kill their fetuses, as mentioned

---

[3] The statistics concerning the number of abortions performed on forced laborers in the two Vienna hospitals and the Innsbruck hospital were gathered in the course of my research project called Austrian Gynecology and National Socialism, which was funded by the Austrian Society of Obstetrics and Gynecology (Österreichische Gesellschaft für Geburtshilfe und Gynäkologie).

[4] In Vienna 670 abortions on forced laborers were performed in a special barrack for foreigners at the Ottakring City Hospital (Czech 2004, pp. 269–273).

[5] Pregnant forced laborers working in Tyrol and Vorarlberg were sent for an abortion to the City Hospital in Hohenems. The number of performed abortions there is not known. The highest number of abortions on forced laborers in an Austrian or German Hospital took place in the "Barrack for Eastern workers" at the Landesfrauenklinik and Hebammenlehranstalt (Governmental Country Hospital and Midwives School) in Linz, where 900 were performed (Hauch 2001, pp. 422–432).

in the report on the meeting, but injected 75 of the women with radiological contrast media and recorded precise data on the fetuses. For example, the baby of 22-year-old Anna H. from Kiev, who he killed by formalin injection shortly before birth, was 51 cm long and weighed 3,000 g (LG für Strafsachen, Vr 4434/1949).

Ehrhardt's last publication on this subject appeared in February 1945 in the last war edition of the *Medical Journal*. The title was, "Does the Child Breathe in the Uterus? A Contribution on the Intrauterine Biology of the Child." In this article, he approached the transition from medical abortion to medical infanticide more closely than in any of his previous texts. He alluded to his six years of previous research, and his remarks revealed a progression to ever more brutal acts. First, he referred to his earlier publications, in which he had shown "that one can depict the fetal gastrointestinal tract and lungs roentgenologically through the intra-amnial injection of colloidal thorium and thereby study the intrauterine functions of drinking and respiration ... active functional processes absent in the dead child, of course." He then formulated his research interest as follows: "What occurs in the child that dies in utero? Does the action of swallowing continue until the onset of death or already cease beforehand? I have pursued this question during 'formalin abortions'" (Ehrhardt 1945, pp. 182–183). These were the very experiments recorded in the protocol, which Ehrhardt had presented to the members of the Medical Society of Styria in March 1944.

The second part of his contribution was again devoted to the "problem of the intrauterine respiratory movement of the child," in this case "particularly of the full-term or nearly full-term child." The research results available "thus far" had been obtained from "fetuses in the first half or the middle of pregnancy," leaving open the question whether these findings also applied to the final phase of pregnancy. To be sure, "the opportunities for extending our studies of the intra-amnial biology and pathology of the fetus to the later months of pregnancy ... are extremely limited." This made "any finding we can obtain towards the end of pregnancy especially valuable." And he attributes it to chance rather than Nazi policy "that I have had two opportunities in recent years to test the matter of intrauterine respiratory movement in a full-term and a nearly full-term child" (Ehrhardt 1945, pp. 182–183).

A live fetus in utero was the precondition for the application of this "biological method," which took one to three days to produce sufficient X-ray images of the fetus. Killing of the fetus, even shortly before birth, was the end of the process. In this article, unlike in his previous publications, Ehrhardt provided no information about his research subjects. Eugenic or therapeutic abortions were only legally permitted until the sixth month of pregnancy, although there are records of later procedures at German hospitals. Abortions performed on "Eastern workers" and Polish women, however, could substantially exceed this limit if the operating physician was willing. Ehrhardt could only conduct his experiments on forced laborers who were admitted to the Graz University Women's Hospital for a termination well past their sixth month. It is worth pointing out, however, that a Frankfurt colleague already mentioned in a 1938 publication concerning "our experience in the drug treatment of intrauterine asphyxia" (Dörr 1938, p. 135) that Ehrhardt had injected Thorotrast through the abdominal walls of women in their sixth to ninth month of pregnancy. He does not say how many patients Ehrhardt experimented on in this manner, but he does announce the forthcoming publication of the results. Ehrhardt himself described only procedures up to the sixth month in his previous publications, although he had conjectured in 1939 that the respiration of the fetus might be different at the end of pregnancy than in earlier stages. Now he was able to confirm this conjecture experimentally, although not in a scientifically adequate manner as he noted. His final sentence read, "Since the number of cases observed thus far (2) is insufficient for an ultimate judgment, the communication of further relevant observations is desirable" (Ehrhardt 1945, p. 183).

Ehrhardt was not the only German gynecologist to experiment on the victims of sterilization and on "Eastern workers." We know that Dr. Hoff was also conducting tests at the Graz University Women's Hospital. Hoff performed physiological experiments on uterine motility in nearly every forced laborer he operated on, approximately 200 women and girls (Czarnowski 2004, pp. 263–273). And we know about experiments with "Euxyl soap abortions" on forced laborers at the Women's University Hospital in Erlangen (Frobenius 2004). Scant research exists on abortions and experiments on forced laborers in the other university gynecological hospitals of the Greater German Reich, and we have only slightly more information about experiments on the victims of forced sterilization (e.g., Czarnowski 2001). There is still a lot of work to do.

Karl Ehrhardt was also not the first or the only "fetographer." According to a review article published in 1942, approximately 20 scientists and research groups worldwide had been experimenting on animals and pregnant women with various contrast media since the late 1920s (Erbslöh 1942). They included six researchers or research groups in Italy alone, three in the United States, two each in Japan and Britain, and individual scientists in Mexico, Turkey, and Hungary. In Germany, two doctors from the medical faculty of Würzburg were testing the fetographic method described by Ehrhardt; the German gynecologist Joachim Erbslöh, the author of the review article, was conducting research at the women's hospital at Bydgoszcz (Bromberg in German) in occupied Poland. The American gynecologists Menees, Miller, and Holly published the first paper on "amniography" in 1930. Which patients these scientists used for their research remains an open question.

Conclusion

The doctors of Graz University Women's Hospital performed numerous abortions on healthy women and girls, in violation of the traditional reproductive ethics of academic gynecology, which up until then had only recognized "purely scientific" indications for abortion. On the other hand, invoking these ethics as well as moral and religious principles, the heads of the University Women's Hospitals in Munich I, Tübingen, Germany, and Innsbruck, Austria, all refused to terminate the pregnancies of forced laborers (Czarnowski 2004, pp. 243–245). Eugenic forced sterilization and abortions authorized by the Law for the Prevention of Genetically Diseased Offspring, however, continued to be performed without further ado at all other university women's hospitals. In Graz, doctors not only performed abortions on forced laborers but continued to do them into the final stages of pregnancy—surgeons were active participants in the Nazi genocidal program to exterminate and selectively incorporate the so-called "Eastern peoples."

The misuse of involuntary patients for scientific experiments continued an old tradition. Poor patients and later those entitled to health insurance by the government were treated as "material" for teaching and research hospitals well into the twentieth century. Nazism added "racial" to social difference, which considerably and specifically expanded the available "material." And yet, we cannot draw a distinct line between the misuse of these and other victims—the transitions were fluid and always dependent on the individual's physical status. Women classified as "genetically diseased" and admitted for eugenic forced sterilization with or without termination, as well as the forced laborers admitted for abortions, were special patients not ordinarily found in university hospitals. It was not just their status as involuntary patients, but also their physical state that made their presence unusual. They were generally healthy, young, and pregnant, and large numbers of them were admitted regularly over a long period for very particular operations. For these reasons, they were especially prized as research subjects, and as a result, they were especially endangered. Could they still bear healthy children after repeated X-rays? Did the radioactive contrast agents injected into their wombs lead to cancer?

None of the women misused by Ehrhardt applied for compensation from the Austrian Fund for Reconciliation (J. Strasser, personal communication, December 2009). Their fate remains unknown.

Acknowledgments I would like to thank Dr. Jürgen Strasser, the office manager of the Zukunftsfonds Österreich, for assistance concerning the Austrian Fund for Reconciliation. I would also like to thank Pamela Selwyn for the translation services she provided.

References

Becker, N., D. Liebermann, H. Wesch, and G. van Kaick. 2006. *Epidemiologische auswertung der mortalität in der Thorotrast-exponierten gruppe und der kontrollgruppe im vergleich zur mortalität in der allgemeinbevölkerung.* Bonn: Bundesministerium für Umwelt, Naturschutz und Reaktorsicherheit (BMU-2006-682).

Bock, G. 1986. *Zwangssterilisation im Nationalsozialismus. Studien zur rassenpolitik und frauenpolitik.* Opladen, Germany: Westdeutscher Verlag.

Czarnowski, G. 1997. Hereditary and racial welfare [Erb- und Rassenpflege]: The politics of sexuality and reproduction in Nazi Germany. *Social Politics* 4(1): 114–135.

———. 1999. Women's crimes – state's crime: Abortion in Nazi Germany. In *Gender and crime in modern Europe*, ed. M. Arnot and C. Usborne, 238–256. London: UCL Press.

———. 2001. Die restlose Beherrschung dieser Materie. Beziehungen zwischen Zwangssterilisation und Sterilitätsforschung im Nationalsozialismus. *Zeitschrift für Sexualforschung* 14: 226–246.

———. 2004. Vom "reichen material ... einer wissenschaftlichen Arbeitsstätte". Zum problem missbräuchlicher medizinischer praktiken an der Grazer Universitäts-Frauenklinik in der zeit des Nationalsozialismus. In *NS-Wissenschaft als vernichtungsinstrument. Rassenhygiene, zwangssterilisation, menschenversuche und NS-euthanasie in der Steiermark*, ed. W. Freidl and W. Sauer, 225–273. Vienna: Facultas.

———. 2008. "Russenfeten".Abtreibung und forschung an schwangeren zwangsarbeiterinnen in der Universitätsfrauenklinik Graz 1943–1935. *VIRUS – Beiträge zur Sozialgeschichte der Medizin* 7: 53–67.

———. 2012. Österreichs "Anschluss"an Nazi-Deutschland und die österreichische Gynäkologie. In *Herausforderungen. 100 Jahre Bayerische Gesellschaft für Geburtshilfe und Frauenheilkunde*, ed. C. Anthuber, M.W. Beckmann, J. Dietl, F. Dross, and W. Frobenius, 138–148. Stuttgart: Thieme.

Czech, H. 2004. Zwangsarbeit, medizin und "rassenpolitik"in Wien. Ausländische arbeitskräfte zwischen ausbeutung und rassistischer verfolgung. In *Medizin und zwangsarbeit im Nationalsozialismus. Einsatz und "behandlung"von ausländern im gesundheitswesen*, ed. A. Frewer and G. Siedbürger, 253–280. Frankfurt: Campus.

de Snoo, K. 1937. Das trinkende kind im uterus. *Monatsschrift für Geburtshilfe und Gynäkologie* 105: 88–97.

Dietl, J. 2005. *1805-2005. 200 Jahre frauenklinik und hebammenschule Würzburg*. Würzburg: Universitätsfrauenklinik Würzburg.

Dörr, H. 1938. Unsere erfahrungen bei der medikamentösen behandlung der intrauterinen asphyxie. *Monatsschrift für Geburtshülfe und Gynäkologie* 107: 129–137.

Duden, B. 1993. Ein falsch gewächs, ein unzeitig wesen, gestocktes blut. Zur geschichte von wahrnehmung und sichtweise der leibesfrucht. In *Unter anderen umständen. Zur geschichte der abtreibung*, ed. G. Staupe and L. Vieth. Berlin: Deutsches Hygiene Museum and Argon.

Ehrhardt, K. 1933. Zur Biologie der intravenösen plazentographie. IV. Mitteilung. *Fortschritte auf dem Gebiete der Röntgenstrahlen* 48: 405–418.

———. 1937. Der trinkende Fötus. Eine röntgenologische Studie. *Münchener medizinische Wochenschrift* 84: 1699–1700.

———. 1939. Atmet das kind im mutterleib? Eine röntgenologische studie. *Münchener Medizinische Wochenschrift* 86: 915–918.

———. 1941. Weitere erfahrungen mit meiner methode der intraamnialen thoriuminjektion (fetale Organographie). *Zentralblatt für Gynäkologie* 64: 114–120.

———. 1944. Schwangerschaftsunterbrechung jenseits des IV. bis V. Monats durch intraamniale injektion von wirkstoffen. *Medizinische Klinik* 40:507.

———. 1945. Atmet das kind im mutterleib? [Does the child breathe in the uterus?] Ein weiterer beitrag zur intrauterinen biologie des kindes. *Medizinische Zeitschrift* 1.1944–1945: 182–183.

Erbslöh, J. 1942. Die methoden der röntgenologischen darstellung der schwangeren gebärmutter mit hilfe von kontrastmitteln (Amniographie, Plazentographie, Fetographie). Ein rückblick und ausblick. *Geburtshilfe und Frauenheilkunde* 4: 349–365.

Franken, I. 2012. Varianten des rassismus. Zwangssterilisierte, Jüdinnen und zwangsarbeiterinnen als patientinnen der Kölner Universitäts-Frauenklinik 1934 bis 1945. In *Schlagschatten auf das "braune Köln". Die NS-Zeit und danach*, ed. J. Dülffer and M. Szöllösi-Janze, 179–201. Cologne: SH-Verlag.

Frobenius, W. 2004. Abtreibungen bei "Ostarbeiterinnen"in Erlangen. Hochschulmediziner als helfershelfer des NS-Regimes. In *Medizin und zwangsarbeit im Nationalsozialismus. Einsatz und behandlung von "ausländern"im gesundheitswesen*, ed. A. Frewer and G. Siedbürger, 283–307. Frankfurt: Campus.

Hauch, G. 2001. Zwangsarbeiterinnen und ihre kinder. Zum geschlecht der zwangsarbeit. In *Zwangsarbeit – sklavenarbeit. Politik-, sozial- und wirtschaftshistorische Studien*, ed. C. Gonsa et al., 355–448. Vienna: Böhlau.

Landgericht (LG) für Strafsachen. 1947. (19 Vr 20/1947). Steiermärkisches Landesarchiv (StLA), Graz, Austria.

———. 1949. (Vr 4434/1949). Steiermärkisches Landesarchiv (StLA), Graz, Austria.

Link, G. 1999. *Eugenische zwangssterilisationen und schwangerschaftsabbrüche im Nationalsozialismus, dargestellt am beispiel der Universitätsfrauenklinik Freiburg*. Frankfurt: Mabuse.

Medizinische Gesellschaft Steiermark (1944). Vereinsberichte. Sitzung vom 12. März 1944. In *Medizinische Klinik* 40.1944:507.

Menees, T.O., J.D. Miller, and L.E. Holly. 1930. Amniography: Preliminary report. *American Journal of Roentgenology and Radiation* 24: 363–366.

Oakley, A. 1993. *Essays on women, medicine and health*. Edinburgh: Edinburgh University Press.

Soemmering, S.T. 1799. *Icones embryonem humanorum*. Frankfurt on Main: Varrentrapp et Wenner.

Spring, C.A. 2009. *Zwischen krieg und euthanasie. Zwangssterilisationen in Wien 1940–1945*. Vienna: Böhlau.

Chapter 9
Abusive Medical Practices on "Euthanasia" Victims in Austria during and after World War II

Herwig Czech

Introduction

In this chapter, I will discuss medical research and abusive medical practices on "euthanasia"[1] victims in Austria during and after World War II. Of particular importance in this context is the research performed by Austrian psychiatrist Dr. Heinrich Gross on brain specimens from the victims at the notorious Spiegelgrund in Vienna, one of the biggest killing centers of the children's euthanasia program. Although the Gross affair has occupied a central place in public debate on the subject in Austria, I will show that the scientific exploitation of victims of medical atrocities during and after the war was much more widespread and implicated many more institutions and individual researchers than the public focus on this case suggests.

Hartheim

The most important site of medical crimes on Austrian territory was the infamous Hartheim Castle near Linz in Upper Austria, which was one of the six killing centers in the Nazis' *Aktion T4,* the 1940 and 1941 gassing campaign targeting mental patients, named for its headquarters located at *Tiergartenstrasse 4* in Berlin.

---

[1] Whereas "euthanasia" usually refers to a voluntary death with dignity requested by patients or their surrogates at the end of life, in the context of National Socialism it meant the systematic selection and killing by the government of approximately 250,00 to 300,00 children and adults whose lives were considered unworthy of living.

H. Czech, PhD (✉)
Recipient of an APART-Fellowship of the Austrian Academy of Sciences; Department of Contemporary History, University of Vienna, Spitalgasse 2-4, 1090 Vienna, Austria
e-mail: herwig.czech@univie.ac.at

S. Rubenfeld and S. Benedict (eds.), *Human Subjects Research after the Holocaust,*   109
DOI 10.1007/978-3-319-05702-6_9, © Springer International Publishing Switzerland 2014

These killing centers were the first institutions in history established for the industrialized murder of human beings, and they provided much of the technology for the extermination of European Jews in death camps in occupied Poland (Friedlander 1995). Of all T4 killing centers, Hartheim was in operation over the longest period of time and accounted for more than a quarter of all T4 victims, 18,269 out of a total of 70,273 ("Hartheim Statistics," ca. 1942–1945; see also Kugler 2003). A high proportion of all victims of *Aktion 14f13*, the murder of sick or weak concentration camp inmates in T4 killing centers, died at Hartheim. According to one eyewitness, the total death toll at Hartheim reached 30,000, making it the most important center of medical mass atrocities in Austria and probably in all of Nazi Germany (Baumgartner 2003). A comparison of the death rate within Hartheim's catchment area with the death rates throughout the rest of the Reich reveals a disproportionate number of victims in the former Austrian territories (Baumgartner 2003, p. 78).[2] In Germany, excluding the Austrian territories, roughly one person for every 1,280 inhabitants was killed during T4, whereas the same figure for Austria and its annexed territories was one person for every 493 inhabitants, or 2.6 times higher (Statistisches Reichsamt 1943, pp. 18–19; Faulstich 1998, p. 262).[3]

Given its importance as a center of medical killing, it is unfortunate that few records of Hartheim's operations exist. Nonetheless, we know that victims were exploited not only for their personal belongings and their gold teeth but also for research purposes. During the admission procedure, the doctors noted individuals of "medical interest" who were then labeled and photographed before being murdered. One of the photographers at Hartheim reported that between 60 and 80 percent of the victims were photographed. After the gassings, the corpses chosen for autopsy were brought to a special dissection room where brains and other organs were removed and preserved for later use. Hermann Wentzel, an assistant pathologist from the neurological clinic Berlin-Buch, was responsible for the conservation of brains and other body parts (Kepplinger 2008, pp. 82–84).

In 1967, Dr. Georg Renno, the former deputy medical director of the Hartheim facility, was indicted in Germany. At his trial, a woman living in a house close to Hartheim Castle, Maria Achleitner née Schuhmann, presented a glass container to the court and gave an interesting testimony. Together with American forces, her father, who became mayor of Hartheim after the war, had retrieved the glass container from the castle's basement. Her father had told her that in the same place, they found a number of such glass containers containing body parts that were seized and taken by the American authorities (Achleitner 1969; Hartheim Collection (n.d.)).

---

[2] In addition to the former Austrian territories, mental patients from Eastern Bavaria and from annexed regions in Slovenia and Czechoslovakia were deported to Hartheim.

[3] According to the census of May 17, 1939, there were 79,375,281 inhabitants in the German Reich as a whole, of which 6,650,306 were living in the former Ostmark (Third Reich's name for Austria from 1938–1942). The calculation is a conservative approximation based on the assumption that out of 70,273 T4 victims mentioned in the "Hartheim Statistics," 13,500 came from the Ostmark.

The fate and whereabouts of these specimens are unknown, but many years after his trial, which had ended in 1975 without a verdict, and shortly before his death in 1997, Renno named Dr. Hans Bertha as one of the recipients of the specimens (Kohl 2000, pp. 203–210). Bertha had received his training in psychiatry with Rudolf Lonauer, who eventually became medical director at Hartheim, at the psychiatric clinic in Graz under Fritz Hartmann (Hubenstorf 2002). Bertha was a medical "expert" in the T4 killing operation and became director of the Viennese Steinhof Clinic in January 1944, where he implemented a policy of systematic starvation and neglect that took the lives of hundreds if not thousands of patients (Schwarz 2002). Bertha's research interest was in epileptic dementia. When patients with this diagnosis were killed in Hartheim, their brains were preserved for him. According to Renno, Bertha visited the institution several times to receive the specimens (Kohl 2000, pp. 203–210). However, Bertha never published the results of this work, and therefore, it is unknown how far it had progressed. All in all, the details of research activities on Hartheim victims remain largely unknown and, given the extensive destruction of the relevant records, they may remain so in the future.

Gugging

During his investigation of German medical crimes during WWII, the psychiatrist and war crimes investigator Leo Alexander tried—ultimately without much success—to establish the neologism "thanatology" for the development of scientific methods of killing (Weindling 2010, p. 142). A striking practitioner of thanatology was Emil Gelny in Gugging, who invented an entirely new method of killing his victims based on the latest medical technology.

Gugging, a psychiatric hospital near Vienna, was a large institution with nearly 1,400 patients at the end of 1937. During the T4 campaign in 1940 and 1941, close to 780 patients, including 106 children, were transferred from Gugging to Hartheim to be gassed. After T4 ended, death rates in the institution increased due to starvation, willful neglect, and unchecked infectious diseases. In all, excess mortality during these years (excluding T4 victims) is estimated at more than 1,420 deaths (Czech 2012, p. 578).

While in the case of starvation and neglect it is often difficult to prove intent, hundreds of patients were directly murdered in the Gugging hospital by doctors and nurses. In the spring of 1943, Hartheim's medical director, Helmut Lonauer, spent several days in the institution and killed more than 100 patients with drug overdoses (Oman 1948). A few months later, another wave of killings was unleashed in Gugging—Dr. Emil Gelny was the driving force.

Gelny was born in Vienna in 1890 and had been a Nazi activist since 1932 (Gelny 1938–1947). Due to his excellent relations with the Gauleiter, he was appointed director of the two psychiatric hospitals in the *Reichsgau Niederdonau*: Gugging and Mauer-Öhling. Gelny began working at Gugging on November 1, 1943, and killed up to 100 people each month with lethal doses of medicine (*Case against Georg Renno*, 1969).

Gelny saw no reason to conceal his activities. In the summer of 1944, between 30 and 50 of Germany's leading psychiatrists, most of them directors of psychiatric hospitals, convened at Gugging. Euthanasia issues were at the very top of the agenda and Gelny chose this forum to present his killing methods. Using a specially modified electroshock device, he killed a patient in front of the audience to demonstrate the effectivity of his invention (Dokumentationsarchiv des österreichischen Widerstandes 1987, p. 656). In April 1945, just before the end of the war in Europe, Gelny personally murdered approximately 150 people, many of them forced laborers, using his shock device in the Mauer-Öhling hospital (*Case against Dr. Emil Gelny* 1948).

Gelny learned of the recently developed electroshock therapy at the Psychiatric University Clinic in Vienna, where he did an internship under Professor Otto Pötzl before assuming control of Lower Austria's psychiatric institutions (Haminger 1948; see also Oman 1948). While research at the clinic included chemically induced states of shock using either insulin or Cardiazol (e.g., Birkmayer 1939), the development and testing of electroshock treatment was of particular importance to the T4 organization, which provided strong support for its dissemination and application (Reichsbeauftragter 1942). Wolfgang Holzer, one of the assistants at the clinic, developed his own prototype of an electroshock device that was competing with Siemens, the leading manufacturer of these devices (Elektroschockapparat, ca. 1944). Holzer was in contact with the medical director of the T4 program, Professor Paul Nitsche, to promote his device and his plans for a research institute in Vienna that would focus on the development of physical methods of therapy in psychiatry (Nitsche 1944). In a planning document submitted to T4, he cited the window of opportunity opened by the ongoing euthanasia killings as the main motive for his project (Holzer 1944).

Spiegelgrund

In Austria, the name of Heinrich Gross, a doctor at the children's euthanasia facility at Spiegelgrund, has over the last decade become synonymous with unethical research on victims of Nazi medical crimes.

Preparations for the systematic registration and extermination of children with mental handicaps began in the spring of 1939 with the establishment of a special front organization in the Führer's Chancellery in Berlin: the Reich Committee for the Scientific Registration of Serious Hereditary and Congenital Ailments. A secret circular in August 1939 obliged doctors and midwives to report all cases of "idiocy" and diverse "deformities" to the public health offices (Reich Minister of the Interior 1939; cited in Klee 2010, pp. 673–676). The Reich Committee then ordered the person classified as such to be committed to one of the so-called "special children's wards," which had been set up either in existing clinics or as separate institutions.

There were at least 30 of these clandestine killing centers in the Reich (Benzenhöfer 2000).

Spiegelgrund, the second killing center of the children's euthanasia program (after Brandenburg-Görden), and one of the largest, was opened in July 1940 on the premises of the psychiatric hospital Steinhof in Vienna (Decree no. 572, 1946). The facility was designed as a permanent part of the city's child and youth care system and had multiple functions, including the concentration, examination, and discreet killing of children in the context of the children's euthanasia program. The aim was to eliminate children who were regarded as economically and biologically worthless because of mental retardation, disability, or severe malformations (Illing 1946). The key criterion in the selection process was "educability" (*Bildungsfähigkeit*), a prognostication of a child's ability to earn a living independently as an adult ("Medical Case Files," 1940–1945).[4] The children were observed and examined, and the results were sent to Berlin where the Reich Committee would reach a final decision on the course of action in individual cases. When the authorization for killing was received in Vienna, children were plied with high doses of barbiturates until they became so weak that they died of pneumonia or other infectious diseases. The *Totenbuch* (Book of the Dead) records the names of 789 individuals who died at the institution between August 25, 1940, and June 3, 1945, mostly of poisoning, neglect, hunger, and various infections (Czech 2002b, pp. 186–187).

Inmates at Spiegelgrund were also subject to various forms of medical torture. Survivors report injections of apomorphine, which caused hours of vomiting, nausea, and feelings of impending doom (*Vernichtungsgefühl*), and of sulfur or its derivatives, which caused excruciating pain and paralysis for several days. Isolation, cold baths, wet sheets, and deprivation of food, sleep, and clothes were other examples of the abuse suffered by the Spiegelgrund inmates (Kaufmann 1991; Gross 2000; Lehmann and Schmidt 2001).

After the deaths of the children, their brains, spinal cords, and other organs were removed and conserved for further scientific use. In this respect, Vienna was, of course, not a unique case. In many of Nazi Germany's euthanasia institutions, the victims were exploited for scientific purposes after their death. For instance, Julius Hallervorden boasted to the war crimes investigator Leo Alexander after the war that he had collected more than 700 brains from euthanasia victims at the Kaiser Wilhelm Institute for Brain Research at Berlin-Buch (Alexander 1945; cited in Seidelman 2012).

Dr. Ernst Illing, the second director of the Spiegelgrund clinic, collected body parts from 21 individuals for his research on tuberous sclerosis, first in the killing center Brandenburg-Görden and then in Vienna. He published his results in 1943 intending to prove that pneumatic encephalography would be a reliable tool for the diagnosis of tuberous sclerosis in living patients (Illing 1943). The possibility of killing research subjects provided a unique opportunity to immediately verify the

---

[4] In the context of T4, children were selected according to the same criteria (Fuchs 2010).

diagnoses through postmortem examinations (Illing 1943).[5] Pneumatic encephalography, the diagnostic method that Illing advocated in his paper, was dangerous, painful, and provided little benefit for the patients because there were no known cures for their ailments. During the procedure, which was routinely performed on the children at Spiegelgrund, including the sick and the weak, the ventricles of the brain were filled with air to render them visible on X-ray ("Medical Case Files," 1940–1945).[6] We know from the postwar trial against the Spiegelgrund doctors that pneumatic encephalography led to the deaths of some victims (Dahl 1998, p. 95). Despite the dangers involved, most of the patients were subjected to this and other diagnostic methods, thereby laying the groundwork for the later scientific exploitation of the victims' dead bodies.

Heinrich Gross was in charge of the ward (Pavilion 15) where the killings took place. His academic achievements during the war were rather modest, but they were the foundation for his later career. His first attempt at exploiting the Spiegelgrund victims dates back to November 1942, when he presented a case study on a two-month-old boy to the Viennese Biological Society. The study, published two years later, was based on a collaboration with the Anatomical Institute of the University of Vienna, headed by Professor Eduard Pernkopf (Gross 1944).

It appears that body parts of euthanasia victims circulated widely among researchers and institutions in the field. For example, according to records from the pathology department of the Steinhof hospital, 24 brains were sent to Professor Carl Schneider at Heidelberg between April and July 1944 (Autopsy record of Otto Wagner-Spital 1944).[7] The details of this cooperation and the fate of the specimens remain unknown. The patients whose brains were sent to Heidelberg had not died at the Spiegelgrund facility, but in the adjacent psychiatric hospital for adults, indicating that research efforts during the war extended beyond the "children's euthanasia" to potential victims of the "decentralized euthanasia" implemented between 1941 and 1945 after the centralized T4 program officially ended (Autopsy records of Otto Wagner-Spital 1944; see also Schwarz 2002).

The Neurological Institute at Vienna University also received specimens, some of which were kept there as part of the institute's collections until the official burial of the Spiegelgrund victims' remains in 2002 (Angetter 1998; Czech 2002a, pp. 160–163).[8] The neuropathological samples obtained from the corpses of the euthanasia victims were used for scientific publications for decades. Although the foundations for this research were laid during the war, the majority of the publications were published in the 1950s and 1960s, as I will describe in the last section of this chapter.

---

[5] In 1978 the same specimens were used in another paper, this time published by Dr. Heinrich Gross and two coauthors (Gross et al. 1978).

[6] Excerpts are published in Häupl 2006.

[7] Carl Schneider was one of the central figures in "children's euthanasia" and in the research efforts in the context of T4 (see Becker-von Rose 1990).

[8] Regarding the burial, see conclusion of this chapter.

Tuberculosis Experiments

Some of the Spiegelgrund patients were used as guinea pigs in lethal experiments. The aim of these experiments was to test the reliability of the Bacillus Calmette-Guérin (BCG) vaccine against tuberculosis, one of the main causes of death in the first half of the twentieth century. Many called it *morbus viennensis,* or *Wiener Krankheit* (Viennese disease) (Dietrich-Daum 2007). The use of the BCG vaccine in Germany had suffered a huge setback in 1930 in Lübeck when 75 children died and many more suffered permanent damage because the attenuated bacillus was mistakenly replaced by a virulent bacillus in the preparation of the vaccine. By the end of the 1930s, new methods of application were recommended, but there was still no certainty about the reliability of the vaccine (Dahl 2002). Doctors found favorable conditions to answer this question in the special facilities of the children's euthanasia program.

Experiments with tuberculosis vaccines took place in various institutions, most importantly in Berlin, Kaufbeuren-Irsee in Bavaria, and Spiegelgrund in Vienna. The initiative in Vienna did not come from the euthanasia doctors, but from the Vienna University Pediatric Clinic. The clinic's director was Franz Hamburger, a committed National Socialist. One of his assistants, Elmar Türk (1907–2005), wanted to take advantage of the research opportunities offered at the Spiegelgrund clinic. The initiative was certainly facilitated by the facts that the first director of the Spiegelgrund clinic, Dr. Erwin Jekelius, was a former assistant of Hamburger, and that the two institutions were already collaborating on different levels (Dahl 1998, pp. 110–112; Dahl 2002; Czech 2011b, pp. 43–49).

In two series of experiments, Türk intentionally infected five children with virulent tuberculosis bacilli, three of them after receiving vaccinations and two without any protection as controls. During the first experiment, three children were transferred from the University Pediatric Clinic to the Spiegelgrund. They were between one and a half and five years of age. One girl had been vaccinated against tuberculosis and then infected; the two other children—a boy and a girl—had been infected as controls without any immunization. They all died within one to three months after their admission to Spiegelgrund in 1943. In 1944, Türk published the results of a second experiment in which he had used two more children, following the same pattern. In the records of these children, some harrowing documents survive. When Türk transferred the children to the Spiegelgrund clinic, he sent along a letter with his detailed wishes concerning their autopsies. The experiments were not only about observing the effects of the vaccination and the infection but also about complementary postmortem examinations. In this case, it was not the euthanasia apparatus in Berlin that decided to kill these children. This decision was made at the University Pediatric Clinic, since preparations for the children's autopsies began before they were admitted to the Spiegelgrund clinic (Dahl 1998, pp. 110–112; Czech 2011b, pp. 43–49; Türk 1942; 1944).

In other cases of experimental research on human beings, the boundaries of the ethically acceptable were clearly tested and sometimes crossed, although it is often impossible to determine with certainty if patients consented and, if not, whether they were harmed. One such example was a study on the temperature regulation in children that was also done at Hamburger's clinic. In the most extreme of the experiments, described in a 1943 doctoral dissertation, a six-month-old child with a mental handicap was exposed to cold until his body temperature fell by several degrees Celsius. The aim of the research was to show that children, in keeping with Nazi ideals of physical toughness, could be exposed to large temperature changes without being harmed (Kawura 1943).

The focus on the role of psychiatry in the euthanasia program has sometimes eclipsed the role of other medical specialties. At Spiegelgrund, practically all children were subjected to various forms of psychological testing and assessment by either psychiatrists or psychologists. Igor Caruso, who later would become one of the best-known psychologists and psychoanalysts in Austria, wrote over 100 reports on children at Spiegelgrund. Some of these assessments were part of the decision-making process that led to the killing of his subjects.[9]

The wider context was a movement to establish child psychiatry, psychology, and therapeutic pedagogy as independent academic disciplines. Therapeutic pedagogy (Heilpädagogik) is an interesting case in point because efforts to establish this new discipline united euthanasia activists from the Spiegelgrund, such as Erwin Jekelius and Heinrich Gross, with staff from the University Pediatric Clinic, such as Franz Hamburger or the much better known Hans Asperger, who was able to greatly advance his career during the war (Hubenstorf 2005; Czech 2011b). Asperger is an important figure in the history of child psychology because of his work on autism, which was acknowledged in the term "Asperger syndrome" in psychiatric nomenclature until recently. Asperger kept a certain distance from the Nazi movement, which may account for the fact that his involvement in children's euthanasia remains practically unknown to this day. In 1942, as part of a seven-member commission, Asperger examined 220 inmates of the Gugging hospital's children's facility. Of these, 35 were sent to the Spiegelgrund at the behest of the commission; all 35 children died within a short period of time (Czech 2011b, p. 27).

Neuropathological Research After WWII

The case of Dr. Heinrich Gross highlights the way Austria dealt with the crimes of National Socialism.[10] In 1948, after returning from Soviet internment as a prisoner of war, Gross was indicted for his part in the killing of children at the Spiegelgrund facility, but the verdict was eventually set aside because of a legal technicality. Gross embarked on a second career as a specialist at the Neurological Hospital at

---

[9] Caruso's role at the Spiegelgrund became known not long ago and led to a heated controversy in the Austrian psychoanalytical community. It was triggered by List (2008).

[10] For a more detailed account in English, see Neugebauer and Stacher (1999).

Rosenhügel in Vienna, and then returned to Steinhof where he rose to the rank of head physician (Czech 2002a; Neugebauer and Schwarz 2004).

Around 1952, he resumed research on the brains of the Spiegelgrund victims, which had been carefully preserved by the institution's pathologist, Dr. Barbara Uiberrak, a Nazi party member since before 1938 who was in charge of the pathology laboratories at the entire Steinhof complex from 1938 into the 1960s (*Case against Dr. Barbara Uiberrak* 1948). In 1946, barely a year after the killings had ended, she testified before the People's Court in Vienna about the children she had autopsied:

> Almost all of the cases are very interesting when you look at them scientifically. We still have all of the 700 brains at Steinhof. Also, in most cases, we have the glands with their inner secretions so they could be ready for scientific pathological examinations anytime. I believe it would be rewarding to take a few cases from each year. (Uiberrak 1946, Vg 2b Vr 2365/45)

Even though she had autopsied almost each body personally, she claimed that she never had seen anything that would have indicated an unnatural death (Uiberrak 1946).

In 1953, Dr. Gross published his first article about interesting neuropathological cases from the Spiegelgrund brain collection (Gross 1953). It was based on the same patient about whom Gross had already published in 1944. The patient, a boy named Günther Pernegger, who was born November 16, 1941, had been placed in the Spiegelgrund facility when he was six weeks old because he had abnormalities of his head and hands. After seven weeks at the institute, where according to his patient file he was eating very little, he developed pneumonia and died six days later (Pernegger 1941–1942).

Over the following 25 years, Gross would use body parts as a basis for at least 38 scientific publications in neuropathology, often with prominent coauthors (Czech 2002a). In 1968, the Ludwig Boltzmann Society established the Ludwig Boltzmann Institute for Research into Malformations of the Nervous System. As head of the institute, Gross systematically exploited the brain specimens preserved during the Nazi period, many of which were kept at the Steinhof hospital until 2002, in some cases until 2012 (Gross 1968).[11]

The Ludwig Boltzmann Institute was not the only scientific institution that profited indirectly from the murders. In the 1950s, Gross gave body parts from approximately 20 Spiegelgrund victims to Vienna University's Institute of Neurology, then under the direction of Hans Hoff (Angetter 1998), which resulted in at least two publications (Seitelberger et al. 1957; Bernheimer and Seitelberger 1968). One of these papers, authored by Franz Seitelberger and two colleagues, was about two sisters, of whom one, Anna Fritz, had died in 1942 at Spiegelgrund ("Book of the Dead," Wiener Stadt- und Landesarchiv [WStLA], 1.3.2.209.10.B4 1940–1945). The specimens from the two sisters were also given to the Max-Planck Institute for Brain Research in Gießen under Julius Hallervorden, and two more

---

[11] Regarding the fate of the human remains, see conclusion of this chapter.

papers were published about them in 1954 (Diezel 1954a, b). The brain of Anna Fritz was found in 2002 among other brain specimens of euthanasia victims at Vienna University's Institute of Neurology.[12]

Among Gross's coauthors (Gross et al. 1968) on the Spiegelgrund specimens was the aforementioned Hans Hoff, one of the few prominent medical refugees from the Nazi regime who returned to Austria after the war. In 1949, he became the head of Vienna University's Neurological Institute. In 1959, Hoff, Gross, and Uiberrak coauthored a long paper based on 29 victims, including 26 who had died at Spiegelgrund, entitled "The Principal Malformations of the Telencephalic Brain Chambers" (Gross et al. 1959). This collaboration clearly boosted Gross's reputation and gave Hoff access to the Spiegelgrund specimens, although it is unclear if he knew about the origin of the collection at the time. If Gross had hoped to further his academic career this way, he was in for a disappointment. According to Michael Hubenstorf, three years later Hoff thwarted Gross's attempt to get his habilitation because of the tainted origin of his research material (2002, p. 371). Hoff did not, however, take a clear public stance on the issue of the euthanasia killings, and in 1965 he prefaced an apologetic book on the euthanasia crimes that largely exonerated psychiatry (Hubenstorf 2002, p. 369).

Hans Hoff's successor at the Neurological Institute was Franz Seitelberger, who also collaborated with Gross on at least two publications in 1962 and 1966 based on specimens from the Spiegelgrund victims (Gross et al. 1962; Gross and Seitelberger 1966). Seitelberger became rector of Vienna University in the 1970s despite his Nazi past, which included membership in the SS (Seitelberger (n.d.): Bundesarchiv Berlin; Hubenstorf 2002, p. 370). He based his habilitation on brain specimens of three sisters who had died between 1942 and 1944 in the Brandenburg-Görden killing facility. In all likelihood, he received the material during a stay with Julius Hallervorden in Gießen in the 1950s (Seitelberger 1954; Professor Herbert Budka, personal communication, 2001).

Andreas Rett, who made important contributions to the treatment of children with mental handicaps while covering up his Nazi past until relatively late, was another of Gross's prominent coauthors (Ronen et al. 2009). In 1968, Rett coauthored a paper on "Infantile Cerebral Disorders" with Gross, Jellinger, and Kaltenbäck in the *Journal of the Neurological Sciences*, which was partly based on data from the Spiegelgrund brain collection.

Although Gross worked as a court expert and brain researcher, he also found time to conduct extensive pharmacological trials. New medications would often come directly from animal experiments for trials on Steinhof patients. According to his own words, Gross tested 83 different psychotropic drugs between 1958 and 1968, in many cases on more than 100 patients (Gross and Kaltenbäck 1968, p. 2). Referring to one such trial carried out in 1973, he made it clear that there was no therapeutic benefit for the patients involved (Arbeitsgemeinschaft Kritische Medizin 1980, p. 25). The drug in question, Clozapin, was forbidden in Finland

---

[12] Witnessed by the author.

in 1975 because nine patients had died from its side effects. Gross had tested it on 500 patients (Gross 1973).

Gross was in high demand as a psychiatric expert in court. In 1976, he was unexpectedly confronted with a Spiegelgrund survivor, Friedrich Zawrel, who had been interned at the institution as an "unreformable" ten-year-old. In his testimony, Gross freely quoted from Zawrel's 1944 Spiegelgrund file and did everything in his power to have Zawrel incarcerated for life as a dangerous criminal (Lehmann and Schmidt 2001).[13] Dr. Werner Vogt and the Arbeitsgemeinschaft Kritische Medizin (Association of Critical Medicine) took up Zawrel's cause and saw to it that Heinrich Gross's career suffered its first setback in 1981. When he lost a sensational libel case against Vogt, the court regarded Gross's participation in the children's murder as proven.[14]

Nevertheless, it took another 20 years until the Viennese Public Prosecution Service chose to indict Gross for his involvement in Nazi euthanasia. In the meantime, another affair sensitized at least some of the medical community and the wider public to research on victims of National Socialism. As became publicly known in the 1990s, the dead bodies of close to 1,400 persons executed in Vienna during the war years had been transported to the Anatomical Institute at Vienna University in order to be used in teaching and research. The head of the institute, Professor Eduard Pernkopf, used part of these bodies for his famous anatomical atlas published in several volumes between 1937 and 1960 (Pernkopf 1937–1960). Pernkopf himself had been an ardent follower of National Socialism and became dean of the medical faculty in 1938 and rector of Vienna University in 1943. Although he was dismissed in 1945, he nevertheless was able to continue his work using the university's resources. Despite its problematic origins, his atlas was regarded as a masterpiece by the medical community and used in many countries for teaching and studying anatomy. In 1996, the university commissioned an investigation into accusations about the origins of some of the source materials for the drawings. In 1998, the commission issued a report of more than 500 pages that unfortunately was never officially published and remains difficult to obtain (Akademischer Senat der Univ. Wien 1998). This so-called Pernkopf Commission was Vienna University's only major effort to shed light on unethical research practices by the medical faculty during National Socialism (e.g., Seidelman and Israel 1997; Malina 1998).

The Gross affair finally culminated in the year 2000, when a belated attempt to bring him to justice failed because an expert witness declared him unfit for trial (Escher 2000). As a consequence, a verdict was never pronounced, and Gross died a free man in 2005. After the unsuccessful trial, in April 2002 the human remains of the Spiegelgrund victims were finally buried in an honorary grave at the Viennese main cemetery. At the same time, the Documentation Center of the Austrian

---

[13] For an extensive online interview with Fritz Zawrel, go to www.gedenkstaettesteinhof.at (Czech 2011a).

[14] For an account by one of the main protagonists, see Vogt (2000).

Resistance prepared a permanent exhibition (The War Against the Inferior) on the premises of the Steinhof hospital, which documents the history of Nazi medicine during World War II in Vienna.[15]

The dead from the Spiegelgrund, however, were not the only victims of Nazi euthanasia whose mortal remains were preserved over decades. Human remains of former patients of the Steinhof psychiatric hospital (as opposed to the Spiegelgrund facility, which was located on the hospital's premises but organized as a separate institution), who had died during the war under conditions that, in many cases, amounted to a discreet form of killing, were known to be stored at Gross's former research institute and at the Steinhof hospital's pathology unit. These human remains from more than 60 individuals were finally documented and individually identified in 2010.[16] It is safe to assume that most of these victims died in the "decentralized euthanasia" that followed the official end of the T4 program. The remains offer clear proof that postmortem research on euthanasia victims in Austria did not occur solely in children's euthanasia, as previously thought (Czech and Mettauer 2011). These specimens were finally buried in a public ceremony in Vienna in May 2012 ("Vienna Buries Victims," 2012).

Many other places and institutions have yet to come under scrutiny. One such example is the former Medical Academy of the SS in Graz, an institution that was affiliated with the local university (Anonymous 1941). We know of its connections to the concentration camps Dachau and Mauthausen, where research material was taken from prisoners' dead bodies. In Dachau students from the SS Medical Academy allegedly operated on hundreds of prisoners without any medical indication for surgery. Still, there has not been a serious study of this academy (Freund 1998, p. 176; Klee 2001, p. 22–23). With the exception of Vienna, no region in Austria has so far undertaken a comparable effort to search for human remains in hospitals, psychiatric institutions, and research institutes.

Acknowledgments The author would like to thank Sheldon Rubenfeld for many helpful comments on an earlier version of this chapter.

References

Achleitner, M. 1969. *Case against Georg Renno and others*, LG Frankfurt/Main, Ks 1/69. [Testimony by Maria Achleitner]. Hessisches Hauptstaatsarchiv 631a/817, [Hessian Main State Archive], Wiesbaden, Germany. Copy in DÖW Hartheim collection no. 261. Dokumentationsarchiv des österreichischen Widerstandes [Documentation Center of the Austrian Resistance], Vienna, Austria.

---

[15] The exhibition's Web site is www.gedenkstaettesteinhof.at.

[16] This identification process was done in a project financed by the City of Vienna and carried out by the Documentation Center of the Austrian Resistance (DÖW), under the responsibility of the author.

Akademischer Senat der Univ. Wien. (Ed.). 1998. *Untersuchungen zur anatomischen Wissenschaft in Wien 1938–1945. Senatsprojekt der Universität Wien.* Vienna: Universität Wien.

Alexander, L., 1945. *Neuropathology and neurophysiology, including electroencephalography, in wartime Germany.* Combined Intelligence Objectives Sub-Committee G-2 Division SHAEF (Rear) APO 413. July 20. National Archives and Records Administration, Washington, DC.

Angetter, Daniela. 1998. Überprüfung der Sammlung des Neurologischen Instituts. In *Untersuchungen zur anatomischen wissenschaft in Wien 1938–1945. Senatsprojekt der Universität Wien*, ed. Akademischer Senat der Univ. Wien, 266–288. Vienna: Universität Wien.

Anonymous. 1941. SS-Ärztliche Akademie in Graz. *Ärzteblatt für die deutsche Ostmark* 4: 157–158.

Arbeitsgemeinschaft Kritische Medizin. 1980. Gross als Forscher und die Ideologie der NS-Hirnforschung. *Eingriffe/Informationen der Arbeitsgemeinschaft Kritische Medizin* 22–25.

Autopsy record of Otto Wagner-Spital. 1944. Medical case files: Deceased girls and boys. Wiener Stadt- und Landesarchiv (WStLA), 1.3.2.209.10.A2. Municipal and Provincial Archives of Vienna.

Baumgartner, A. 2003. "Die Kranken sind dann vergast worden." Die Ermordung von KZ-Häftlingen in Hartheim. In *Der Wert des Lebens: Gedenken/lernen/begreifen. Begleitpublikation zur Ausstellung des Landes OÖ in Schloss Hartheim 2003*, ed. Institut für Gesellschafts- und Sozialpolitik an der Johannes Kepler-Universität Linz u. a., 74–79. Trauner Verlag: Linz.

Becker-von Rose, P. 1990. Carl Schneider, wissenschaftlicher Schrittmacher der Euthanasieaktion und Universitätsprofessor in Heidelberg 1933–1945. In *Von der Heilkunde zur Massentötung. Medizin im Nationalsozialismus*, ed. G. Hohendorf and A. Magull-Seltenreich, 192–217. Heidelberg: Wunderhorn.

Benzenhöfer, U. 2000. *"Kinderfachabteilungen" und NS-Kindereuthanasie.* Wetzlar: GWAB.

Bernheimer, H., and F. Seitelberger. 1968. Über das Verhalten der Ganglioside im Gehirn bei 2 Fällen von spätinfantiler amaurotischer Idiotie. *Wiener Klinische Wochenschrift* 9: 163–164.

Birkmayer, Walter. 1939. Motorische Erscheinungen im Cardiazol-Krampf [Motor symptoms in Cardiazol shock treatment]. *Archiv für Psychiatrie und Nervenkrankheiten* 2/3: 291–313.

Book of the Dead of the Viennese Neurological Clinic for Children at Spiegelgrund. 1940–1945. Wiener städtische Nervenklinik für Kinder. Wiener Stadt- und Landesarchiv (WStLA), 1.3.2.209.10.B4. Municipal and Provincial Archives of Vienna.

Case against Dr. Barbara Uiberrak. 1948. Landesgericht Wien (LG Wien), Vg 3b Vr 474/48. Copy in DÖW E 22719. Dokumentationsarchiv des österreichischen Widerstandes [Documentation Center of the Austrian Resistance], Vienna, Austria.

Case against Dr. Emil Gelny and accomplices. 1948. Landesgericht für Strafsachen Wien (LG Wien),Vg 8a Vr 455/46. Copy in DÖW 18860. Dokumentationsarchiv des österreichischen Widerstandes [Documentation Center of the Austrian Resistance], Vienna, Austria.

Czech, H. 2002a. Forschen ohne Skrupel. Die wissenschaftliche Verwertung von Opfern der NS-Psychiatriemorde in Wien. In *Von der Zwangssterilisierung zur Ermordung. Zur Geschichte der NS-Euthanasie in Wien Teil II*, ed. E. Gabriel and W. Neugebauer, 143–163. Vienna: Böhlau.

———. 2002b. Selektion und Kontrolle. Der "Spiegelgrund" als zentrale Institution der Wiener Jugendfürsorge zwischen 1940 und 1945. In *Von der Zwangssterilisierung zur Ermordung. Zur Geschichte der NS-Euthanasie in Wien Teil II*, ed. E. Gabriel and W. Neugebauer, 165–187. Vienna: Böhlau.

———. 2011a. Interview with Friedrich Zawrel. [video]. Retrieved from Dokumentationsarchiv des österreichischen Widerstandes (DÖW) website at http://gedenkstaettesteinhof.at/de/interviews/videos/Friedrich-Zawrel Accessed 23 Dec 2013.

———. 2011b. Zuträger der Vernichtung? Die Wiener Universitäts-Kinderklinik und die NS-Kindereuthanasieanstalt "Am Spiegelgrund". In *Festschrift 100 Jahre Wiener*

Universitätsklinik für Kinder- und Jugendheilkunde, ed. A. Pollak, 23–54. Vienna: Universitätsklinik für Kinder- und Jugendheilkunde, Wien.

————. 2012. Nationalsozialistische Medizinverbrechen in der Heil- und Pflegeanstalt Gugging. In *Update! Perspektiven der Zeitgeschichte. Zeitgeschichtetage 2010*, ed. L. Erker, A. Salzmann, L. Dreidemy, and K. Sabo, 573–581. Innsbruck: Studienverlag.

Czech, H., and P. Mettauer. 2011. *Überprüfung von Präparaten im Otto Wagner Spital der Stadt Wien auf einen möglichen NS-Hintergrund* (unpublished report). Vienna

Dahl, M. 1998. *Endstation Spiegelgrund. Die Tötung behinderter Kinder während des Nationalsozialismus am Beispiel einer Kinderfachabteilung in Wien*. Vienna: Erasmus.

————. 2002. "... deren Lebenserhaltung für die Nation keinen Vorteil bedeutet." Behinderte Kinder als Versuchsobjekte und die Entwicklung der Tuberkulose-Schutzimpfung. *Medizinhistorisches Journal* 37: 57–90.

Decree no. 572. 1946. Wiener Stadt- und Landesarchiv (WStLA), 1.3.2.209.A1. Municipal and Provincial Archives of Vienna.

Dietrich-Daum, E. 2007. *Die "Wiener Krankheit". Eine Sozialgeschichte der Tuberkulose in Österreich*. Vienna: Oldenbourg.

Diezel, P.B. 1954a. Histochemische Untersuchungen an primären Lipoidosen. Amaurotische Idiotie, Gargoylismus, Niemann-Picksche Krankheit, Gauchersche Krankheit, mit besonderer Berücksichtigung des Zentralnervensystems. *Virchows Archiv*, 326: 89–118.

————. 1954b. Histochemischer Nachweis des Gangliosids in Ganglien- und Gliazellen bei amaurotischer Idiotie und Isolierung der lipoidspeichernden Zellen nach der Methode von M. Behrens. *Deutsche Zeitschrift für Nervenheilkunde und deren Grenzgebiete*, 172: 344–350.

Dokumentationsarchiv des österreichischen Widerstandes (DÖW) (ed.). 1987. *Widerstand und Verfolgung in Niederösterreich. Band 3*. Vienna: Österreichischer Bundesverlag.

Elektroschockapparat "Elkra I" and "Elkra II." (ca. 1944). [Advertising brochures]. Bundesarchiv Berlin (BAB), R 96 I-12. [German Federal Archives Berlin].

Escher, R. 2000. Gross-Prozess geplatzt [Gross trial has failed, 21.3.2000]. *Salzburger Nachrichten*. p. 6.

Faulstich, H. 1998. *Hungersterben in der Psychiatrie 1914–1949. Mit einer Topographie der NS-Psychiatrie*. Freiburg: Lambertus, Akademischer Senat der Universität Wien.

Freund, F. 1998. Zum Umgang mit Leichen im KZ Mauthausen und dem Nebenlager Gusen. In *Untersuchungen zur anatomischen Wissenschaft in Wien 1938–1945. Senatsprojekt der Universität Wien*, ed. Akademischer Senat der Univ. Wien, 163–176. Vienna.

Friedlander, H. 1995. *The origins of Nazi genocide: From euthanasia to the final solution*. Chapel Hill: University of North Carolina Press.

Fuchs, P. 2010. Zur Selektion von Kindern und Jugendlichen nach dem Kriterium der "Bildungsunfähigkeit". In *Die nationalsozialistische "Aktion T4" und ihre Opfer. Von den historischen Bedingungen bis zu den ethischen Konsequenzen für die Gegenwart*, ed. M. Rotzoll, G. Hohendorf, P. Fuchs, C. Mundt, and W.U. Eckart, 287–296. Paderborn, Germany: Schöningh.

Gelny, E. 1938–1947. Gauakt [NSDAP personnel file] Dr. Emil Gelny. Österreichisches Staatsarchiv, Archiv der Republik (ÖStA, AdR), Austrian State Archives, Vienna.

Gross, H. 1944. Ein Fall von Akrocephalosyndaktylie. *Wiener klinische Wochenschrift* 57: 493.

————. 1953. Zur Morphologie des Schädels bei der Acrocephalosyndaktylie. *Morphologisches Jahrbuch* 92: 350–372.

————. 1968. Institut zur Erforschung der Mißbildungen des Nervensystems. In *Geschäftsbericht 1968* (unveröffentlicht), ed. L. Boltzmann-Gesellschaft, 5–6. Vienna.

————. 1973. *Vortrag beim Clozapin-Symposium der Psychiatrischen Universitätsklinik Wien am 25. 5. 1973. Typoskript, 5 Seiten*. [unpublished paper]. Copy at Dokumentationsarchiv des österreichischen Widerstandes [Documentation Center of the Austrian Resistance], Vienna, Austria.

————. 2000. *Spiegelgrund. Leben in NS-Erziehungsanstalten*. Vienna: Ueberreuter.

Gross, H., and E. Kaltenbäck. 1968. The clinical position of moperone among the butyrophenons. *Nordic Journal of Psychiatry* 23(1): 4–9.

Gross, H., and F. Seitelberger. 1966. Die pathologische Anatomie der zerebralen spastischen Paresen. *Wiener Medizinische Wochenschrift* 116: 756–760.

Gross, H., H. Hoff, and E. Kaltenbäck. 1959. Über die wichtigsten Fehlbildungen der telencephalen Hirnkammern. *Wiener Zeitschrift für Nervenheilkunde* 1–34.

Gross, H., E. Kaltenbäck, and F. Seitelberger. 1962. Über eine systemisierte Fehlbildung des Rautenhirns. *Wiener Klinische Wochenschrift* 74: 705–708.

Gross, H., K. Jellinger, E. Kaltenbäck, and A. Rett. 1968. Infantile cerebral disorders. *Journal of the Neurological Sciences* 7: 551–564.

Gross, H., E. Kaltenbäck, and M. Godizinski. 1978. Tuberöse Sklerose. Neuropathologischer Befund und klinisches Korrelat bei 21 Fällen. In *Aktuelle Probleme der Neuropathologie 4*, ed. K. Jellinger and H. Gross, 75–87. Vienna: Facultus.

Hartheim Collection. (n.d.). (Copies of material concerning "euthanasia crimes" in Austria). Collection no. 261. Dokumentationsarchides österreichischen Widerstandes (DÖW) [Documentation Center of the Austrian Resistance]. Vienna, Austria.

"Hartheim Statistics." [ca. 1942–1945]. Microcopy no. T-1021, Record Group 242/338, item no. 000-12-463, Exhibit 39, roll no. 18, frame no. 91. National Archives and Records Administration, Washington, DC. Copy in DÖW 22862. Dokumentationsarchiv des österreichischen Widerstandes [Documentation Center of the Austrian Resistance], Vienna, Austria.

Häupl, W. 2006. *Die ermordeten Kinder vom Spiegelgrund: Gedenkdokumentation für die Opfer der NS-Kindereuthanasie in Wien*. Vienna: Böhlau.

Holzer W. Vorschlag zur Gründung einer Forschungsanstalt für active Therapie der Nerven- und Geisteskrankheiten, Sept 1944; BAB, R 96 I-18.

Hubenstorf, M. 2002. Tote und/oder lebendige Wissenschaft. Die intellektuellen Netzwerke der NS-Patientenmordaktion in Österreich. In *Von der Zwangssterilisierung zur Ermordung. Zur Geschichte der NS-Euthanasie in Wien Teil II*, ed. E. Gabriel and W.G. Neugebauer, 237–420. Vienna: Böhlau.

———. 2005. Pädiatrische Emigration und die "Hamburger-Klinik" 1930–1945. In *90 Jahre Universitäts-Kinderklinik am AKH in Wien. Umfassende Geschichte der Wiener Pädiatrie*, ed. K. Widhalm and A. Pollak, 69–220. Vienna: Literus-Universitätsverlag.

Illing, E. 1943. Pathologisch-anatomisch kontrollierte Enzephalographien bei tuberöser Sklerose. *Zeitschrift für die gesamte Neurologie und Psychiatrie*, 178: 160–171.

———. 1946. Interrogation of the defendant Dr. Ernst Illing. Wiener Stadt- und Landesarchiv (WStLA), Landesgericht Wien (LG Wien), Vg 2b Vr 2365/45. Municipal and Provincial Archives of Vienna. Copy in DÖW 19542/2. Dokumentationsarchiv des österreichischen Widerstandes [Documentation Center of the Austrian Resistance], Vienna, Austria.

Kaufmann, A. (ed.). 1991. *Totenwagen. Kindheit am Spielgrund*. Vienna: Mandelbaum.

Kawura, H.D. 1943. *Beobachtungen der Wärmeregulation der Kinder bei Abkühlung* (Diss. Med.), Universität Wien, Wien, Österreich.

Kepplinger, B. 2008. Die Tötungsanstalt Hartheim 1940–1945. In *Tötungsanstalt Hartheim*, ed. B. Kepplinger, G. Marckhgott, and H. Reese, 63–116. Linz: Oberösterreichisches Landesarchiv.

Klee, E. 2001. Sichten und vernichten. In *Medizin und Nationalsozialismus in der Steiermark*, ed. W. Freidl, A. Kernbauer, and R. Noack, 10–26. Innsbruck: Studien, Verlag.

———. 2010. *"Euthanasie" im Dritten Reich. Die "Vernichtung lebensunwerten Lebens"*. Frankfurt: Fischer Taschenbuch.

Kohl, W. 2000. *"Ich fühle mich nicht schuldig". Georg Renno, Euthanasiearzt*. Vienna: Zsolnay.

Kugler, A. 2003. Die "Hartheimer Statistik". "Bis zum 1. September 1941 wurden desinfiziert: Personen: 70273". In *Der Wert des Lebens: Gedenken/lernen/begreifen. Begleitpublikation zur Ausstellung des Landes OÖ in Schloss Hartheim 2003*, ed. Institut für Gesellschafts- und Sozialpolitik an der Johannes Kepler-Universität Linz u. a, 124–131. Trauner: Linz.

Lehmann, O., and T. Schmidt. 2001. *In den Fängen des Dr. Gross. Das misshandelte Leben des Friedrich Zawrel.* Czernin: Vienna.

List, E. 2008. "Warum nicht in Kischniew?" - Zu einem autobiographischen Tondokument Igor Carusos. *Zeitschrift für psychoanalytische Theorie und Praxis* 23(1/2): 117–141.

Malina, P. 1998. Eduard Pernkopf's atlas of anatomy or: The fiction of "pure science." *Wiener Klinische Wochenschrift* 110: 193–201.

Medical Case Files: Deceased girls and boys. 1940–1945. Wiener Stadt- und Landesarchiv (WStLA), 1.3.2.209.10.A2. Municipal and Provincial Archives of Vienna.

Neugebauer, W., and P. Schwarz. 2004. *Der Wille zum aufrechten Gang. Offenlegung der Rolle des BSA bei der gesellschaftlichen Reintegration ehemaliger Nationalsozialisten. Herausgegeben vom Bund sozialdemokratischer AkademIkerinnen, Intellektueller und KünsterInnen (BSA).* Vienna: Czernin.

Neugebauer, W., and G. Stacher. 1999. Nazi child "Euthanasia" in Vienna and the scientific exploitation of its victims before and after 1945. *Digestive Disease* 17(5–6): 279–285.

Nitsche, P. 1944. Letter to Wolfgang Holzer. (R 96 I-18). Bundesarchiv Berlin (BAB) [German Federal Archives Berlin].

Oman, K. 1948. *Case against Dr. Emil Gelny and accomplices.* Landesgericht für Strafsachen Wien (LG Wien), Vg 8a Vr 455/46. Testimony by Dr. Karl Oman, 19.6.1948. Copy in DÖW 18860/4. Dokumentationsarchiv des österreichischen Widerstandes [Documentation Center of the Austrian Resistance], Vienna, Austria.

Pernegger, G. 1941–1942. Medical Case Files: Deceased girls and boys. [patient file]. Wiener Stadt- und Landesarchiv (WStLA), 1.3.2.209.10.A2. Municipal and Provincial Archives of Vienna.

Pernkopf, E. 1937–1960. *Topographische Anatomie des Menschen, Band 1 bis 4.* Berlin: Urban und Schwarzenberg.

Reich Minister of the Interior. 1939. *Meldepflicht für missgestaltete usw. Neugeborene* [Obligatory registration of newborns with deformities etc.]. Decree by the Reich Ministry of the Interior, RdErl.d.RMdI. IVb 3.088/39-1079/Mi, 18.8.1939.

Reichsbeauftragter Heil- und Pflegeanstalten Linden to Landesregierungen. 1942. [Circular]. (R 96 I-12), Bundesarchiv Berlin (BAB) [German Federal Archives Berlin].

Ronen, G.M., B. Meaney, B. Dan, F. Zimprich, W. Stögmann, and W. Neugebauer. 2009. From eugenic euthanasia to habilitation of "disabled'" children: Andreas Rett's contribution. *Journal of Child Neurology* 24(1): 115–127.

Schwarz, P. 2002. Mord durch Hunger. "Wilde Euthanasie" und "Aktion Brandt" in Steinhof in der NS-Zeit. In *Von der Zwangssterilisierung zur Ermordung. Zur Geschichte der NS-Euthanasie in Wien Teil II*, ed. E. Gabriel and W. Neugebauer, 113–141. Vienna: Böhlau.

Seidelman, W.E. 2012. Dissecting the history of anatomy in the Third Reich—1989–2010: A personal account. *Annals of Anatomy.* doi:10.1016/j.aanat.2011.1011.1013.

Seidelman, W.E., and H. Israel. 1997. Nazi origins of an anatomy text: The Pernkopf atlas. *Journal of the American Medical Association* 276(20): 1633.

Seitelberger, F. (n.d.). BDC [composite file on person of interest]. Bundesarchiv Berlin (BAB) [German Federal Archives Berlin]. Copy in DÖW. Dokumentationsarchiv des österreichischen Widerstandes [Documentation Center of the Austrian Resistance], Vienna, Austria.

———. 1954. Die Pelizaeus-Merzbachersche Krankheit. *Wiener Zeitschrift für Nervenheilkunde* 9: 228–289.

Seitelberger, F., G. Vogel, and H. Stepan. 1957. Spätinfantile amaurotische Idiotie. *Archiv für Psychiatrie und Zeitschrift für die gesamte Neurologie* 154–190.

Statistisches Reichsamt. Ed. 1943. *Volkszählung. Die Bevölkerung des Deutschen Reichs nach den ergebnissen der Volkszählung 1939. Heft 1: Stand, Entwicklung und Siedlungsweise des Deutschen Reichs.* Berlin: Verlag für Sozialpolitik, Wirtschaft und Statistik, Paul Schmidt

Türk, E. 1942. Über BCG-Immunität gegen kutane Infektion mit virulenten Tuberkelbazillen. *Medizinische Klinik* 36: 846–847.

————. 1944. Über die spezifische Dispositionsprophylaxe im Kindesalter (Tuberkulose-Schutzimpfung). *Deutsches Tuberkulose-Blatt* 18: 23–28.

Uiberrak, B. 1946. *Case against Dr. Ernst Illing* et al. [Testimony]. Landesgericht Wien (LG Wien), Vg 2b Vr 2365/45). Copy in DÖW 19542. Dokumentationsarchiv des österreichischen Widerstandes [Documentation Center of the Austrian Resistance], Vienna, Austria.

Vienna buries victims of Nazi medical experiments. 2012. *The Telegraph*. Retrieved from http://www.telegraph.co.uk/history/world-war-two/9255267/Vienna-buries-victims-of-Nazi-medical-experiments.html. Accessed 8 Aug 2013.

Vogt, W. 2000. Euthanasiearzt und Gerichtsgutachter. Zwei Möglichkeiten der Ausübung von Gewalt gegen Menschen. *Wespennest*, 120: 89–104.

War Against the Inferior. (n.d.). Dokumentationsarchiv des österreichischen Widerstandes [Documentation Center of the Austrian Resistance]. [exhibit website]. http://www.gedenkstaettesteinhof.at. Accessed 23 Dec 2013.

Weindling, P.J. 2010. *John W. Thompson: Psychiatrist in the shadow of the Holocaust*. Rochester: University of Rochester Press.

Wien LG. Case against Dr. Emil Gelny and accomplices, testimony by Dr. Otto Haminger, (23.6.1948) Vg 8a Vr 455/46, (copy in DÖW 18860/4).

Chapter 10
Medical Research and National Socialist Euthanasia: Carl Schneider and the Heidelberg Research Children from 1942 until 1945

Gerrit Hohendorf and Maike Rotzoll

Euthanasia and the Holocaust

The systematic murder of unwanted "inferior" groups of people began under National Socialism in 1939 with the centrally organized part of the policy to murder psychiatric patients, the T4 program, named after its location at *Tiergartenstrasse* 4. Shortly after the attack on Poland in September 1939, the perpetrators registered, evaluated, and gassed residents of the mental hospitals and nursing homes of the German Reich and its annexed regions, including Austria. The extermination centers were located in the middle of the Reich, usually in or near smaller cities. Over 70,000 patients were murdered in gas chambers in 1940 and 1941. This centralized, bureaucratic killing of institutionalized patients was called "euthanasia," suggesting that incurable patients were saved from suffering and that society was saved from the burden of their care (Klee 1983; Aly et al. 1985; Friedlander 1995; Schmuhl 1987). In a comprehensive study of the available patient files from the victims of the T4 program, we established that the patients' economic productivity played the crucial role in the selection of victims (Hohendorf 2010).

In August 1941, when it became impossible to keep the T4 program secret, Adolf Hitler gave a verbal directive to discontinue the gassing of psychiatric patients, but not their registration. This tactical retreat was by no means the end of the murder of patients. As a result of decentralized euthanasia in the home territory of the Reich alone, another 90,000 residents perished from either intentional medication overdoses or deliberate food deprivation (Faulstich 1998, p. 660).

G. Hohendorf, PD Dr (✉)
University Lecturer; Institute for the History and Ethics of Medicine, Technical University of Munich–TUM, Ismaninger Str. 22, 81675 Munich, Germany
e-mail: hohendorf@gesch.med.tum.de

M. Rotzoll, Dr
Institute for the History and Ethics of Medicine, University of Heidelberg, INF 327, Heidelberg 69120, Germany
e-mail: maike.rotzoll@histmed.uni-heidelberg.de

S. Rubenfeld and S. Benedict (eds.), *Human Subjects Research after the Holocaust*,
DOI 10.1007/978-3-319-05702-6_10, © Springer International Publishing Switzerland 2014

Eventually, the euthanasia measures were extended to other groups of people, including physically and/or mentally ill forced laborers, residents of foster homes and institutions for corrective training, as well as concentration camp inmates selected for their inability to work (Schmuhl 1987, pp. 224–229). As a consequence of the close organizational relationships between the T4 program and the concentration and extermination camps, some of the staff trained to kill under the guise of euthanasia were transferred to extermination camps in occupied Poland, such as Sobibor, Belzec, and Treblinka (Schmuhl 1987, pp. 248–255).

Euthanasia and Pathological-Anatomical Research

Medical and scientific criteria, along with economical considerations, played an important role in the selection of euthanasia victims. The large number of corpses generated by the euthanasia program in a short period of time also offered the participating doctors unimagined research possibilities for pathological-anatomical examination. Neuropathologist Julius Hallervorden (1882–1965), who was head of both the department of histopathology at the Kaiser-Wilhelm-Institute for Brain Research and the department of pathology of the Brandenburg asylums, approached the organizers of the euthanasia program with a request, as he explained in 1946:

> I had heard something there, that that was going to be done, and then I went to them and told them, now, for goodness sake, if you're going to kill them all at least take out the brains, so that the material is used. They asked, well, how many can you examine, then I told them an unlimited amount—the more the better (quoted in Peiffer 1997, p. 44)

Hallervorden ultimately examined about 700 brains from victims of the T4 program in the Kaiser–Wilhelm-Institute for Brain Research (Peiffer 2002, p. 166). The medical experts affiliated with the T4 head office were also aware of the opportunity for extensive fundamental pathological-anatomical research. On January 23, 1941, they adopted a research plan that included anatomical studies of the victims of the T4 program in 14 German anatomical institutes, but it is not clear if this plan was fully realized (Anatomical Research Program 1941, p. 127, 140). In any case, on September 18, 1941, shortly after the discontinuation of the T4 program, Paul Nitsche, the acting medical director of the T4 head office, noted somewhat more modestly, "that one should first of all move on now to a suitable institution as a research facility to exhaustively examine the still existing cases of congenital idiocy and epilepsy before the disinfection" (Nitsche 1941, pp. 127, 149–150; Hohendorf et al. 1996, p. 941).

This important research facility was eventually established in the state asylum Görden near Brandenburg, which was headed by the child and youth psychiatrist Hans Heinze (1895–1983) from the fall of 1938 until 1945 (Beddies 2002, pp. 131–134, pp. 145–148). Within the framework of the T4 program, numerous children and adolescents already had been selected in the Brandenburg-Görden state asylum and transferred to the T4 gas chambers (Beddies 2002, p. 136). In the summer of

1940, moreover, the first special pediatric unit (*Kinderfachabteilung*) in the German Reich for the evaluation and killing of "unfit" children by overdose with the antiepileptic medicament Luminal had been established in this asylum, and it had served and continued to serve as the model for additional units (Beddies 2002, pp. 140–145). Midwives and doctors had to report all children with a mental or physical disability to the Reich's Committee for the Scientific Registration of Serious Hereditary and Constitutional Illnesses, which controlled the specialized pediatric units. If the expert panel, which included Hans Heinze, chose a child as a candidate for euthanasia, the child was committed to one of the specialized pediatric units. After a period of observation, a decision was made to kill or not to kill the child (Topp 2004, pp. 17–54). In contrast to the T4 program, which formally ended in 1941, the killing of children and adolescents continued in these units until 1945, even after the end of the war (Proctor 1990). Heinze's senior physician, Dr. Friederike Pusch (1905–1980) became head of the Görden special pediatric unit in 1942 and, just as Heinze had done since the 1930s, cooperated closely with Hallervorden at the Kaiser-Wilhelm-Institute for Brain Research in Berlin-Buch.[1]

On October 28, 1940, a group of 59 children was transferred from the Brandenburg-Görden state asylum to the Brandenburg killing institution's gas chamber, in order to examine their brains immediately after killing them. The sole motivation for the selection of these Görden "research children" most probably was the scientific interest in the causes of the forms of "idiocy" in childhood. These children, who predominantly came from the Brandenburg state asylums in Lübben and Potsdam, were probably already marked as interesting for science when they were transferred to Görden in 1938. In contrast to the scientific exploitation of the euthanasia victim's bodies, which was a common practice, this group was chosen for its scientific interest and killed "on order" (Beddies 2002, pp. 137–138; Reicherdt 2010, p. 148). For example, it has been verified that Julius Hallervorden removed the brains from 40 of these children immediately after they were killed for research and scientific evaluation (Beddies 2002, p. 137; see also Aly 2013, pp. 131–133).

The Görden children's files contain additional evidence of the doctors' interest in their research potential. Most of the medical histories of the Görden children end with an extensive discharge summary, which usually includes the results of a thorough neurological examination, followed by the final entry, the handwritten date of "transfer." The possible hereditary transmission of each illness, primarily congenital or acquired idiocy, and the "educability" or prognosis for intellectual development of each child were considered. Repeated encephalograms and intelligence tests gave additional evidence of the research value of these children.

---

[1] Friederike Pusch has only recently entered the historical field of vision in research on the history of National Socialist euthanasia. For more about her and her participation in the murder of children in the framework of the Reich's Committee Proceedings and the Nazi T4 euthanasia program, see Falk and Hauer (2002, p. 100f) and Schwoch (2004, pp. 195–198).

In January 1942, several months after the end of the T4 program, its head office opened an internal research department in Görden as previously proposed by Nitsche (Beddies 2002, p. 145). In addition to investigating idiocy, dementia, organic brain syndromes, and the so-called epileptic character changes, "the question of euthanasia in individual cases of illness or certain groups of diseases (athetosis for example) [should be] clarified. . ." (Heinze 1942, pp. 127, 055). The research resulted not only in the killing of selected patients but also in the development of selection criteria for killing.

The main focus of the research activity was a thorough clinical evaluation of the patients, including a determination of the heritability of the patient's illness, which could eventually be correlated with a postmortem examination of the brain. The research was hindered by wartime's limited availability of encephalograms and other laboratory tests. Heinze specially reported to the Reich's Working Committee on Mental Hospitals and Nursing Homes on the "social fate" and personality changes of those patients suffering from organic brain diseases such as athetosis. One can assume that the prognosis of a "destroyed personality" would entail "practical conclusions concerning the problem of euthanasia" (Heinze 1942, pp. 127, 054).

The Görden research department had 160 available beds. The research patients were chosen from the Brandenburg asylum and some other nearby psychiatric facilities. On September 9, 1943, Heinze reported on 135 patients (Heinze 1943, pp. 127, 059a–062a). Ninety-eight of these children and adolescents have already been identified (Beddies 2002, p. 146). Thirty died in the Görden asylum before the end of the war, others were transferred to institutions, such as Hadamar and Meseritz-Obrawalde, where patients died of starvation at an especially high rate or were killed during decentralized euthanasia (Beddies 2002, p. 147).

The Görden research department was closely connected to the research activities of the Heidelberg psychiatrist Carl Schneider, who, like Heinze, pursued interests in child and adolescent psychiatry, including epilepsy and idiocy. Schneider was one of the psychiatric experts affiliated with the T4 head office and visited Heinze's research department in March 1942 (Beddies 2002, p. 145). He developed a master plan for the research financed and facilitated by the head T4 office including a periodic exchange of resident doctors between Görden and Heidelberg/Wiesloch and the possibility of writing dissertations (Beddies 2002, p. 146).

Carl Schneider

Carl Schneider came from impoverished circumstances, began his medical studies in Würzburg in 1911, and took part in the World War I as a military doctor's assistant (Teller 1990, pp. 465–467). His experience of World War I as a disgrace and shame for Germany may have motivated Schneider's scientific and psychiatric philosophy: weakness, fear, chaos, and indecency in humans must be overcome by order and middle-class virtue. In his 1939 essay "Degenerative Art and Lunatic's

Art," Schneider revealed his image of humans. He thought it essential to eliminate all that was instinctual and pathological, so that "the faithful, diligent, disciplined, decent, discerning, willing to sacrifice, honor-loving, and honorable person" can come into being (Schneider 1939b, p. 162; see also Rotzoll et al. 2002).

In 1926, Schneider became a civil servant and medical officer in the Saxon asylum in Arnsdorf. In 1926 and 1927, he was sent for one year to the German Psychiatric Research Institute. In the 1920s, his interest in abnormal psychology was mainly directed toward schizophrenia. So Schneider can be seen as a highly regarded psychiatrist with a strong inclination to research at the beginning of the 1930s. He presented a number of noted scientific papers about schizophrenia, epilepsy, and organic psychosis (Rotzoll and Hohendorf 2012, p. 311). In 1930, he became head doctor at the Bethel asylum in Westphalia, the largest Protestant facility for epileptic and mentally disabled people. The ascendancy of National Socialism opened other, more promising opportunities for Carl Schneider, and although he previously had not been politically active, he joined the Nazi party in May 1932.

In November 1933, Schneider became chair of the Heidelberg Department of Psychiatry, succeeding Karl Wilmanns, who was driven from his post for political reasons (Rotzoll and Hohendorf 2012, pp. 311–312). Schneider reorganized the Heidelberg Psychiatric Hospital by introducing work therapy on a wide scale, in keeping with the ideas of the "new era" and wrote about it extensively (Schneider 1933, p. 234; 1939a, pp. 91–220). Work therapy was not, however, a humane treatment option for a sick patient. Since productivity was the *only* option for a person to regain his or her place in a totalitarian society, work therapy exerted considerable psychological pressure on patients (Rotzoll and Hohendorf 2012, pp. 313–314). The ability to work eventually became the most influential criteria for the selection of euthanasia victims. In a 1943 memorandum about reform of German psychiatry, Schneider wrote:

> But also the measures of euthanasia will find even more understanding and approval, once it is established and known, that in every case of mental illness, all possibilities for healing the patients, or at least improving their conditions enough so that they are brought to economically valuable work, be it in their profession, or another form, will be exhausted.[2] (Rüdin et al. 1943, p. 126, 424)

Schneider's unrelenting scientific ambitions persisted even as World War II was coming to an end. He may have been driven by his belief that modern psychiatry could both develop criteria of productivity and heredity to assess the value of life, and create new humans and a new society in which human frailty, inferiority, and abnormality were eliminated or minimized. Indeed, Schneider was involved in the euthanasia program from the very beginning, first as expert of the T4 head office, then at Bethel as a member of a medical commission to select patients for extermination, and as a consultant on a euthanasia law and euthanasia propaganda films.

---

[2] See also a draft of this document by Carl Schneider (1943, pp. 126, 437–442).

Carl Schneider's Research Program

Within the context of the euthanasia program, Schneider developed an extensive research proposal to modernize psychiatric care. He applied for 15 million Reichsmark over 15 years for this research with the far-ranging goal of "... dispell[ing] once and for all the old ideas about humans [and] really consider the organism as a biological unit in the development and evolvement of its functions" (Schneider 1942, pp. 127, 127). Using his biological concepts of human psychology, he intended to develop the prerequisites for therapy and for the prevention of transmission of undesirable hereditary conditions. For example, Schneider proposed four long-term studies in 1942:

1. Charting the development of 30 healthy boys and 30 healthy girls.
2. Parallel studies on children with idiocy of extrinsic and of hereditary origin.
3. Examining the pathophysiological effects of shock therapy and work therapy on schizophrenics in the framework of *Symptomverbandslehre* (theory of symptom groups).
4. Systematically examining the biology of the schizophrenic *Sprunghaftigkeits-verbands* (symptom group of volatility).[3]

Concerning the fate of those examined, he laconically wrote, "It is a self-evident that histopathological and pathological-anatomical examinations must and can also be conducted for numerous studies especially in the framework of the [euthanasia] program" (Schneider 1942, pp. 127, 128).

Although the large-scale research plan was not funded in 1941 due to the war, the T4 head office granted Schneider 5,000 Reichsmark in 1942 to investigate idiocy and epilepsy. He received at least 187 brains for pathological–anatomical examination from different institutions in the Reich.[4] In addition to the one at Görden, a second research department was opened in the Wiesloch asylum near Heidelberg in December 1942. Although the Wiesloch research department was closed in March 1943, 35 children, adolescents, and young adults from the Wiesloch asylum with the diagnoses of epilepsy and idiocy were examined here. Nine died during the war in different institutions, one child as the result of an encephalogram performed in the Heidelberg Hospital (Hohendorf et al. 1996, p. 942; Peschke 2012, pp. 636–663). After the Wiesloch research department was closed, Schneider continued his research at the Heidelberg Psychiatric Hospital itself, located in the middle of a university town, together with Julius Deussen (1906–1970), an employee of Ernst Rüdin at the German Psychiatric Research Institute in Munich (Hohendorf et al. 1996, pp. 942–944; Hohendorf et al. 1999;

---

[3] Schneider's concept of the psychopathology of schizophrenia is characterized by three groups of symptoms called *Symptomverbände*. One of them was the *Sprunghaftigkeitsverband*; *Sprunghaftigkeit* means incoherence in thought and speech.

[4] After the war, the police discovered 187 brains from euthanasia victims in the Heidelberg Psychiatric Hospital (State Attorney Heidelberg I Js 1698/47, 1947–1948).

Roelcke et al. 1994, pp. 525–532; Roelcke et al. 1998; Rotzoll and Hohendorf 2012). At the same time that he coordinated Carl Schneider's research project, Deussen was also assigned to the military hospital of the Heidelberg Psychiatric Hospital as a medical officer.

As far as we can reconstruct the history of the Research Department of the Heidelberg Psychiatric Hospital, the euthanasia program offered unlimited possibilities to investigate the different forms of hereditary and nonhereditary mental disabilities in childhood: Clinical, radiological, and genealogical data could be directly correlated with the pathological–anatomical findings of the brain. Schneider and Rüdin, who supported the research program in Heidelberg, believed their research was of decisive importance for the outcome of the war, both to provide a scientific basis for the euthanasia of mentally disabled children and to either prevent conception by parents of children with presumed hereditary forms of idiocy or to encourage procreation by parents of children with presumed exogenous forms of idiocy. In a letter to the *Reichsdozentenführer* on October 23, 1942, Rüdin, stressing the "supreme importance for racial hygiene," wrote that conclusive identification and differentiation of these children would be meaningful both in their own interest as well as in that of the German *Volk*, and that euthanasia, "a humane and safe counteraction," could then be recommended with full conviction. (Rüdin quoted in Roelcke 2000, p. 136; see also Schneider 1944, pp. 127, 878–883).

From August 1943 through December 1944, 52 children and adolescents were examined in the Heidelberg Psychiatric Hospital. Twenty-one of these "research children" were killed by medication (phenobarbital and/or morphine-scopolamine) in the Eichberg asylum in 1944.[5] Schneider and Deussen were responsible for the transfer of these children to the Eichberg asylum (Frankfurt District Court 4 Kls 15/46 1946; State Attorney Heidelberg I Js 1698/47, 1947–1948; State Attorney Heidelberg 10 Js 32/83, 1983–1986). Preserved correspondence conclusively indicated that Deussen spoke to parents about a possible "release" for their children in the Eichberg asylum (Heidelberg Psychiatric Hospital, Research files F 11, 15, 21, 35, 36, and 43). For that purpose, the children had to be transferred to the Eichberg asylum.

The children examined in Heidelberg came from different institutions including a large number from Schwarzacher Hof, a church institution in northern Baden (Scheuing 2004), and some from private homes. Professor Duken transferred seven children from the Heidelberg Pediatric Hospital. Nursing staff especially hired by the T4 head office and medical collaborators of Carl Schneider subjected the children to neurological and psychiatric evaluations, genealogical surveys,

---

[5] In 1942 one child was already killed at the Eichberg asylum and was added to the research program after his death. Another child was transferred from the Heidelberg hospital to the Eichberg asylum in the beginning of 1943 without being formally included in the research program. Nonetheless, his brain was examined in Heidelberg after he was killed. We count this child as one of the murdered "research children" here.

anthropometric measurements, photographs, laboratory tests, and observation of behavior and suitability for work therapy trials during their usual six-week hospital stay. After the tests were finished, most of the children were sent back to the Schwarzacher Hof or released to their relatives. Some were transferred directly from the Heidelberg Psychiatric Hospital to the Eichberg Asylum. Those children sent back to the Schwarzacher Hof were then transferred from there to the Eichberg asylum on July 28, 1944. The children were to be killed there and their brains sent back to Heidelberg. In the end, however, "only" four of the brains of all "research children" were examined in Heidelberg, mostly because of the difficulties that the war presented.[6]

The Heidelberg Research Children

Given below is a typical example of the fate of the Heidelberg research children. Two-and-a-half-year-old Klaus, son of a merchant from the Ruhr area, was lagging in his development since his seventh month but eventually learned to sit and walk with some support. After the child's examination in the Essen city pediatric hospital, his father approached the Heidelberg Pediatric Hospital and asked that his son be admitted for testing. Apparently the father had great hopes for the examination and possible treatment of the child in Heidelberg. The admission note included this laconic statement, "child seems idiotic." Professor Schneider consulted a few days later, and, on April 8, 1944, he wrote his colleague Duken:

> It certainly is a case of an idiotic child, who couldn't be closely examined in detail because of his behavior. He was unruly and reluctant to such an extent that I barely gained a connection to the child. Nothing could arouse his interest. As he neither speaks nor plays properly, one must expect that it has to do with mental standstill already in the very first stages of development. ... I would like nothing better than to include him in our idiocy research, if the parents agree. The parents would accumulate no special costs, on the contrary, we would be willing to bear the expenses for accommodation on our research budget. I am most likely to believe that it is a generalized abnormality, perhaps a congenital cerebellar agenesis or something like that. (Medical Record Klaus B. Prot, Nr. 936/1944)

Klaus was transferred to the Heidelberg Psychiatric Hospital for examination in the research department that same day for the prescribed examination procedure.[7] Three people held him down for his physical examination and photographs. Klaus also had laboratory tests, a chest X-ray, and an encephalogram, which revealed moderate hydrocephalus. Deussen requested a comprehensive family tree from the health department in Klaus's hometown. On May 16, 1944, Klaus's father came to

---

[6] See the list of brains from euthanasia victims compiled by the Heidelberg Criminal Police Department in 1947 in State Attorney Heidelberg, I Js 1698/47, 1947–1948.

[7] All the following information is from Heidelberg Psychiatric Hospital, Research file F 35.

Heidelberg where Dr. Deussen gave him an extremely bad prognosis for his son. On June 2, 1944, he wrote the father:

> After an in-depth examination we must unfortunately inform you that your son Klaus suffers from severe idiocy and in the medical discretion no improvement of his condition can be reached. Based on our discussion on the occasion of your visit we would now like to ask you to address an application to both the board of the Eichberg state asylum near Eltville/Rheingau as well as to the hospital here that your son will be transferred to the Eichberg institution. We ask you to resolve to take this step as soon as possible, since the question of Klaus's further accommodation must be resolved soon after the conclusion of the examination. (Heidelberg Psychiatric Hospital, Research file F 35)

It can be concluded by the tone of this letter and from similar correspondence in the files of other children examined in Heidelberg, that Deussen advised a prompt "release" of the child in the Eichberg asylum. But the father informed Dr. Deussen on May 15, 1944, that his wife wished to place Klaus in a home near his hometown; therefore, he wanted to pick up Klaus at the end of the month in Heidelberg. But, for unknown reasons, this did not happen. Klaus was transferred to the nearby asylum Schwarzacher Hof where Dr. Deussen had regular access to the children he had previously examined. After the Schwarzacher Hof was vacated to set up an arms factory, Klaus and 16 other children were transferred from there to the Eichberg asylum on July 28, 1944. Thirteen of these 17 children were "research children" examined in the Heidelberg Psychiatric Hospital (Scheuing 2004, pp. 402–426). Klaus was killed immediately after his arrival on July 29, 1944 (Durchgangsbuch 1941–1944).

The only known scientific presentation of the data from the Heidelberg Research Department was a dissertation on the metabolism of idiotic children at the University of Leipzig in 1946 by Schneider's daughter-in-law, Monika Schneider. The context of the studies was not mentioned.

After the war, information about research in the euthanasia program, including the death of children, was totally repressed. The criminal investigation by the Heidelberg state attorney was abandoned in 1947, and Schneider's colleague Hans Joachim Rauch, who was responsible for the neuropathological examination of the brains, continued to work in the hospital until the 1990s (Rotzoll and Hohendorf 2007). It was not until 1998 that a memorial was placed in front of the Heidelberg Psychiatric Hospital to remind us of the 21 children who were victims of criminal medical research.

On the one hand, the history of the Heidelberg research unit demonstrates that highly esteemed psychiatrists and researchers like Schneider, Rüdin, and Deussen were involved in the euthanasia program. On the other hand, the existence of such a research unit in the middle of a small university city like Heidelberg illustrates that human subjects research was not limited to concentration camps.

References

Aly, G. 2013. Die Belasteten. 'Euthanasie' 1939–1945. Eine Gesellschaftsgeschichte. Frankfurt: S. Fischer.

Aly, G., A. Ebbinghaus, M. Hamann, F. Pfäfflin, and G. Preissler (eds.). 1985. *Aussonderung und Tod. Die klinische Hinrichtung der Unbrauchbaren*. Berlin: Rotbuch.

Anatomical Research Program. 1941. U.S. Army, Europe. Records of the Judge Advocate General, War Crimes Branch. Records Relating to Medical Experiments ("Heidelberg Documents"), p. 127, 140. Record Group 549 (location 290/59/17/05). National Archives and Records Administration, College Park, MD.

Beddies, T. 2002. Kinder und Jugendliche in der brandenburgischen Heil- und Pflegeanstalt Görden als Opfer der NS-Medizinverbrechen. In *Brandenburgische Heil- und Pflegeanstalten in der NS-Zeit. Schriftenreihe zur Medizin-Geschichte des Landes Brandenburg*, vol. 3, ed. K. Hübener, 129–154. Berlin: be.bra Wissenschaft.

Durchgangsbuch. 1941–1944. Psychiatrisches Krankenhaus Eltville/Rheingau. In possession of Eichberg Psychiatric Hospital, Eltville, Germany.

Falk, B., and F. Hauer. 2002. Erbbiologie, Zwangssterilisation und "Euthanasie" in der Landesanstalt Görden. In *Brandenburgische Heil- und Pflegeanstalten in der NS-Zeit. Schriftenreihe zur Medizin-Geschichte des Landes Brandenburg*, vol. 3, ed. K. Hübener,79–104. Berlin: be.bra Wissenschaft.

Faulstich, H. 1998. *Hungersterben in der Psychiatrie 1914–1949. Mit einer Topographie der NS-Psychiatrie*. Freiburg: Lambertus.

Frankfurt District Court 4 Kls 15/46. 1946, February 21. [Court case]. (430/32442). Hessen State Public Records Office, Wiesbaden, Germany.

Friedlander, H. 1995. *The origins of Nazi genocide: From euthanasia to the Final Solution*. Chapel Hill: University of North Carolina Press.

Heidelberg Psychiatric Hospital [Universitätsklinikum Heidelberg, Klinik für Allgemeine Psychiatrie]. (n.d.). Research files F 11–F 43. Heidelberg Psychiatric Hospital, Historical Archives [Universitätsklinikum Heidelberg, Klinik für Allgemeine Psychiatrie, Historisches Archiv], Heidelberg, Germany.

Heinze, H. 1942, September 9. *Report on the Görden research facility by Hans Heinze*. U.S. Army, Europe. Records of the Judge Advocate General, War Crimes Branch. Records Relating to Medical Experiments ("Heidelberg Documents"), pp. 127, 050–055. Record Group 549 (location 290/59/17/05). National Archives and Records Administration, College Park, MD.

———. 1943, September 9. *Report on the Görden research facility by Hans Heinze*. U.S. Army, Europe. Records of the Judge Advocate General, War Crimes Branch. Records Relating to Medical Experiments ("Heidelberg Documents"), pp. 127, 059a–062a. Record Group 549 (location 290/59/17/05). National Archives and Records Administration, College Park, MD.

Hohendorf, G. 2010. Die Selektion der Opfer zwischen rassenhygienischer "Ausmerze", ökonomischer Brauchbarkeit und medizinischem Erlösungsideal. In *Die nationalsozialistische "Euthanasie"-Aktion "T4" und ihre Opfer, Geschichte und ethische Konsequenzen für die Gegenwart*, ed. M. Rotzoll, G. Hohendorf, P. Fuchs, P. Richter, C. Mundt, and W.U. Eckart, 310–324. Paderborn: Schoeningh.

Hohendorf, G., V. Roelcke, and M. Rotzoll. 1996. Innovation und Vernichtung—Psychiatrische Forschung und "Euthanasie" an der Heidelberger Psychiatrischen Klinik 1939–1945. *Der Nervenarzt* 67(11): 935–946.

———. 1999. Innovation without ethical restriction: Remarks on the history and ethics of psychiatric research at the University of Heidelberg, 1941–1945. In *Knowledge and power: Perspectives in the history of psychiatry. Selected papers from the Third Triennial Conference of the European Association for the history of psychiatry, 11–14 September 1996, Munich, Germany*, ed. E.J. Engstrom, M.M. Weber, and P. Hoff, 173–179. Berlin: VWB.

Klee, E. 1983. *"Euthanasie" im Dritten Reich. Die "Vernichtung lebensunwerten Lebens"*. Frankfurt: Fischer.

Medical Record Klaus B. Prot. (Nr. 936/1944). Pediatric Hospital Inventory Acc. 15/01 L-II. Universitätsarchiv Heidelberg [University Archives Heidelberg], Germany.

Nitsche, H. 1941, September 18. *Memorandum concerning research* [Aktennotiz betr. Forschung]. U.S. Army, Europe. Records of the Judge Advocate General, War Crimes Branch. Records Relating to Medical Experiments ("Heidelberg Documents"), pp. 127, 149–150. Record Group 549 (location 290/59/17/05). National Archives and Records Administration, College Park, MD.

Peiffer, J. 1997. *Hirnforschung im Zwielicht: Beispiele verführbarer Wissenschaft aus der Zeit des Nationalsozialismus Julius Hallervorden—H. J. Scherer—Berthold Ostertag. Abhandlungen zur Geschichte der Medizin und der Naturwissenschaften*, vol. 79. Husum: Matthiesen.

———. 2002. Die Prosektur der brandenburgischen Landesanstalten und ihre Einbindung in die Tötungsaktionen. In *Brandenburgische Heil- und Pflegeanstalten in der NS-Zeit. Schriftenreihe zur Medizin-Geschichte des Landes Brandenburg*, vol. 3, ed, K. Hübener, 155–177. Berlin: be.bra Wissenschaft.

Peschke, F. 2012. *Ökonomie, Mord und Planwirtschaft. Die Heil- und Pflegeanstalt Wiesloch im Dritten Reich. Aspekte der Medizinphilosophie* vol. 10. Bochuml Freiburg: Project verlag.

Proctor, R. 1990. *Racial hygiene: Medicine under the Nazis*. Cambridge, MA: Harvard University Press.

Reicherdt, B. 2010. "Gördener Forschungskinder". NS-"Euthanasie" und Hirnforschung. In *Die nationalsozialistische "Euthanasie"-Aktion "T4" und ihre Opfer. Geschichte und ethische Konsequenzen für die Gegenwart*, ed. M. Rotzoll, G. Hohendorf, P. Fuchs, P. Richter, C. Mundt, and W.U. Eckart, 147–151. Paderborn: Schöningh.

Roelcke, V. 2000. Psychiatrische Wissenschaft im Kontext nationalsozialistischer Politik und "Euthanasie". Zur Rolle von Ernst Rüdin und der Deutschen Forschungsanstalt für Psychiatrie/ Kaiser-Wilhelm-Institut. In *Geschichte der Kaiser-Wilhelm-Gesellschaft im Nationalsozialismus. Bestandsaufnahme und Perspektiven der Forschung*, vol. 1/1, ed. D. Kaufmann, 114–148. Göttingen: Wallstein.

Roelcke, V., G. Hohendorf, and M. Rotzoll. 1994. Psychiatric research and "euthanasia." The case of the psychiatric department at the University of Heidelberg, 1941–1945. *History of Psychiatry* 5: 517–532.

———. 1998. Erbpsychologische Forschung im Kontext der "Euthanasie": Neue Dokumente zu Carl Schneider, Julius Deussen und Ernst Rüdin. *Fortschritte der Neurologie Psychiatrie* 66: 331–336.

Rotzoll, M., B. Brand-Claussen, and G. Hohendorf. 2002. Carl Schneider, die Bildersammlung, die Künstler und der Mord. In *Wahn Welt Bild. Die Sammlung Prinzhorn. Beiträge zur Museumseröffnung. Heidelberger Jahrbücher*, vol. 46, ed. T. Fuchs, I. Jádi, B. Brand-Claussen, and C. Mundt, 41–64. Berlin: Springer.

Rotzoll, M., and G. Hohendorf. 2007. Zwischen Tabu und Reformimpuls. Der Umgang mit der nationalsozialistischen Vergangenheit der Heidelberger Psychiatrischen Universitätsklinik nach 1945. In *Vergangenheitspolitik in der universitären Medizin nach 1945. Institutionelle und individuelle Strategien im Umgang mit dem Nationalsozialismus. Pallas Athene Beiträge zur Universitäts- und Wissenschaftsgeschichte*, vol. 22, ed. S. Oehler-Klein and V. Roelcke, 307–330. Stuttgart: Franz Steiner.

———. 2012. Krankenmord im Dienst des Fortschritts? Der Heidelberger Psychiater Carl Schneider als Gehirnforscher und "therapeutischer Idealist". *Nervenarzt* 83: 311–320.

Rüdin, E., M. de Crinis, C. Schneider, H. Heinze, and H. Nitsche. 1943. *Reflections and propositions concerning the further development of psychiatry* [Gedanken und Anregungen betr. die künftige Entwicklung der Psychiatrie]. U.S. Army, Europe. Records of the Judge Advocate General, War Crimes Branch. Records Relating to Medical Experiments ("Heidelberg Documents"), pp. 126, 420–427. Record Group 549 (location 290/59/17/05). National Archives and Records Administration, College Park, MD.

Scheuing, H.-W. 2004. *"...als Menschenleben gegen Sachwerte gewogen wurden". Die Geschichte der Erziehungs- und Pflegeanstalt für Geistesschwache Mosbach/Schwarzacher Hof und ihrer Bewohner 1933–1945* (Ed. 2). Heidelberg: Winter Universitätsverlag.

Schmuhl, H.-W. 1987. *Rassenhygiene, Nationalsozialismus, Euthanasie. Von der Verhütung zur Vernichtung 'lebensunwerten Lebens', 1890–1945. Kritische Studien zur Geschichtswissenschaft*, vol. 75. Göttingen: Vandenhoeck & Ruprecht.

Schneider, C. 1933. Die Auswirkungen der bevölkerungspolitischen und erbbiologischen Massnahmen auf die Wandererfürsorge. *Der Wanderer* 50: 233–240.

———. 1939a. *Behandlung und Verhütung der Geisteskrankheiten. Allgemeine Erfahrungen Grundsätze Technik Biologie. Monographien aus dem Gesamtgebiete der Neurologie und Psychiatrie*, vol. 67. Berlin: Julius Springer.

———.1939b. Entartete Kunst und Irrenkunst. *Archiv für Psychiatrie* 110: 135–164.

———. 1942, March 12. Research proposal by Carl Schneider. U.S. Army, Europe. Records of the Judge Advocate General, War Crimes Branch. Records Relating to Medical Experiments ("Heidelberg Documents"), pp. 127, 127–129. Record Group 549 (location 290/59/17/05). National Archives and Records Administration, College Park, MD.

———. 1943, January 28. *Comments on the further development of psychiatry* [Bemerkungen über die zukünftige Ausgestaltung der Psychiatrie]. U.S. Army, Europe. Records of the Judge Advocate General, War Crimes Branch. Records Relating to Medical Experiments ("Heidelberg Documents"), pp. 126, 437–442. Record Group 549 (location 290/59/17/05). National Archives and Records Administration, College Park, MD.

———. 1944, January 24. *Report on the Research on Idiots and Epileptics in the Framework of the [Euthanasia] Action*. U.S. Army, Europe. Records of the Judge Advocate General, War Crimes Branch. Records Relating to Medical Experiments ("Heidelberg Documents"), pp. 127, 878–883. Record Group 549 (location 290/59/17/05). National Archives and Records Administration, College Park, MD.

Schneider, M. 1946. *Stoffwechselbelastungsproben bei schwachsinnigen Kindern* (Medizinische Dissertation Leipzig). Library of Leipzig University, Leipzig, Germany. Photocopy in possession of authors.

Schwoch, R. 2004. Ärztinnen in der Landesanstalt Görden. 1936–1947. Anpassung, Unterordnung oder Karriere?. In *Kinder in der NS-Psychiatrie. Schriftenreihe zur Medizin-Geschichte des Landes Brandenburg*. vol. 10, eds., T. Beddies and K. Hübener, 185–202. Berlin: be.bra Wissenschaft.

State Attorney Heidelberg, I Js 1698/47. (1947/1948). [Criminal investigation]. (309 Zug. 1992/34 No. 4). Baden-Württemberg General State Archives, Karlsruhe, Germany.

State Attorney Heidelberg, 10 Js 32/83. 1983/1986. [Criminal investigation]. Central Office of the State Departments of Justice. (439 AR-Z 40/83). Federal Archives Ludwigsburg, Ludwigsburg, Germany.

Teller, C. 1990. Carl Schneider. Zur Biographie eines deutschen Wissenschaftlers. *Geschichte und Gesellschaft* 16: 464–478.

Topp, S. 2004. Der "Reichsausschuß zur wissenschaftlichen Erfassung erb- und anlagebedingter schwerer Leiden". Zur Organisation der Ermordung minderjähriger Kranker im Nationalsozialismus 1939–1945. In *Kinder in der NS-Psychiatrie. Schriftenreihe zur Medizin-Geschichte des Landes Brandenburg*, vol. 10, eds., T. Beddies and K. Hübener, 17–54. Berlin-Brandenburg: be.bra Wissenschaft.

Chapter 11
Victims of Human Experiments and Coercive Research under National Socialism: Gender and Racial Aspects

Paul Weindling

When concentration camps were liberated, groups of former prisoners documented the medical experiments carried out on them at Auschwitz, Buchenwald, Dachau, Mauthausen, and the predominantly women's camp of Ravensbrück. On March 4, 1945, a declaration by The Prisoner Doctors of Auschwitz to the International Public declared that prisoners had been treated as experimental animals; that the Allies and neutral states should bring to trial those responsible; and that prosecuting perpetrators should prevent coerced human experiments and medical atrocities in the future (International Scientific Commission [TNA WO 309/470] 1946–1948; Weindling 2004). These actions to document the research atrocities were of far-reaching significance in terms of investigation, prosecution, ethical safeguards, and the struggle for compensation. In addition to seeing the revelations of experiments as a matter of war crimes prosecutions and courtroom proceedings, the victims and firsthand prisoner observers impressed on the Allied military liberators, war crimes investigators, and prosecutors not only their suffering at the hands of Nazi doctors but also the wider importance of fully documenting the human subjects research as violations of human rights.

At some concentration camps, documents were already collected during the war. Lists were kept by the Dutch prisoner-nurse Henrik Nales for victims of Eugen Haagen at Natzweiler and of the Luxemburgian Eugene Ost for malaria subjects at Dachau (Alexander 1947b; Ost 1988). At Ravensbrück this effort to document the experiments involved solidarity among prisoners in the camp (notably, the taking of clandestine photographs of the wounds) and efforts to alert international opinion during the war. Prisoners intervened to save the lives of victims, as with the Ravensbrück "Rabbits" (the Polish survivors of the sulfonamide experiments) and the Jewish children experimented on at Sachsenhausen. The efforts to alert the

P. Weindling (✉)
Wellcome Trust Research Professor in the History of Medicine; Department of History, Philosophy and Religion; Oxford Brookes University, Gypsy Lane, Headington, Oxford, Oxon OX3 0BP, UK
e-mail: pjweindling@brookes.ac.uk

S. Rubenfeld and S. Benedict (eds.), *Human Subjects Research after the Holocaust*,
DOI 10.1007/978-3-319-05702-6_11, © Springer International Publishing Switzerland 2014

Allies about the Ravensbrück experiments were successful. In October 1944, the International Council of Women in London demanded that the International Committee of the Red Cross give "all possible protection" to the women subjected to experiments. The Council expressed horror at the "barbarous experiments under the guise of scientific research" (Zimmern 1944). The British and Polish Red Cross forwarded messages of concern to the Red Cross in Geneva but without apparent effect (British Red Cross 1944; Polish Red Cross 1944; see also "An Experiment in Co-Operation," n.d; TNA WO 309/470, 1946–1948).

These efforts to document Nazi medical experiments had a profound impact on the Allied scientific intelligence and war crimes investigation teams during the immediate postwar aftermath. The British liberators of Bergen-Belsen encountered survivors of Auschwitz experiments. The scientific intelligence officer John Thompson interrogated the Belsen doctor Fritz Klein, who had conducted experiments with mescaline and Rutenol in Auschwitz ("War Criminal, Dr. Klein," 1946; Gries 1945). Concerned about the criminality of Nazi experiments, Thompson as head of the British branch of the Field Information Agency Technical (FIAT) scientific organization first identified the experiments as "Medical War Crimes" in November 1945, and he alleged that 90 percent of German medical research by leading scientists and clinicians was criminal (Thompson 1945; Weindling 2010, p. 115). Thompson had immense empathy and compassion for the victims; he rightly saw that the experiments were not restricted to the concentration camps but that there were important links to major university and research institutes.

The British military pathologist, Keith Mant, investigated medical war crimes at Ravensbrück as part of a brief from the British Special Operations Executive to determine the circumstances of the women killed at Ravensbrück (Scientific and Technological Branch [TNA FO 1031/74] 1945–1946; Somerhough to FIAT 1946; Thompson to Somerhough 1946; Weindling 2001). Mant broadened his mission to the totality of medical conditions at Ravensbrück, including inducements to potential research subjects, such as a day without work or extra food. He extended the investigation from Ravensbrück to Auschwitz, including sterilization and infertility experiments by Carl Clauberg and twin experiments by Mengele, again drawing upon the testimonies of concentration camp survivors (TNA FO 1031/74, draft minutes, May 15, 1946; Mant to Thompson 1946; Mant 1949, 1961; "Testimonies," BB/35/263). Mant compiled two important reports. The first was "On the Medical Services, Human Experimentation and Other Medical Atrocities committed in Ravensbruck Concentration Camp" (1947–1949, RW 2/5, 31). His second report was "Experiments in Ravensbruck Concentration Camp Carried out under the Direction of Professor Karl Gebhardt" (TNA WO 309/469, 1946).

Mant interviewed several of the Rabbits with both humane sympathy and with clinical precision, taking details of their wounds and arrest histories. He took affidavits from Janina Iwanska and Helena Piasecka from June 24 to 28, 1946 (Iwanska and Piasecka 1946; Mant notebook 1945–1946; Mant 1950); from Sofia Macza—this document was used at the Nuremberg Medical Trial—on April 15–16, 1946, in Stockholm; and from Irena Stanislawa Suchon in Lund on July 11, 1946 (Dörner et al. 1999). These investigations provided evidence for the Ravensbrück trial administered by the British in Hamburg from December 5, 1946, to February 3, 1947. When the British handed over the "Hohenlychen group"—Gebhardt, his assistant Fritz Fischer, and the former camp doctor Herta Oberheuser (the only woman prosecuted at the Nuremberg Trials)—to the Americans for prosecution for their role in the sulfonamide experiments at Ravensbrück, Mant and Thompson briefed the US prosecutors at Nuremberg from December 1946 to August 1947. The Ravensbrück trial defendants included camp doctors, nurses, and prisoners accused of complicity in medical atrocities. Polish victims Helena Dziedziecka, Helena Piasecka, Zofia Sokulska, and Maria Adamsk gave evidence. The camp doctors Percival Treite, who was accused of conducting sterilization experiments, and Gerhard Schiedlausky were among the accused (Procès de Ravensbrück 1946–1947; "Ravensbrück War Crimes Trial," 1946–1948).[1]

The Medical Trial was the only one of the US-mounted successor trials at Nuremberg that called victims to give evidence, rather than wholly relying on captured documents (Weindling 2012). The "Polish girls," as the prosecutors called them, were expected to make a profound impact on the Nuremberg judges. Other victims of experiments were called to testify. A Polish-Jewish victim of X-ray sterilization at Auschwitz presented his deep burns and scarring to the court. The German Roma Karl Höllenreiner, when asked to testify against the internal medicine specialist Beiglböck, who administered the seawater drinking experiments at Dachau in 1944, sprang into the dock and punched the defendant (Dörner et al. 1946–1947, microfiche 2/10229–10234). Höllenreiner's liver had been punctured, he had been forced to drink putrid yellow water, and he had lost his child, sister, and both of her children at Auschwitz. This dramatic confrontation between a victim and his erstwhile perpetrator showed the strain felt by victims (Alexander 1947a; Dörner et al. 1946–1947, microfiche 6/1612, 6/1622–3).

At the same time, the scientific intelligence officer Thompson set out to document all "medical war crimes." As a British trial observer at Nuremberg, he pointed out that the flaw in such a trial was that when a perpetrator was missing or no longer alive, the experiment could pass unnoticed. The International Scientific Commission instigated by Thompson, who then acted as secretary, and supported by the British and French worked parallel to the trials to gather documentation for a full reconstruction of all Nazi human experiments. The Commission met from 1946 until 1948 (Weindling 2002). Although extensive documentation was collected at Nuremberg, Dachau, Berlin, and Vienna, and then sorted, indexed, copied, and evaluated, there was only a brief concluding report by the British Foreign Office, *Scientific Results of German Medical War Crimes: Report of an Enquiry by a Committee under the Chairmanship of Lord Moran M.C., MD*, published in London in 1949.

Whereas the prosecutors at Nuremberg alleged a high number of victims of the experiments and a large number of involved doctors, the defense at the trials claimed that there were only a small number of experiments. Typical was SS

---

[1] For a Treite victim deposition, see Kormornicka (1952).

administrator Oswald Pohl's affidavit of March 26, 1947, stating that there were at most 350–400 victims in eight experiments (Pohl, NARA 1019/54). Leading German medical scientists, such as the physiologist Hermann Rein, claimed that there were a small number of perpetrators whose crimes were detached from mainstream German science and whose activities were condemned as "pseudoscience." The West German Chambers of Medicine alleged that those on trial in Nuremberg represented only "a minute part of the medical profession" (Weindling 2004, p. 332). The claim of no more than 350 perpetrators, which was made by the *Westdeutschen Ärztekammern* in November 1949, was added to the account of the medical trial by Alexander Mitscherlich and Fred Mielke in *Wissenschaft ohne Menschlichkeit*. Overall, the position was that of marginalization, which obscured linkages between the experiments and mainstream academic medicine (*Westdeutschen* Ärztekammern 1949). There was no known figure of the number of victims, and international health agencies, lawyers, and administrators concerned with establishing compensation assumed that only a few survived a small number of potentially fatal experiments.

During the 1960s, the journalist Günther Schwarberg, working for the illustrated magazine *Stern*, became concerned with the fate of the 20 children selected in Auschwitz for research by Kurt Heissmeyer for a TB immunity experiment at the concentration camp of Neuengamme, where the children arrived on December 15, 1944. The children and the accompanying French prisoner-nurses were killed on April 20, 1945. By the 1980s, Schwarberg identified these tragic victims, long known as "the twenty children," and reconstructed their life histories (1996).

What Schwarberg achieved in microcosm for the 20 children is inspirational for those aiming to identify as far as possible all victims and to reconstruct their life histories. In 2007, a project called Victims of Medical Experiments and Coercive Medical Research under National Socialism began at Oxford Brookes University under my direction. The goal has been to identify the victims of the coerced human experiments and other types of unethical research under National Socialism.[2] We have examined available records from the period 1933–1945, postwar compensation, and the International Tracing Service and have correlated publications on individual experiments. To date, we have identified 27,759 victims overall. Fuller documentation exists on at least 12,121 persons (Weindling 2014). The project covers:

I. Experiments

- Physiological experiments
- Comparative pathology
- Genetic research

---

[2] I acknowledge here Anna von Villiez (database design), Aleksandra Loewenau (Polish victims), Marius Turda (Greek victims), and Nichola Farron (Russian/ Soviet victims).

II. Body Parts/Fluids

- Body parts as anatomical specimens: executed prisoners
- Brains for dissection from "euthanasia" victims
- Draining of blood as a culture medium for infectious diseases research

III. Vaccine Research

- Comparing effects of vaccines and sera and infected but non-vaccinated control groups
- Human reservoirs of infectious pathogens

IV. Sterilization Research

- Experimental X-ray sterilization and operations

V. Observations in Camps and Ghettoes

- Anthropological measurements, face masks, hand and footprints
- Psychological observations on racial groups

The victims project involves all locations where coerced experiments and research occurred, that is, not only concentration camps but also psychiatric hospitals and clinics. This approach includes the terrifying experiments supported by the SS, but also seeks to go beyond these often deadly experiments and to reconstruct all coerced research. By way of contrast, in responding to Thompson's first memorandum on the criminal experiments, the Allied authorities decided to focus only on the concentration camps, as high-level officials at the British Foreign Office believed that to investigate and prosecute more widely would be too disruptive. The interactions between the Hohenlychen orthopedic and sports clinic and Ravensbrück demonstrate the importance of medical institutions taking a role in experiments because there were orthopedic and bone transplantation experiments done at both locations. Including all locations where coerced experiments and research occurred allows academic institutions and mainstream German medicine to be linked to the concentration camps and the SS. The project, furthermore, allows agency to the researchers who opportunistically exploited victims as resources, as well as agency to victims who endured considerable pain and suffering, but struggled to survive, and at times resisted and sabotaged the experiments.

The methodology is based on linking as many biographical, administrative, and medical records concerning the victims and survivors as possible. Names, whenever available, ensure that victims are uniquely counted because when using multiple sources, there is a risk of double counting. The researching of life histories allows the process of atrocity and its impact to be understood, permits the identification of the wider networks including senior scientists authorizing the experiments, and takes into account transitional locations. This approach also allows groups of victims to be considered by gender, religion, ethnicity, and nationality, effectively reconstructing victims' life histories. Finally, this approach allows a number of perspectives: first, a structural cohort analysis in terms of the age, gender, nationality, ethnic identity, and religion of the victims; second, the documentation of the biographical and subjective aspects of victims; and third, a consideration of the different medical priorities shaping the experiments.

The Dimensions of the Experiments

One finding is quite striking: Overall, males outnumbered females by approximately 2 to 1. This gender imbalance is clearly the case in the largest national cohort of victims, those from Poland, a group including both Roman Catholics and Jews. In certain cohorts, such as the Russians and Yugoslavians, males also predominated. Among other less sizeable cohorts such as Czechoslovakians, Dutch, French, Hungarians, and especially Greeks, it was women who predominated. Whereas male prisoners predominated in the concentration camps on German territory, there were at least equal if not higher numbers of women in the death camps instituted later. Most death camps (like those of the *Aktion Reinhardt*) were not locations for experiments. In contrast, Auschwitz became a major center of human experiments as well as a death camp, a forced labor camp, and a holding location for Sinti and Roma. Here, our project data shows that there were higher numbers of women who were victims of the experiments, whereas overall men predominated by a ratio of 2:1 as already noted (Weindling 2014).

The experiments became more frequent, and victims rose steeply beginning in mid-1941. The infectious disease experiments relating to epidemics, notably malaria and murine typhus (*Fleckfieber*), involved several hundred victims, and gender was not of medical significance. Thus men were used as experimental subjects because of their availability (Weindling 2000, pp. 237–238, 352–363). Indeed, the large numbers of prisoners of war, such as Soviet and Yugoslavian prisoners, increased the available pool of male victims. Gender only became relevant in research relating to race, reproduction, and heredity.

Although religion was not always given for victims, the project has identified 3,864 Jewish victims as the largest religious group. At first, Jews were measured and racially studied with a sense that here was a group destined for extinction. Then, figures like Kremer and Rascher would select a Jew when they wanted to dissect a dead specimen. Only with the establishment of Auschwitz did Jews become used for generally nonfatal experiments. In Auschwitz Sephardic, as opposed to Ashkenazi, Jewish Greek women were of especial interest to the German medical researchers for racial reasons. This can be seen in that Greeks, both men and women, were the largest group selected for the Jewish skeleton collection. Of the women victims among this tragic group, the 20 Greek Jewesses were the largest in number. They were transported from Auschwitz to the concentration camp of Natzweiler-Struthof in Alsace, where they were killed on October 11 and 13, 1943. The bodies were transported to the Strassburg Anatomical Institute for an intended Jewish skeleton collection. The identification of the victims by the journalist Hans-Joachim Lang—they had been previously a known but nameless group—makes this aspect of the high proportion of Sephardic Jews apparent (Lang 2004).

The military-related experiments required victims comparable to German Wehrmacht soldiers in order to test vaccines and pharmacological remedies against infectious diseases, notably typhus and malaria. Soviet prisoners of war were used, but the experimenters encountered resistance from these tough prisoners. Prisoners of war were utilized, for example, at Sachsenhausen or on Crete, for the transmissibility of hepatitis, where British prisoners were among those experimented on (Leyendecker and Klapp 1989). It can be conjectured that the shift from males to females was in part an expectation of females being more docile, but also due to availability and the convenience of proximity to the SS sanatorium of Hohenlychen.

Jewish children were used increasingly in 1943–1944. In July 1943, the geneticist Hans Nachtsheim experimented on at least six children with epilepsy from the Landesanstalt Görden psychiatric institution in a low-pressure chamber (Schmuhl 2008, pp. 331–338; Schwerin 2004, pp. 397–417; Weindling 2003). We see child research with Mengele's experiments on twins, the September 1944 hepatitis experiments by Arnold Dohmen on Jewish children selected from Auschwitz and brought to Sachsenhausen, and the Heissmeyer experiments in Neuengamme. Astrid Ley suggests that adult male prisoners were needed for slave labor later in the war and, therefore, children were substituted for them (Eschebach and Ley 2012). This switch from adult males to children shows how scientific exploitation of prisoner bodies was part of a war economy mobilized to use all human resources to the maximum.

Moreover, laboratory animals were in increasingly short supply. Gebhardt explained that the kidnapping of Gibraltar apes was considered, but deemed too risky (Weindling 2004, p. 23). The geneticist Hans Nachtsheim lost his rabbit strains bred to show similar symptoms of tremor for genetic significance. Hungry Berliners stole the rabbits for food toward the close of the war.

For the sulfonamide experiments on the 20 men and the 74 Polish Rabbits, the proximity of Ravensbrück to the Hohenlychen orthopedic clinic of the SS surgeon Karl Gebhardt was crucial (Mant 1946). It is known that tests were first carried out on 20 men (Ebbinghaus et al. 1946–1947, 2001). Christl Wickert (personal communication, August 18, 2010) suggests that these men came from the small men's camp at Ravensbrück, which differs from defendant and Gebhardt assistant Fritz Fischer's testimony at the Nuremberg Medical Trial that the men were transferred from Sachsenhausen. Starting July 24, 1942, the 74 Polish women were selected for further experiments, and they were cruelly wounded so that their legs became infected (Loewenau 2012). In this case, it appears that women were available in greater numbers in a convenient location. The anonymity of the male prisoners set against the articulate female Rabbits is striking. This difference may be because the procedures were far worse in terms of the severity of the wounds for the women. There was a remarkable solidarity among the prisoners, which assisted the survival of many of the Rabbits, despite the fact that five died from the effects of the operation and six were executed (Jurkowska 1952). We do not know if the men prisoners survived or not, although it is certain from the camp records that there were male fatalities in that period.

As Astrid Ley has suggested, the 20 children killed at the Bullenhuser Damm, another iconic group of victims, were selected to free up adult prisoners for slave labor (Eschebach and Ley 2012). The perpetrating doctor, Kurt Heissmeyer, who lived in close proximity to Gebhardt at Hohenlychen, carried out experiments previously on Soviet and Serb prisoners of war at Neuengamme before substituting children (Schwarberg 1996). Children could be manipulated in a way that the Polish women prisoners could not be. Children were highly vulnerable to being killed as evidenced by what happened. In the case of 18 boys and an adult selected for hepatitis experiments at Sachsenhausen, there was a last-minute reprieve from execution through the intervention of a Norwegian prisoner (Oren-Hornfeld 2005).

After the war, the assumption among politicians and administrators was that most coerced research subjects died in the experiments. The United Nations Human Rights Division, which took up the case of the survivors of the experiments largely due to the plight of the Rabbits, at first believed that "Possibly less than one hundred or only a couple of hundred..." (UNOG SOA 417/3/01). The International Military Tribunal at Nuremberg heard about experiments in which the subject's death was part of the experiment. For example, at Dachau, Rascher killed living persons for physiological and anatomical reasons as part of his experiments ("Trials of War Criminals," 1944–1949, p. 101; Records of Judge Advocate [RG 153] 1944–1949, p. 24; Neff, ca. 1946).

Iwan Ageew, a Russian survivor of the freezing water experiments at Dachau, described vividly how the experiment took him to the point of loss of consciousness. The context is highly medicalized, and the presence of medical students (likely from the SS Medical Academy at Graz) should be noted.

> Two warders pushed me to a bathroom. 3 doctors and about 10 students were already gathered there. After a heart examination I was injected with some red stuff and put in to a bath-tub with a thermometer. They switched on a ventilator. I was covered in water all but head and hands. Two of the physicians took my wrists, controlling my pulse and making notes. I was not able to describe the agony I felt being completely helpless in the hands of the so unscrupulous tormentors to whom the life of a concentration camp inmate meant less than nothing. The last thing I remember before I lost consciousness was that a slight ice-covering began to appear on the surface of the water. (Ageew 1951, UNOG SOA 417/3/01)

Some victims were selected for killing after they were examined, especially in psychiatric experiments, as illustrated by Ernst Rüdin and Carl Schneider's assistant Julius Deussen who selected 21 "idiot" children for killing after a detailed physical examination (Roelcke 2008). The brains of many euthanasia victims were removed for study after they died. Executed victims of Nazi "justice" increased enormously, thereby making high numbers of cadavers available for research and teaching (Hildebrandt 2009). Even so, our project found that 11,489 persons survived to make claims as victims of experiments, which means approximately half the victims survived not only the experiments but also the camps, clinics, and prisons (Weindling 2014).

Many survivors suffered severe physical disabilities with life-long consequences. Here again, the Rabbits exemplify how survival was often synonymous with chronic pain. Others had damaged reproductive organs and had to live with the devastating consequences. Carl Clauberg, one of the perpetrating gynecologists, caused both sterility and enduring pain. One survivor wrote:

> The experiments actually started 2 or 3 days later but they never let us know what it was all about. I think the experiment was on sterilization of women. We had to go downstairs into a special room, you had to stretch out on a table, they strapped you down and they started first with 24 injections in many parts of our body. We were terribly sore after this. Another day they injected into the womb and ovaries some substance, we didn't know what, that burned like hell and our pains were beyond endurance. Of course all this was done always without anaesthetic. Oh, were we ever miserable!
>
> The victim wished to know what substance caused her injuries.
>
> After the war, my husband wanted to find out what was the substance they injected us with, for I had so many health related troubles and problems. Through the German Ambassador, in a roundabout way, we did find out that it was formaldehyde the Nazi doctors injected into us. What else they tried to find out other than measure our endurance or methods for sterilization, I really don't know. Certainly they knew well how to torture women. Those very long injection needles left open sores on our bodies. (de Jong, n.d., para 22)

Clauberg's research falls into the category of gender-related research with a strongly racial motivation. He carried out chemical injection experiments at Auschwitz-Birkenau starting December 28, 1942, and then moved to Block 10 in the main camp of Auschwitz for his research from April 1, 1943, until December 1944. Out of 456 women, 341 were Jewish. According to Danuta Czech (1989), approximately 550 victims went through Block 10 and only 273 were left on November 28, 1944. Women were periodically sent to the gas chambers from Block 10, so it can be presumed that a few hundred of Clauberg's victims were killed.

The Auschwitz camp doctor and SS officer, Horst Schumann, carried out radiation experiments, mostly on males, alongside Clauberg in Block 10. Clauberg and Schumann were seeking to develop a procedure to sterilize large numbers of people rapidly, using a less complex method than surgical sterilization. Of the 420 victims of X-ray sterilization at Auschwitz, all were Jewish and nearly all were male. Of these, records show that 188 males were castrated after being radiated. Among X-ray sterilization victims, ten Greek female Jews were subsequently subjected to oophorectomies and then died. We have identified 500 Jewish female victims of Clauberg's intrauterine chemical injections (Weindling 2014).

The nationalities of Clauberg's victims were mainly Dutch, Greek, and Polish. At Ravensbrück, our project found claims of surgical sterilization—out of ten victims, two were Jewish; chemical sterilization—out of 377, 64 were Jewish (61 were Hungarian and three were of Czechoslovak origins); and sterilization by methods unspecified—out of 144 victims, 34 were Jewish and nearly all of them Hungarian.[3]

The Clauberg method used chemical injection of a milky substance causing a painful sensation of burning. Individual victims' testimonies show that extensive additional research was involved. Rene Molcho, a victim of Schumann, testified to the elaborate procedures. These involved the taking of a sperm sample, and then the surgical removal of one and, in most cases, the second irradiated testicle (Molcho 1963).

In some cases, men were the subjects of reproductive research, such as the homosexuals subjected to hormone experiments at Buchenwald by the Danish doctor, Carl Peter Jensen, known as "Carl Vaernet" (Hackett 1995, p. 79; Høiby et al. 2002; "Gay Holocaust," 2002). One of his techniques was the implantation of a testosterone-filled tube into a victim's right groin, where, theoretically, the device would slowly release the hormone over many years (Rooke 2012). Our database has details of 35 victims of Vaernet's experimental implantation on September 16 and December 8, 1944. They were born between 1884 and 1921. While 19 were German, three formerly Austrians,[4] one Czechoslovak, and one Pole, 11 were Jewish. The numbers of claimants are higher than in the standard literature.

Mengele's research appears complex. He conducted research on a variety of victims from a genetic point of view, including growth (hence the research on dwarves), physical traits like eye color, and differential responses to infections. In certain cases, body parts were sent to the Kaiser Wilhelm Institute for Anthropology, Human Heredity, and Eugenics, such as the heterochromic eyes of a Sinti family by the name of Mechau, who were selected for killing. Blood samples were also collected and sent to Berlin. All of Mengele's research fit with the program on hereditary pathology of Otmar von Verschuer at the Kaiser Wilhelm Institute for Anthropology, Human Heredity, and Eugenics. Race and heredity rather than gender appear to be the motivating characteristics of Mengele's diverse and extensive research. While many victims were killed as part of the research, others survived because the research was not brought to a conclusion. The Victims of Coerced Research project has identified 598 Jewish twins by name or camp number (there were 12 for whom only a camp number is known) to date.[5] There was also the extended Ovitz family of dwarves. However, no Sinti and Roma victim of Mengele appears to have survived the gassing of the Gypsy camp at Auschwitz (Heese 2001; Schmuhl 2005, pp. 478–481). The documentation indicates a far higher number of female than male victims of Mengele, certainly among the Jewish twins.

The protest by the Ravensbrück Rabbits that their rights as prisoners were violated, and their subsequent refusal to undergo further experiments were part of

---

[3] My thanks to Aleksandra Loewenau for this data from the Victims of Human Experiments project. For the victims of Schumann and Clauberg, see Weinberger (2009).

[4] The Nazis abolished Austrian nationality in March 1938. Therefore, our project takes March 1938 as the date for when nationality is given.

[5] As of October 8, 2013.

a pattern of resistance and sabotage of Nazi experiments and, indeed, more generally by prisoner research assistants. The French prisoner Claudette Bloch sabotaged Auschwitz research on developing a plant for producing rubber by deliberately mixing up seeds (Claudette Bloch Kennedy, personal communication, March 1997). Ludwik Fleck, among other prisoners involved in the vaccine research at Buchenwald, delivered fake vaccine to the Germans. The fragmented instances that have been recorded in human subjects research indicate a wider pattern of sabotage and manipulation. These instances include fake twins such as Gyorgy Kuhn (born on January 1, 1932) and his younger brother Istwar Kuhn (born on December 12, 1932) (see "List of Twins," n.d.), and partial operations by prisoner surgeons such as Dr. Samuel at Auschwitz to prevent the surgical removal of ovaries.

At times the prisoner assistants conducted sabotage. For example, some of them manipulated the water temperature in the freezing experiments. French prisoner-nurses, who were also to be killed, attenuated vaccines for the Bullenhusen children by tampering with the cultures that were to be injected. It was said that during the seawater drinking experiment on Sinti and Roma at Dachau, drinking water was smuggled in. The prisoner assistants told the Swiss conscientious objector Wolfgang Furrer, one of Claus Schilling's malaria research subjects at Dachau, "We prisoners do whatever is necessary to ensure no one dies... " (n.d., Yad Vashem 0.33/4000). Prisoner-nurses reduced doses of highly toxic experimental drugs (Majdański 2009, p. 80; Zámečnik 2007, p. 290).

The postwar position of the German medical profession was denial of responsibility. Rather than offering refertilization to victims who could benefit from it, their position was to leave victims sterilized. Similarly, care and therapy were not offered, even though victims wanted medical attention to assist them in recovering both their health and fertility. High medical expenses are a recurrent theme in victims' files (e.g., Gerrites-DeBoer, UNOG SOA 417/3/01). The situation was similar to victims of the 1933 Sterilization Law who also agitated for operative refertilization. Victims of the Clauberg and Schumann experiments had hormonal treatments in an attempt to restore their health and fertility at considerable personal expense (Molcho 1963). The Ravensbrück Rabbits also were in need of treatment for their atrocious wounds.

After the UN Human Rights Division took up the matter, the Federal Republic's observer informed the UN on July 26, 1951, that it would compensate all victims (Baumann 2009). The UN wanted to compensate pain and suffering, but the Germans insisted on the narrower medical criteria of damage to health and loss of earning capacity. Herein lay a bone of contention. Victims wrote about their state of mind and nerves, but that did not influence the German Finance Ministry in their calculations. The German bureaucrats required documentation that an experiment, which they interpreted in narrow terms, had taken place—unless an experiment was cited at the Nuremberg trials, a claim might be rejected. For example, the claims of the survivors of Mengele's experiments on twins were initially rejected in the 1950s because the experiments were not noted by Mitscherlich and Mielke, and again in the 1960s because it was decided that Mengele's anthropological research did not constitute an experiment. Officials also demanded that only victims who suffered permanent damage to their health should be compensated (Gerrites-DeBoer, UNOG SOA 417/3/01). They mainly viewed financial compensation as covering damage to earning capacity (UNOG SOA 417/3/01, Part E). Claims were rejected when applicants were not in financial need or had not suffered any damage to health because of the experiments. Therefore, if the damage to health was due to the conditions within the concentration camp rather than the experiment, no compensation was paid.

When dealing with victims of experiments that left them sterile, such as men whose testicles were surgically removed, the Germans asked to what extent the victim was currently in pain and unable to work, meaning that pain in the past was not compensated. The inability to have children was compensated at the lowest rate. After all, where was the physical pain or damage to health in it? So the sterilization victims—among the most numerous—received only a minimal, token amount, with victims of X-ray sterilization receiving damages (*Schmerzensgeld*) at the lowest rate for their pain and suffering.

Those who were purposely infected in order to develop vaccines constituted another category of victims. Malaria victims were subject to recurrent bouts of the disease, and typhus victims could be left temporarily paralyzed after a bout of delirium. Both diseases—malaria with about 1,000 victims, and typhus with about 1,500 victims— also caused many fatalities. In the case of malaria, the German official also argued that the disease wore off after nine years. For example, in fixing the amount of the grant to the Polish priest Francis Bielicki, consideration was given to the facts that he was in good health and that medical experience demonstrated the disappearance of the effects of malaria before nine years from the date of infection (later relapses are extremely rare even in the worst form of malaria). He was awarded 3,000 DM in April 1953 ("Victims of Experiments," UNOG SO 262/2, GEN B).

Time and time again, the ministerial officials attributed the victim's physical and psychological problems to the general conditions in the concentration camp. In the case of Mrs. Yvonne Allouche, the Finance Ministry reached the conclusion that "according to expert medical opinion, her 75 percent loss of earning capacity is to be attributed not primarily to the effects of this sterilization but to the shock she experienced during the period of persecution" (Allouche, UNOG SOA 417/3/01).

Victims found it difficult to separate the experiments from the rest of their persecution. The Countess Yvonne Komornicka, a Ravensbrück prisoner who was operated on by Treite, was outraged at the misunderstanding of the long-term effects of the operation. Another survivor understood experiments as "experiences" because he wanted to file a claim for "physical and moral experiences" rather than experiments. It meant the victims' representative organizations themselves turned down many claims. Considerable detail was required. Victims said that the pain blocked their memory. One claimant, Herzlik Adler, wrote, "I do not remember what kind they were as the pains I suffered got me almost out of my mind" (Adler, UNOG SOA 417/3/01).

Victims felt single, lump sum payments were of only limited value in covering long-term medical costs and, by August 1952, victims complained about the meager amounts of their reparations. The survivors felt further pain concerning the procedures. The Polish priest Leo Michaelowski, who had undergone freezing and malaria experiments at Dachau and then testified in Nuremberg, felt that the requirement to submit to a further medical examination by the German government was prejudicial when he could cite court testimony and provide a certificate on his state of health from the expert witness Leo Alexander (Michaelowski, UNOG SOA 417/3/01). The UN lawyers felt let down and undermined by German bureaucracy (United Nations Human Rights Division, UNOG SOA 417/3/01). This situation was even more acute for political reasons. No one from a country not recognized by Germany—notably Poland—could receive payment until special agreements were negotiated with the Federal Republic. Thus, the country with the largest victim group could not at first receive compensation (Baumann 2009).

In other cases, monthly earnings were considered too high, or a loss of under 30 percent of monthly earnings was considered non-compensatable (Eisemann, UNOG SOA 417/3/01). Pain at the time, long-term chronic injuries to internal organs, and enduring psychological distress were not compensated. Thus, by 1958 there were 34 priests who had undergone malaria experiments in Dachau who merited no compensation on the basis of loss of earnings ("Victims of Experiments," UNOG SO 262/2 GEN B). Housewives were similarly regarded as not suffering a loss of earnings from the result of experiments, despite enduring disabilities and, in some cases, being rendered sterile.

John Thompson joined forces with other psychiatrists in New York, who were concerned about the lack of recognition of psychological effects of persecution. At the Forum on Late Consequences of Massive Traumatization with Particular Emphasis on the Problems of Nazi Victims held on March 14, 1965, he commented:

> There is an irony that we ask for data to convince us that someone who has been brutalized, mauled, tortured, humiliated, spat upon, mercilessly thrown into the pit of death yet kept alive only to be tortured more by ever deepening wounds—we ask for data to convince us that such a one has been permanently damaged. (Thompson Papers 1906–2014)

Conclusions

The Allied prosecutions were very much based on the victims' self-documentation. Bringing the human subjects researchers to trial represented a considerable achievement, especially in contrast to the shameful neglect of prosecutions in the German Federal Republic (West Germany) and the German Democratic Republic (East Germany). The Rabbits remained a major voice in the struggle for compensation for victims of human experiments. Although West Germany expected only a small number of survivors, in fact, there were large numbers of claimants, most of whom had severe physical injuries as well as the psychological characteristics associated with what became recognized as "survivor syndrome." Because Mengele's twin experiments were not documented by the Mitscherlich and Mielke volume on the

Nuremberg Doctors' Trial, claims by the twins were rejected by West Germany in the 1950s and 1960s; their claims were finally recognized in the 1970s.

Race was a complex and disputed category under National Socialism, as were the nature of the human subjects research and the authority directing them. The analysis presented in this chapter shows that multiple types of unethical human subjects research occurred under National Socialism. Not only were large numbers of victims affected, but the number of surviving victims was also far higher than anticipated. The complex data will be further augmented and refined, and the narratives of survivors will be analyzed in order to understand more fully the consequences of coerced research. This analysis will provide a basis in historical evidence for discussions of the ethics of coerced human subjects research.

Acknowledgements Wellcome Trust Grant No 096580/Z/11/A on research subject narratives. AHRC GRANT AH/E509398/1 Human Experiments under National Socialism. Conference for Jewish Material Claims Against Germany Application 8229/ Fund SO 29. My thanks also to CMATH for the opportunity of visiting archives in Majdanek and Auschwitz.

References

Adler, H. 1950–1956. United Nations human rights division. Compensation for injuries, 1950–1956 (UNOG SOA 417/3/01). Herzlik Adler file. United Nations Archive, Geneva, Switzerland.

Ageew, I. 1951. United Nations human rights division. Compensation for injuries, 1950–1956 (UNOG SOA 417/3/01). Iwan Ageew file. Letter from Iwan Ageew to Egon Schwelb, 15 October 1951. United Nations Archive, Geneva, Switzerland.

Alexander, L. 1947a, June 28. Alexander to McHaney. Neuro-psychiatric examination of the witness, Karl Höllenreiner, 28 June 1947. (4/33). Alexander Papers Medical Archives, Durham, NC.

———. 1947b, July 1. Letter to Phyllis Alexander on testimony of Nales. Leo Alexander family papers (HOLLIS 13079113). Harvard Law School Library, Harvard University, Cambridge, MA.

Allouche, Y. 1950–1956. United Nations human rights division. Compensation for injuries, 1950–1956 (UNOG SOA 417/3/01). Yvonne Allouche file. United Nations Archive, Geneva, Switzerland.

Ärztekammern, Westdeutschen. 1949. Vorwort der Westdeutschen Ärztekammern. In *Wissenschaft ohne Menschlichkeit*, ed. A. Mitscherlich and F. Mielke. Heidelberg: Lambert Schneider.

Baumann, S.M. 2009. *Menschenversuche und Wiedergutmachung*. Munich: Oldenbourg.

Office, British Foreign. 1949. *Scientific results of German medical war crimes: Report of an enquiry by a committee under the chairmanship of Lord Moran M.C., MD*. London: HMSO.

British Red Cross Society and Order of St. John. 1944, November 15. Correspondance générale avec Croix-Rouge 1.12.1945-30.5.1945, Otages, détenus politiques, Allemagne (microfilm G44.01, G 44/13-11) International Committee of the Red Cross Archives, Geneva, Switzerland.

Czech, D. 1989. *Kalendarium der Ereignisse im Konzentrationslager Auschwitz-Birkenau 1939–1945*. Hamburg: Rowohlt Verlag.

de Jong, Elisabeth. (n.d.). Auschwitz survivors' essays. Personal reflections. http://www.luketravels.com/auschwitz/essay-b.htm. Accessed 23 Jan 2014.

Dörner, K., A., Ebbinghaus, and K., Linne. (eds.). 1999. *The Nuremberg medical trial 1946/47. Transcripts, material of the prosecution and defense. Related documents.* (English Edition), on behalf of the *Stiftung für Sozialgeschichte des 20. Jahrhunderts* in cooperation with K. Roth and P. Weindling. Microfiche Edition. microfiches 8/ 440. Munich: Saur.

Eisenmann, W. 1950–1956. United Nations human rights division. Compensation for injuries, 1950–1956 (UNOG SOA 417/3/01). Yvonne Wilhelm Eisenmann file. United Nations Archive, Geneva, Switzerland.

Eschebach, I., and A. Ley (eds.). 2012. *Geschlecht und Rasse in der NS-Medizin.* Berlin: Metropol Verlag.

Furrer, W. (n.d.). Menschen als Versuchskaninchen [unpublished typescript]. Chapter 8. Call no. 0.33/4000. Yad Vashem Archive, Jerusalem, Israel.

An experiment in co-operation 1925–1945. The history of the liaison committee of women's international organisations 1925–1945. (n.d.). London: no publisher.

Gay Holocaust. 2002. The hunt for Nazi concentration camp doctor Carl Vaernet. Retrieved from International Homosexual Web Organization (IHWO). http://users.cybercity.dk/~dko12530/hunt_for_danish_kz.htm. Accessed 1 Nov 2011.

Gerrites-DeBoer, M. (n.d.). United Nations human rights division. Compensation for injuries, 1950–1956. (UNOG SOA 417/3/01). Translation of Netherlands Red Cross statement of Marianna Gerrites-DeBoer. United Nations Archive, Geneva, Switzerland.

Gries, L. 1945, November 15. Note concerning Leo Gries evidence against Klein and Rutenol. War Office: Judge Advocate General's Office, British Army of the Rhine War Crimes Group (North West Europe) and predecessors. Bergen Belsen Concentration Camp: First Trial. (WO 309/484). The National Archives, Kew, London, UK.

Hackett, D.A. (ed.). 1995. *The Buchenwald report.* Boulder, CO: Westview Press.

Heese, H. 2001. *Augen aus Auschwitz—Ein Lehrstück über nationalsozialistischen Rassenwahn und medizinische Forschung—Der Fall Dr.* Karin Magnussen, Essen: Klartext Verlag.

Heger-de Koning, D. 1950–1956. United Nations human rights division. Compensation for injuries, 1950–1956 (UNOG SOA 417/3/01). Debora Heger de Koning file. United Nations Archive, Geneva, Switzerland.

Hildebrandt, S. 2009. Anatomy in the Third Reich: An outline, part 2. Bodies for anatomy and related medical disciplines. *Clinical Anatomy* 22(8): 894–905.

Høiby, N., J. Rubin, H.D. Nielsen, and N.B. Danielsen. 2002. *Værnet, den danske SS-læge i Buchenwald.* Copenhagen: JP Bøger.

International Scientific Commission for Investigation of Medical War Crimes. 1946, July 1–1948, January 31. [Records created or inherited by the War Office, Armed Forces, Judge Advocate General, and related bodies]. General Correspondence (WO 3O9/470). The National Archives, Kew, London, UK.

Iwanska, J. and H. Piasecka. 1946, June 28. Deposition de Janina Iwanska faite par devant le Major Arthur Keith Mant, RAMC à la Commission d'investigation des crimes de guerre à Paris. Deuxième Deposition d'Helena Piasecka. Fonds Lépine, Archives Institut Pasteur, Paris, France.

Jurkowska, A. 1952. United Nations human rights division. Compensation for Injuries (1950–1956). Alicia Jurkowska de Serafin, 13 January 1952 (SOA 417/3/01). United Nations Archives, Geneva, Switzerland.

Komornicka, Y. 1952, November 25. Déposition of Comtesse Yvonne Komornicka. Survivante des experiemces soi-disant medicales [UN dossier]. United Nations Archive, Geneva, Switzerland.

Lang, H.-J. 2004. *Die Namen der Nummern.* Hamburg: Hoffman & Campe.

List of Twins. (n.d.). About the survivors. Retrieved from CANDLES Holocaust Museum and Edication Center. http://www.candlesholocaustmuseum.org/learn/about-survivors.htm. Accessed 10 Jan 2014.

Loewenau, A. 2012. *The impact of Nazi medical experiments on Polish inmates at Dachau, Auschwitz and Ravensbrück.* [PhD thesis]. Oxford Brookes University: Oxford, UK.

Leyendecker, B., and B.F. Klapp. 1989. Deutsche Hepatitisforschung im Zweiten Weltkrieg. In *Der Wert des Menschen. Medizin in Deutschland 1918–1945*, eds. C. Pross & G. Aly, 261–293. Berlin: Hentrich.

Majdański, K. 2009. *You shall be my witnesses: Lessons beyond Dachau*. Garden City Park, NY: Square One Publishers.

Mant, A.K. 1944–1946. *Experiments in Ravensbruck concentration camp carried out under the direction of Professor Karl Gebhardt* [report]. War Office: Judge Advocate General's Office, British Army of the Rhine War Crimes Group (North West Europe) and predecessors: Registered Files (BAOR and other series). German medical experiments: general correspondence. (WO 309/469). The National Archives, Kew, London, UK.

————. 1946. On the medical services, human experimentation and other medical atrocities committed in Ravensbrück. John da Cunha Papers: Ravensbrück War Crimes Trials (RW 2/ 5, 31). The National Archives, Kew, London, UK

————. 1949. The medical services in the concentration camp of Ravensbruck. *The Medico-Legal Journal* 18: 99–118.

————. 1950, June 15. Letter to Telford Taylor. (NARA RG 153/ 86-3-1 book 3, box 10 letter). National Archives and Records Administration, Washington, DC.

————. 1961. Medical war crimes in Nazi Germany. *St Mary's Hospital Gazette*.

Mant notebook. 1945–1946. Original with the Mant family. Photocopy held by P.J. Weindling, Department of History, Philosophy and Religion, Oxford Brookes University, Oxford, UK.

Mant to Thompson. 1946, June 26. Scientific and Technological Branch policy on unethical medicine and medical war crimes. (1945–1946). [Papers]. Foreign Office and Predecessors: Control Commission for Germany (British Element), T Force and Field Information Agency Technical: Enemy Personnel Exploitation Section. (FO 1031/74). The National Archives, Kew, London, UK.

Michaelowski, L. 1950–1956. United Nations human rights division. Compensation for injuries, 1950–1956 (UNOG SOA 417/3/01). Leo Michaelowski file. United Nations Archive, Geneva, Switzerland.

Molcho, R. 1963, November. Rene Molcho testimony. (03.2486). Yad Vashem Archive, Jerusalem, Israel.

Neff, W. ca 1946. Interrogations N5. Walter Neff. Staatarchiv Nürnberg, Nuremberg, Germany.

Oren-Hornfeld, S. 2005. *Wie brennend Feuer. Ein Opfer medizinischer Experimente im Konzentrationslager Sachsenhausen erzählt*. Berlin: Metropol.

Ost, E. 1988. Die Malaria-Versuchsstation im Konzentrationslager Dachau. In *Dachauer Hefte 4*, S.174–189.

Pohl. 1947, March 26. Pohl affidavit. (NARA 1019/54). National Archives and Records Administration, Washington, DC.

Polish Red Cross. 1944, November 16. Letter to Max Huber 16 Nov. 1944, J. E. Schwarzenberg reply 28 Nov. 1944. Correspondance générale avec Croix-Rouge 1.12.1945-30.5.1945, Otages, détenus politiques, Allemagne (microfilm G44.01, G 44/13-11) International Committee of the Red Cross Archives, Geneva, Switzerland.

Procès de Ravensbrück. 1946–1947. (AJ 3633, p. 132, d 6087). Centres des Archives de l'Occupation française en Allemagne et en Autriche de Colmar (AOF), Province de Palatinat, Colmar, France.

Ravensbrück war crimes trials. 1946–1948. War Office: Judge Advocate General's Office, British Army of the Rhine War Crimes Group (North West Europe) and predecessors: Registered Files (BAOR and other series). (WO 309/1655-63). The National Archives, Kew, London, UK.

Records of the Office of the Judge Advocate General (RG 153). 1944–1949. Vol.1. Dachau trial, Trial record box 185. National Archives and Records Administration, Washington, DC.

Roelcke, V. 2008. *Lebensläufe schreiben—Die diversen curriculae vitae des Psychiaters Julius Deussen vor und nach 1945* In: K. Grundmann & I. Sahmland (Eds.), Concertino: Ensemble aus Kultur- und Medizingeschichte. Marburg: Universitätsbibliothek Marburg.

Rooke, T.W. 2012. *The quest for cortisone*. East Lansing, MI: Michigan State University Press.

Schmuhl, H.-W. 2005. *Grenzüberschreitungen. Das Kaiser-Wilhelm-Institut für Anthropologie, menschliche Erblehre und Eugenik 1927–1945*. Göttingen: Wallstein.
———. 2008. *The Kaiser Institute for Human Anthropology, Heredity, and Eugenics, 1927–1945*. New York: Springer.
Schwarberg, G. 1996. *Meine zwanzig Kinder*. Göttingen: Steidl.
Schwerin, A.V. 2004. *Experimentalisierung des Menschen. Der Genetiker Hans Nachtsheim und die vergleichende Erbpathologie 1920–1945*. Göttingen: Wallstein.
Scientific and Technological Branch policy on unethical medicine and medical war crimes. 1945–1946. [Papers]. Foreign Office and Predecessors: Control Commission for Germany (British Element), T Force and Field Information Agency Technical: Enemy Personnel Exploitation Section. (FO 1031/74). The National Archives, Kew, London, UK.
Somerhough to FIAT, BAOR. 1946, March 24. International Scientific Commission for Investigation of Medical War Crimes. War Office: Judge Advocate General's Office, British Army of the Rhine War Crimes Group (North West Europe) and predecessors: Registered Files (BAOR and other series). (WO 309/471). The National Archives, Kew, London, UK.
Testimonies of Leo Eitinger, Simon Umschweif. (n.d.). Medical Experiments in Concentration Camps. (n.d.). [Mant]. Camps de Concentration. (BB/ 35/263). Archives de France, Paris.
The Nuremberg Medical Trial 1946/47. Transcripts, material of the prosecution and defense. Related documents. English ed. On behalf of the Stiftung für Sozialgeschichte des 20. Jahrhunderts. Klaus Dörner, Angelika Ebbinghaus, Karsten Linne, editors. In cooperation with Karlheinz Roth and Paul Weindling. Microfiche ed. Munich: Saur; 1999.
Thompson, J. Papers. 1906–2014. Papers of John Thompson [Papers , transcripts, and audiotapes]. Held by P. J. Weindling, Department of History, Philosophy and Religion, Oxford Brookes University, Oxford, UK.
Thompson to War Crimes Branch c/o Judge Advocate. 1945, November 29. War Office: Judge Advocate General's Office, British Army of the Rhine War Crimes Group (North West Europe) and predecessors. German medical experiments (WO 309/468). The National Archives, Kew, London, UK.
Thompson to Somerhough. 1946, April 5. International Scientific Commission for Investigation of Medical War Crimes. War Office: Judge Advocate General's Office, British Army of the Rhine War Crimes Group (North West Europe) and predecessors: Registered Files (BAOR and other series). (WO 309/471). The National Archives, Kew, London, UK.
Trials of war criminals before the Nuernberg Military Tribunals under control council law no. 10. 1946–1949. The medical case. (Vol. 1). Washington, DC: U. S. Government Printing Office. http://www.loc.gov/rr/frd/Military_Law/pdf/NT_war-criminals_Vol-I.pdf. Accessed 23 Jan 2014.
United Nations Human Rights Division. 1953–1980. Compensation for injuries 1950–1956. (UNOG SOA 417/3/01). United Nations Archive, Geneva, Switzerland.
Victims of experiments. 1956–1974. Plight of survivors from Concentration Camps. (UNOG SO 262/2 GEN Parts A to G) General 01.1958–06.1958. United Nations Archive, Geneva, Switzerland.
War Criminal, Auschwitz, Dr. Klein. 1946. Foreign office: Political departments: General correspondence from 1906–1966 (FO 371/57641). The National Archives: Kew, London, UK.
Weinberger, R.J. 2009. *Fertility experiments in Auschwitz-Birkenau: Perpetrators and their victims*. Saarbrucken: Sudwestdeutscher Verlag für Hochschulschriften.
Weindling, P.J. 2000. *Epidemics and genocide in Eastern Europe*. Oxford: Oxford University Press.
———. 2001. Auf der Spur von Medizinverbrechen: Keith Mant (1919–2000) und sein Debut als forensischer Pathologe, 1999. *Zeitschrift f. Sozialgeschichte des 20. und* 21: 129–139.

———. 2002. Die Internationale Wissenschaftskommission zur Erforschung medizinischer Kriegsverbrechen. In A. Ebbinghaus & K. Dörner (Eds.), *Vernichten und Heilen. Der Nürnberger Ärzteprozess und seine Folgen,* ed. paperback, 439–451. Berlin: Aufbau Taschenbuch.

———. 2003. Genetik und Menschenversuche in Deutschland 1940–1960. Hans Nachtsheim, die Kaninchen von Dahlem und die Kinder vom Bullenhuser Damm. In *Rassenforschung an Kaiser-Wilhelm-Instituten vor und nach 1933,* ed. H.-W. Schmuhl, 245–274. Göttingen: Wallstein.

———. 2004. *Nazi medicine and the Nuremberg trials.* Basingstoke: Palgrave Macmillan.

———. 2010. *John Thompson—Psychiatrist in the shadow of the Holocaust.* Rochester: Rochester University Press.

———. 2012. Victims, witnesses and the ethical legacy of the Nuremberg medical trial. In *Reassessing the Nuremberg military tribunals,* ed. K. Priemel and A. Stiller, 74–103. New York: Berghahn Books.

———. 2014. *Science and suffering: Victims of Nazi human experiments.* London: Bloomsbury.

Zámečnik, S. 2007. *Das war Dachau.* Frankfurt: S. Fischer.

Zimmern, E. 1944. Resolution of the Liaison Committee of Women's international Organisations, 23 Oct 1944. Correspondance générale avec Croix-Rouge 1.12.1945–30.5.1945, Otages, détenus politiques, Allemagne (microfilm G44.01, G 44/13-11) International Committee of the Red Cross Archives, Geneva, Switzerland.

Chapter 12
The White Rose: Resisting National Socialism–with an Introduction by Susan Benedict

Traute Lafrenz Page and Susan Benedict

Introduction

National Socialism was at its peak in 1941, and Munich was at the epicenter. Jewish professors had been dismissed from the Maximilian University in Munich like other universities throughout the Reich. Jewish students were not permitted to attend, and Jewish physicians had either fled the country or restricted their practices to other Jews as required by law. It was within this environment that a brave student group, the White Rose, coalesced and decided to oppose the Nazis to preserve the values of pre-Hitler Germany that they so revered.

Traute Lafrenz arrived at the Maximilian University from Hamburg in May 1941 to continue her medical studies. As was the custom in medical education at the time, students would enroll at various medical schools to take particular courses with renowned faculty. Traute began her medical education in Hamburg in the summer of 1939, completed four semesters, and then moved to Berlin for an additional semester before moving to Munich to continue her studies (T. Lafrenz Page, personal communication 2011). In June 1941, at a concert in Munich, Traute recognized Alexander Schmorell, who also had been a student at the University of Hamburg, accompanied by his friend, Hans Scholl. Hans, too, was attending the Maximilian University. Traute and Hans became friends, studied together at each other's apartments, and eventually developed a romantic relationship. Traute became a part of Hans' circle of friends who would often come together to cook meals and discuss philosophy, literature, and current events. Among the members

T.L. Page, MD
Surviving member of the White Rose, 4277 Highway 165, Hollywood, SC 29449-6011, USA

S. Benedict, RN, PhD, FAAN (✉)
Assistant Dean and Department Chair: Acute and Continuing Care; Professor and Director of Global Health; Co-director Campus-wide Program in Interprofessional Ethics; The University of Texas Health Science Center at Houston School of Nursing, 6901 Bertner Ave., Houston, TX, USA
e-mail: Susan.C.Benedict@uth.tmc.edu

S. Rubenfeld and S. Benedict (eds.), *Human Subjects Research after the Holocaust,*   157
DOI 10.1007/978-3-319-05702-6_12, © Springer International Publishing Switzerland 2014

of the group were Sophie Scholl, younger sister of Hans and a biology-philosophy student, as well as medical students Willi Graf, Alexander Schmorell, and Christoph Probst (Sachs 2005).

As 1941 turned to 1942 and the war wound on, the members of the group became increasingly distressed over National Socialism and decided to take action by writing and distributing anti-Nazi leaflets. They wrote the first leaflet in the summer of 1942, quoting from Goethe and describing the shame the German people should feel about their government and the need to resist the crimes of the government (White Rose Society 1942). Three additional leaflets were written and distributed before Hans, Alex, and Willi were shipped off to the Russian front to serve as medics. En route, they saw firsthand the atrocities against the Poles and the Jews, including the Warsaw Ghetto (Sachs 2005).

Meanwhile, in Munich, Sophie Scholl, Christoph Probst, and Traute Lafrenz continued efforts to expand the group, at which point Professor Karl Huber joined their cause. By the end of January 1943, 10,000–12,000 leaflets had been distributed by mail or by placing them in public locations such as telephone booths and on trains. In February 1943, Hans, Willi, and Alex dared to write messages such as "Down with Hitler" on over 70 walls of buildings throughout Munich. Later, Hans would add the word "Freedom" on buildings including the main university building (Sachs 2005).

Although Dr. Kurt Huber had written the sixth leaflet, he broke with the students over a divergence of opinion about the group's further course of action. All agreed that Hitler had to go, but Huber thought Hitler could only be removed by a military coup, whereas Hans believed that the general population would bring Hitler down once they became aware of the atrocities committed by the National Socialists. Hans refused to print the sixth leaflet as written by Professor Huber because of a section about joining the military to overthrow the government (Sachs 2005).

Just two days before the fateful day of February 18, Josef Söhnngen, a shop owner who had been helping the group obtain forbidden books, warned Hans that he was taking too many risks and endangering his friends in the process; however, the group persevered with printing leaflets and posting graffiti (Sachs 2005).

On February 18, 1944, Sophie cut classes. Hans and she, carrying one briefcase and one suitcase filled only with leaflets, went to the university where they placed the leaflets in plain sight on the stairs and by the doorways to the classrooms. Students were leaving the lectures as Hans and Sophie ran into Traute and Willi. When Jakob Schmid, a maintenance man at the university, spotted Hans and Sophie distributing leaflets, they threw the leaflets over the railing into the atrium of the classroom building. Sophie ran into a room to hide, but Schmid apprehended Hans. In Hans' pocket was Christoph Probst's draft of the seventh leaflet (Sachs 2005).

Hans' most recent girlfriend, Gisela Schertling, a loyal Nazi who was aware of the group's activities, was immediately arrested and, in order to obtain leniency for herself, named the members of the group including Professor Huber (Sachs 2005). On February 22, 1943, Nazi lawyer and judge Roland Freisler, President of the People's Court, tried and convicted Hans, Sophie, and Christoph; all three were executed by guillotine at 5:00 P.M. the very same day. On April 19, Traute and other group members were also tried and found guilty. Willi, Alex, and Professor Huber were sentenced to death. Professor Huber and Alexander Schmorell were executed on July 13, 1943 (Scholl 1983, p. 68). Willi Graf's parents sent a letter to Hitler asking for clemency. The request, however, was denied and Willi Graf was executed on October 12, 1943 (Sachs 2005).

Traute was sentenced to one year in prison for distributing the leaflets. She was rearrested immediately upon the completion of her prison term and was finally liberated from prison by US forces. She completed her medical degree and, in 1947, immigrated to the United States, where she completed a residency in internal medicine (T. Lafrenz Page, personal communication 2011). Her life's work was the establishment and administration of a school for indigent children with developmental disabilities—the Esperanza School in Chicago. The nature of this school is especially remarkable when one considers that Traute received her medical education during the time that people with developmental and psychiatric conditions were considered to be "useless eaters" and were being killed at the request of the Nazi government.

On December 2, 2012, the Center for Medicine after the Holocaust (CMATH) presented the first CMATH award to Traute Lafrenz Page, MD, and the White Rose "for their courage and dedication to the highest standards of medical ethics in the gravest of circumstances."

The following section is a summary of her speech prior to the presentation of the award (see also Page 2012, video).

Traute Lafrenz Page's Speech

It is a beautiful thing that we can do: Give people a name and place.

I have been invited to talk about the White Rose, a resistance group of university students in Munich. I wonder why this group has been so popular throughout the years, appearing in books and movies. Is it because of the brother and sister alliance? All of the members were beautiful people striving to find what is good and right.

Hans Scholl: Hans had a restless mind, forever searching for the truth in philosophy and religion.

Sophie Scholl: Sophie was the younger sister of Hans and his housemate. She was beautiful and upright, very young but willing to serve.

Alex Schmorell: Alex was sensitive and artistic. He loved his homeland of Russia for its writers and poetry.

Christoph Probst: Christo had a love of nature and family.

Professor Huber: Professor Huber was a profound searcher with a touch of genius and was deeply musical. Leibniz, Fichte, and Goethe were very important to him and his area of specialization. Professor Huber had a burning desire to see a Germany that could be identified with these values.

Willi Graf: Willi was thoughtful and religious.

In addition to these well-known members of the White Rose, there were other resisters whose names are hardly known. I wish to give them a place in our memory. They, too, gave their lives for what they knew was good and right.

Werner Scharff: Werner was Jewish and thus was not allowed to study at the university. He founded the Community for Peace and Rebuilding in Berlin, helping Jewish families and alerting them to dangers.

Hellmut Hübener: Hellmut was very young and worked as an apprentice in the youth offices in Hamburg. He spread news that he heard over the BBC (British Broadcasting Company) by printing pamphlets. At his trial, he was asked by the judge, "Do you think news from the BBC is true?" To this, knowing that he was going to get the death sentence, Hellmut replied, "Sure, don't you?"

Walter Klingen Beck and Daniel V. Recklinghausen: Walter and Daniel were young, 18–19 years old, and big fans of the radio. They constructed their own and tried to send out news over it. They also wrote pamphlets. They lived in Munich.

Ignatz Schumann: Ignatz was a carpenter who worked near Hartheim in Linz, Austria. When he found out what was causing the horrible smell [from the killing and gassing of patients in Hartheim], he spread the facts both verbally and by newsletters.

Maria Theresa Kreschbauer: Maria was from Vienna. She had a Communist background. She distributed pamphlets on the resistance by hiding them in her baby carriage.

Bernhard Lichtenberg: Father Lichtenberg was a Catholic priest who came from a small town in Silesia to Berlin. He wrote letters of protest to Hitler. For about three years, he openly led prayers from the pulpit of St. Hedwig Cathedral in Berlin for all of the Jewish and Catholic sufferers in the concentration camps.

Joseph Schultz: Joseph was a member of the 171 Infantry Regiment of the German Army. In retaliation for some partisan activities, the Germans lined up 30 or more villagers against a haystack to execute them. Joseph took off his helmet and walked over to the haystack saying, "If you do this, you will have to shoot me, too," which was done.

Many others I do not know about except that they were sentenced to death by a court in Vienna for their resistance activities:

Dr. Alred Haschel
Theresa Horacek
Elfriede Merkel
Heinrich Raab
Karl Teufel

All of the people mentioned above have names, names that need a place in our memory.

The White Rose

The group that came to be known as the White Rose was simply a group of university students who became good friends. They were young, beautiful, and very verbal. They kept diaries and wrote letters. Hans and Sophie Scholl's older sister, Inge, wrote a book about them after the war.

I first met Hans Scholl in the spring of 1941. I had gone to Munich with a friend, Ulla Claudius, to begin my clinical studies. Ulla's father was a poet, writer, and Social Democrat.

One of my earliest influences was a teacher who emphasized that "Your father should protest the right of authorship." I was impressed because at my house, little thought was given to politics. My mother, from Vienna, was influenced by Emperor Franz Joseph, whereas my father was Prussian and influenced by Bismarck.

I remember the burning of the Reichstag [German parliamentary building, burned in 1933]. On the way to school, I saw a display window with pictures that portrayed Jews as being ugly and carriers of inherited diseases. My favorite teachers left the school, not only the Jewish ones. Jewish classmates left. All that I knew emigrated. Books were forbidden so I listened to the radio. I went to an art exhibit featuring fascist Franco in Spain. Thus, by the time I got to Munich, I arrived with the conviction that things were going very wrong.

Soon after arriving in Munich to begin my studies, I met Hans Scholl through my friend Alex Schmorell. We started *Leseabende* [evening reading]. We read books together and discussed them. We studied the German classics, Aristotle, and French and Russian authors. We read on religion written by Paul Claudel. There was a neo-Catholic movement. Hans was very eager to find a religious orientation. All along, we were listening to the news, listening to the lies.

Soon I met the entire Scholl family. There were five children in the family who all read the same books and discussed them. Mr. Scholl, an accountant, was denounced by his secretary and sent to prison. The oldest child, Inge, was working with him in his office. When it was tax season, I used my semester vacation to work with her in the office.

In the summer months of 1942, Hans and Alex wrote the first four pamphlets. About 100 were sent out by mail. They were full of direct criticism of the government and of quotations from Goethe, Schiller, and Plato. These were just the things that we'd been reading as we thought, "We should do something. What can we do?" In the early summer, Sophie started her studies in Munich. Hans and Sophie moved to a very different part of town. I did not see them daily because I was busy with my job.

Although our meetings would start as reading sessions, they became more political: *Dieser Staat ish kein Staat*! [This state is not a state!]

Things accelerated. Professor Huber joined our meetings and two more pamphlets were written, even more direct than the preceding ones. Hans and Alexander Schmorell wrote slogans on houses and university buildings: "Down with Hitler" and "End the war." Hans and Sophie put copies of the last leaflet on the doors of classrooms at our university, the 470-year-old Maximilian University in Munich. Afterward, they threw additional leaflets over the balcony, from the second floor to the first floor courtyard. They were arrested on Thursday, the 18th of February 1944, interrogated on Friday, the 19th, and were kept in prison over the weekend. On Monday, they were tried for treason and beheaded the same day.

Man stands between good and evil. One can find the good, but one must work to do so.

References

Page, T.L. 2012, December 5. *The story of the White Rose* [video]. Presented at Human Subjects Research after the Holocaust Conference. The Methodist Hospital Research Institute in Houston. Available at http://hsrah.org/speaker-recordings/.

Sachs, R.H. 2005. *White rose history, volume II: Journey to freedom.* Los Angeles: Exclamation! Publishers.

Scholl, I. 1983. *The White Rose.* Hanover, NH: University Press of New England.

White Rose Society (The). 1942. The first leaflet. The leaflets of the White Rose. Retrieved from http://www.whiterosesociety.org/WRS_pamphlets_home.html. Accessed 23 Nov 2013.

Chapter 13
The Origins and Impact of the Nuremberg Doctors' Trial

Howard Brody

The Nuremberg Doctors' Trial has been thoroughly described by previous authors (Annas and Grodin 1992). My intent here is to reflect upon a few aspects of the trial and its aftermath that have been less explored. I do so from the vantage point of one whose formal training is in philosophical ethics rather than in history.

The main questions I address are as follows: What were the sources of the ethical concepts that became enshrined in the Nuremberg Code? What immediate or short-term impact did the Trial and the Code have on human subjects research ethics in the United States? Finally, could we envision a set of circumstances in which the Trial might never have occurred? To address the latter question, I will suggest comparisons between the US response to medical war crimes in postwar Germany and Japan. While medical war crimes in Germany were not restricted to unethical experiments on human subjects, I will focus here on that aspect of Nazi medicine, as that discussion is most directly relevant to my list of questions.

Background

When the Allies occupied Germany in 1945, they found evidence of a series of experiments conducted on concentration camp inmates. Some were aimed at military applications, such as exposure to low temperature and atmospheric pressure and the treatment of wound infections. Others were related to Nazi ideas of racial hygiene and eugenics. In the latter group, one could include experiments to refine efficient techniques for eugenic sterilization, studies of brain pathology of psychiatric patients, the Auschwitz twin studies conducted by Joseph Mengele, and

H. Brody, MD, PhD (✉)
John P. McGovern Centennial Chair in Family Medicine; James Wade Rockwell Distinguished Professor in Medical History; Director, Institute for the Medical Humanities University of Texas Medical Branch, 301 University Boulevard, Galveston, TX 77555-1311, USA
e-mail: habrody@utmb.edu

S. Rubenfeld and S. Benedict (eds.), *Human Subjects Research after the Holocaust*, DOI 10.1007/978-3-319-05702-6_13, © Springer International Publishing Switzerland 2014

the skeletons of Jews and other "inferior races" collected by August Hirt (Proctor 1988; Lifton 2000).

At last count, the Oxford Brookes database of identified victims of these experiments included approximately 25,000 individuals (Weindling 2012). While many died in agony from the experiments, others survived, and some of the self-described "rabbits" of the Ravensbrück camp (women who had been subjected to deliberate wounding with infection to test antibiotics) were able to testify at the Doctors' Trial and exhibit their scars to the judges.

The Nazis justified these experiments on a number of grounds, some of which were introduced as part of their defense at the Trial. Generally, they viewed the experimental subjects as subhuman and the experiments ethically comparable to animal studies. The Nazi program of racial hygiene was predicated on the idea that non-Aryan races represented a public health threat to the German *Volk* (people), and so eliminating those inferior races was medically akin to excising a dangerous tumor. These justifications received heightened attention as soon as World War II began; if "good" Aryan youths were being killed in battle, it seemed silly to quibble over the lives or the suffering of inferior non-Aryans. As in other nations, the pressing national interest of winning the war, and hence preserving the lives of soldiers and airmen, seemed to trump all other ethical considerations regarding research. Finally, the experimenters employed crass utilitarian considerations—as all the victims were effectively sentenced to death anyway by virtue of having been confined in concentration camps, why not salvage something of scientific value rather than simply burning the bodies?

The Allied response to these discoveries was the Nuremberg Doctors' Trial, conducted between December 1946 and August 1947. It was the first of 12 trials of specific groups of Nazis accused of war crimes by an American military tribunal, following the main trial of the major leaders of the Third Reich by the International Military Tribunal. Of the 23 defendants at the Doctors' Trial, 16 were found guilty and 7 were hanged. As part of the final ruling, the judges issued what later came to be called the Nuremberg Code, which formed the first major modern document regulating research on human subjects.

Sources of the Code

Two American physicians played key roles in the Doctors' Trial. Leo Alexander (1905–1985), a Viennese-born psychiatrist and neurologist, had been a member of the US Army Medical Corps since 1942 and, in 1945, was assigned to the intelligence unit gathering evidence for the Nuremberg trials. He wrote two key memos to the chief prosecutor, General Telford Taylor, in which he implicated a lack of informed consent as the critical issue in condemning the German experiments. Andrew C. Ivy (1893–1978) was an internationally recognized physiologist who was nominated by the American Medical Association as medical adviser to the

Nuremberg prosecution. He testified for the prosecution at the trial, also stressing informed consent as an absolute requirement for human research.

Shuster (Shuster 1997), reviewing the Trial and the roles of Ivy and Alexander, noted that Ivy was forced to admit on cross-examination that there were no existing written principles of research ethics, let alone any that specified informed consent as a requirement. She characterized the inclusion of an informed consent requirement in the Nuremberg Code as "new" (1997). These sorts of observations might lead some to conclude that the Doctors' Trial and the Nuremberg Code were relatively cynical examples of the victors rewriting history and making up rules for the vanquished out of whole cloth. Indeed, if there was no precedent for the Nuremberg Code and its insistence on informed consent in either Germany or the United States prior to the war, then we could better understand why US research failed to conform to Nuremberg standards both during and after the war.

This view, however, is effectively challenged by Vollman and Winau (1996). They described a scandal in Prussia in 1898 involving the celebrated scientist Albert Neisser (1855–1916), after whom the bacterium causing gonorrhea is named. Hoping that he had discovered a vaccine against syphilis, Neisser injected cell-free serum from syphilis patients into prostitutes, who had been admitted for other conditions, without informing them of the nature of the substance or obtaining their consent. When some of the prostitutes subsequently developed syphilis, he admitted that this showed that his "vaccine" had not worked, but denied that the serum could have caused their disease, blaming instead their status as prostitutes (Vollman and Winau 1996).

The government of Prussia felt obliged to investigate and take action. The panel of physicians they consulted, including the celebrated pathologist Rudolf Virchow (1821–1902), advised that human experimentation was unethical without informed consent. Legal experts similarly advised the government that informed consent would be mandatory for any non-therapeutic research. In 1900, the Prussian ministry responsible for health directed all hospitals and clinics within its jurisdiction that no research was permissible on anyone who had not given "unambiguous consent" after a "proper explanation of the possible negative consequences" (Vollman and Winau 1996, p. 1446). They noted that as consent was mandatory, neither minors nor incompetent adults could be employed as experimental subjects. The directive, however, was not legally binding.

Under the Weimar Republic, the Reich Interior Ministry took similar action in 1931. The ministry was apparently prompted both by a pending reform in the German laws, and also by criticism in the political press of more recent unethical experiments. The new rules, which did have the force of law, specified that informed consent was required for both therapeutic and non-therapeutic research. (An emergency exception was made for therapeutic research only.) The law anticipated, and rejected, the exploitation of some classes of people as research subjects due to their socioeconomic status. The 1931 legal code was never rescinded under the Nazi regime, and so remained officially the law on the books throughout World War II. Vollman and Winau do not explain why, if this was so, neither Andrew Ivy, nor the German expert in medical history consulted by the Nuremberg prosecutors, made any mention of this precedent at the Trial.

If one finds Vollman and Winau's account compelling, then two important conclusions follow. First, the concept of informed consent as a requirement for ethical research was not an invention of the Nuremberg prosecutors. Second, the ideas contained in the Nuremberg Code were as much German as American. Claims that respect for informed consent antedated the Trial are reinforced by David Rothman's historical survey of American World War II medical research (1991). Rothman claimed that the idea of informed consent was clearly present among US physicians and scientists of the period, but was applied in a spotty fashion. Where well-educated subjects were employed, and where failure to obtain consent could most easily have resulted in negative publicity, consent requirements were most scrupulously followed. US prisoners were used as experimental subjects in military-related research, and the popular press used the fact that they had voluntarily consented as evidence of their patriotic support of the war effort (Rothman 1991).

American Exceptionalism in Research Ethics

Even though informed consent was obviously understood by at least some American physician-scientists during the War, the impact of the Nuremberg Code on US human subjects research was, from today's point of view, disappointing. As we now know, Harvard research anesthesiologist Henry K. Beecher (1904–1976) wrote an expose in 1966, published in the prestigious *New England Journal of Medicine*, listing more than 20 unethical experiments conducted by some of the most prominent academic physicians in major US academic medical centers. In the vast majority of cases, the absence of informed consent was the principal ethical violation alleged. Beecher's (1966) revelations were followed in 1972 by the Associated Press story, by Jean Heller, about the Tuskegee Syphilis Study in which the US Public Health Service studied uninformed and unconsenting black sharecroppers in Alabama for 40 years, all the while denying them treatment for their syphilis (Heller 1972/2000).

Americans, it seemed, generally drew the wrong conclusions from the Nuremberg Trial. The reasoning appeared to proceed as follows:

1. The Nazis committed these ethical atrocities.
2. The Nazis were evil people.
3. I, the American physician-scientist, am a good person.
4. Therefore, the Nuremberg Code and similar rules, necessary as they are when evil people are involved, simply do not apply to me or my research.

Students of the history of the Trial note specific causes for such reasoning. Shuster, for example, agreed with the US Belmont Report of 1979 (post-Beecher and -Tuskegee) in clearly distinguishing between an ethical code suitable for therapeutic medicine and an ethical code designed to regulate human subjects research (Shuster 1998; see also National Commission, 1979). By contrast, Shuster argued that neither Alexander nor Ivy could separate the traditional Hippocratic ethic relevant to patient care from the ethics of human subjects research. In the traditional therapeutic ethics of the time (when informed consent as a requirement *for treatment* was largely unknown), the principal protection of the rights and interests of the patient was the physician's beneficent intent. Alexander and Ivy (and to a large extent Beecher) believed that similar beneficent intentions should govern the relationship between investigator and subject, and, therefore, legalistic approaches such as the Nuremberg Code should be avoided in American practice. They could not see, as the authors of Belmont did, that research is fundamentally a different sort of activity from therapy, aimed at generating scientifically valid knowledge rather than at individual patient benefit; and that protecting the rights of human subjects demanded a different ethical approach (Miller and Brody 2003).

Marrus (1999) argued further that the limitations of the Nuremberg Trial also led to misunderstandings of its applicability to US research practice. From the Allied command viewpoint, Germany's crime was waging aggressive warfare against other nations; what Nazis had done to their own citizens was of much lesser concern. The war crimes trials accordingly should focus on Germany as a wartime aggressor. Legal experts were also concerned about the international tribunal's lack of jurisdiction over peacetime activities of the Third Reich.

The limitations thereby imposed on the Doctors' Trial insured that only a restricted slice of Nazi medicine would be placed on public view and documented. What happened to non-Germans in concentration camps received primary attention. Germany's forced sterilization of its own citizens and the racial hygiene theories of eugenics, which prompted these actions and were eagerly embraced by so many within the German medical community, were almost totally ignored. Marrus thus argued that the historical record, as painted by the Nuremberg prosecution, had serious gaps and was very likely to be misread by later generations. For example, if Nazi eugenic theories were omitted as a major motivator, one would not have to address the embarrassing prominence of American and English eugenicists among those whom the Nazis sought to emulate (Marrus 1999; see also Kevles 1985; Proctor 1988).

The Role of John W. Thompson

So far, I have tried to show that Germany itself gave rise to the theory of informed consent as part of ethical research and that, for several reasons, US scientists felt themselves exempt from the Nuremberg Code at least until the late 1970s. I now want to turn to the question of how likely or unlikely were the events leading up to the Doctors' Trial—put crudely, how easy would it have been for this trial never to have occurred at all? One way to explore this question is by way of analogous events in Japan during the same time period; and a further way of illuminating the comparison between Japan and Germany is to investigate the career of John W. Thompson, as told by the historian Weindling (2010).

Thompson (1906–1965) was an intriguingly cosmopolitan figure. He was born in Mexico of US parents, received most of his education in the United States, and then graduated from the Edinburgh Medical School. He then established a combined career in psychiatry and physiology, and during the 1930s, had contacts with both Ivy and Alexander. During the war, he gravitated to Canada and became an officer in the Royal Canadian Air Force (RCAF). He did experiments on the affects of high altitude on pilot safety but, unlike the Germans at Dachau, was scrupulously ethical, including self-experimentation as a central feature of his studies.

In May 1945, Thompson arrived in Germany as part of a RCAF contingent assigned to oversee the disarming of the Luftwaffe. For unclear reasons, he ended up among the Allied medical team at the Bergen-Belsen concentration camp, and thus gained firsthand experience among the camp survivors. Next, he was assigned to assess the Germans' scientific results related to aeronautic stress. As he reviewed the available documents, he soon realized that unethical experiments had been conducted, which he was able to confirm through interviews of surviving German physicians and scientists.

Thompson was eventually appointed Secretary-General of the International Scientific Commission (War Crimes), and the intelligence team that he headed was instrumental in preserving and analyzing captured German experimental records. Thompson's influence on the Nuremberg Trial was, however, mostly indirect. Through his personal relationships with people like Alexander and Ivy, he exerted considerable influence on the proceedings while remaining in the background. Nonetheless, Weindling, writing in an era when the "great man" theory of history is held mostly in disfavor, has no hesitation in labeling Thompson as "Godfather of both the Nuremberg Code and informed consent" in research (Weindling 2010, p. 148).

As recorded in communications with his Canadian government superiors, Thompson quickly reached several key conclusions that helped to shape the Doctors' Trial. While some Allied physicians had advocated the destruction of all the records of the unethical Nazi experiments, Thompson believed that it was essential to record and preserve all scientific findings; he rejected the idea that one could best honor the memory of the victims by ignoring any valid data that resulted from their exploitation. He agreed fully that the means by which these data had been collected were both unethical and criminal; indeed, Thompson seems to have originated the idea of a "medical war crime." It was, therefore, essential that the perpetrators of these experiments be tried and punished. Exposure and punishment were required not because Nazis were uniquely evil—Thompson here seems to have precisely anticipated, and rejected, the later American exceptionalism response. Thompson realized fully that wherever science was practiced, scientists would be tempted to abuse vulnerable research subjects and take advantage of wartime fervor, just as had happened in Germany. The only defense against similar ethical abuses in the future was to painstakingly record the Nazi medical war crimes and to call for firm ethical boundaries (Weindling 2010, p. 125).

Thompson, therefore, had a singularly clear vision of what needed to be done in postwar Germany and set out to bring about the results he sought. The events that followed in Germany stand in marked contrast to the US record in Japan.

Japanese Medical War Crimes

Unlike the haphazard Nazi experiments, the Japanese research on human subjects between 1936 and 1945 was centrally directed by a military unit, of which the chief headquarters was Unit 731 in Ping Fan, near Harbin, Manchuria. This unit was founded by Major Shiro Ishii, a physician with additional training in bacteriology and preventive medicine, who eventually rose to the rank of Lieutenant General. Unit 731 ultimately boasted 150 buildings and 3,000 personnel and could accommodate 600 prisoners at any one time (Williams and Wallace 1989; Harris 2002; Nie et al. 2010).

While the Japanese military did some experiments on extreme environmental conditions similar to those at Dachau, the main thrust of Unit 731 was the development and testing of bacteriological weapons. Japanese scientists grew massive quantities of disease bacteria and tested bombs and other methods for dispersing them to infect enemy troops and civilians. Most of the victims were Chinese nationals, whom the Japanese referred to as *maruta* ("logs"), reflecting their alleged subhuman status. None of the estimated 10,000 victims of these experiments survived. The standard experimental protocol called for the victim to be brought to autopsy, and when Russian troops approached Ping Fan in 1945, all remaining prisoners were killed to assure that no witnesses remained (Harris 2002).

The Japanese resembled the Nazis in attributing racial inferiority to non-Japanese, especially Chinese and Koreans as well as Caucasians. They had, however, no theories of a public health threat from racial mixing. They simply sought to depopulate territories inhabited by inferior peoples to allow expansion of their own empire (Nie et al. 2010).

After the war ended, Unit 731 maintained its cohesion, and General Ishii was able to direct his subordinates' contacts with the American occupation forces. As a result, the first two American investigators to look into the biological warfare program, in 1945–1946, failed to turn up evidence of experiments on human subjects, though they remained suspicious that the Japanese were holding back information. It was not until April 1947, when the Nuremberg Doctors' Trial was well underway, that a third US biological warfare expert finally was approached by Ishii's representatives. The expert, Dr. Norman Fell, was told that the Japanese were now prepared to be more forthcoming with data, so long as the interviews were handled "from a scientific point of view"—that is, with assurances that the Japanese would not be prosecuted for war crimes (Fell 1947a, pp. 1–2).

When Fell issued his final report in June 1947, he stated, ". . . that all information obtained in this investigation would be held in intelligence channels and not used for 'War Crimes' programs." He attributed this decision to "the recommendations of the CinC, FEC," that is, the Commander in Chief of the Far East Command, General Douglas MacArthur (Fell 1947b, p. 2). By then, MacArthur was primarily focused on the anti-communist war that he felt sure would soon break out, and he wanted to return Japan as quickly as possible to peaceful status as a major US ally. Further punishment of Japanese physicians and scientists for events in the past war was, for MacArthur, a dangerous distraction and impediment to his chief goal.

The American biological-warfare scientists, based at Camp Detrick, Maryland, had no problem accepting MacArthur's recommendations in this instance because their own priority was not the punishment of Japanese scientists for medical war crimes. Rather, they sought data that the Japanese *might* have obtained that the Americans, barred from doing similar experiments, *could not* have obtained. A later report by Camp Detrick scientists spoke of the "scruples attached to human experimentation" that (presumably) prevented American scientists from obtaining the data that the Japanese had gathered (Hill and Victor 1947). The potential military and national security value of these data seemed to outweigh any human-itarian or justice concerns regarding the abuse of the experimental subjects. By playing coy in releasing any data, the Japanese shrewdly increased the Americans' hopes that the data would eventually prove valuable. Unlike John Thompson in Germany, none of these American scientists ever came face to face with the victims of Japanese racist policies.

The group that eventually decided the fate of the investigation into the Japanese biological warfare program was a coordinating committee made up of high officials representing the US cabinet departments of State, War, and Navy (SWNCC). In a document dated August, 1947, a SWNCC subcommittee agreed, first, that Unit 731 scientists had "violate[d] the rules of land warfare," and second, that what they had done was substantially similar to what German physicians were then on trial for at Nuremberg (SWNCC 1947a, Appendix A, p. 4). But the SWNCC subcommittee went on to assert, "The value to the U.S. of Japanese data is of such importance to national security as to far outweigh the value accruing from 'war crimes' prosecu-tion" (SWNCC 1947b, p. 1). The background concern running through the discus-sion was that any war crimes trial would mean that important data on the Japanese biological warfare program would be made public, and hence shared with the Soviet Union. Keeping these data out of communist hands was as high a priority to the United States as obtaining the data for themselves (Harris 2002).

While the decision had effectively been made in August 1947, it was not until November 26, 1948, that the investigation into possible war crimes by Japanese scientists was officially closed. This appears to be a form of delaying tactic as the Tokyo War Crimes Trials of Japanese leaders thought most responsible for prose-cuting the war in the Pacific took place between April 1946 and December 1948. By dragging out the final decision whether to pursue war crimes investigations against Unit 731 personnel, American military leaders assured that the Tokyo trial would be winding down before any action could be taken (Harris 2002).

As a result of what amounted to a US cover-up of medical war crimes, Japanese leaders of the biological warfare program achieved prominence in medicine and science in the postwar decades (Nie 2006). Between 1947 and 1983, all but one head of the Japanese equivalent of the National Institutes of Health had served in a biological warfare unit. Others became wealthy in private practice or in the pharmaceutical industry. As a final irony, when Camp Detrick experts finally were able to collate all the data obtained from Ishii and his subordinates, they discovered nothing of importance that the United States had not already discovered (Harris 2002).

Japan and Germany

The US victors in World War II brought many of the perpetrators of unethical Nazi experiments to trial and punishment, but actively covered up similar and, in some ways, even more egregious crimes in Japan (Brody et al. 2014). How are we to account for this difference?

If Weindling's account is plausible, the personal influence of John Thompson, who had no counterpart in Japan, was significant in guiding events in Germany (Weindling 2010). Thompson was clearly worried that failing to aggressively prosecute what he labeled as "medical war crimes" would lead to exactly the mindset that ended up prevailing in the US response to Japan—that the combination of scientific fascination with the forbidden fruit of unethical experiments, coupled with the pressure of national security in time of war or impending war, would lead future scientists in other nations to replicate the wrongdoing of the German physicians. We now know that these predictions were borne out particularly in the radiation experiments conducted by US scientists during the Cold War (Advisory Committee, 1995).

Another key factor appeared to be the timing of events. On June 24, 1948, the Soviet occupation forces instituted their blockade of Berlin, which led the United States to respond with the Berlin airlift. From this moment forward, if not before, US leadership was far more concerned about the Cold War against the Soviet communists than about events that had occurred during World War II. In fighting communism in the Soviet Union and China, the United States desired strong allies in the newly pacified nations of Germany and Japan. A critical aspect of returning these nations to sovereign status as reliable US allies was rebuilding their professional and academic institutions. The willingness of German and Japanese scientists and physicians to cooperate with the United States in the future now seemed much more important than any previous wrongdoing under the old regimes.

By the time of the Berlin blockade and airlift, the Doctors' Trial was nearly ended. While "what ifs" are of questionable value in studying history, one might still wonder what would have happened if the uncovering of the Nazi experiments and the decision of whether to prosecute the perpetrators had been delayed as long in Germany as it had been in Japan.

Conclusion

Superficial accounts of the ethics of human subjects research trace a simple straight line: Nazi atrocities led to the Doctors' Trial, which led to the Nuremberg Code, which highlighted the ethical primacy of informed consent, which led to today's apparatus for regulating research and protecting the rights of participants. I have tried to show that the story is messier than this straightforward, linear account. At least some level of concern for informed consent predated the Nuremberg Code. Even after the Nuremberg Code was promulgated, American investigators often failed to accept any responsibilities regarding informed consent and research subject protection. And had circumstances been altered just a little, it is conceivable that the Doctors' Trial would never have occurred at all.

Is any of this important except as a set of historical footnotes? I would suggest that there is a lesson here for juggling two important ethical insights. These insights at first appear logically incompatible, and yet on reflection both appear to be true and informative. The first insight is that something supremely and uniquely evil happened in Germany during the Nazi era. It is not by accident that popular culture treats the terms "evil" and "Nazi" as virtual synonyms. The second insight is that in many ways, the German physician-scientists were not that different from the rest of us. We ignore the reasons why they acted as they did at our peril. Many of us would also like to think today that had we been in the situation of the Camp Detrick scientists, we would never have participated in a cover-up and would have demanded that the Japanese war criminals be brought to justice. These are comforting, but ultimately misleading thoughts. The historical record suggests rather what John Thompson most feared—that under the right circumstances, most of us are prone to behave in ways that will not later withstand dispassionate ethical scrutiny. Coming to understand both insights at once—that Nazi Germany was a sort of immoral universe unto itself, but that it can teach important lessons for all of us—remains the task at hand.

Acknowledgments I am grateful to Sarah Leonard for research assistance on the Japanese experiments and to Paul Weindling and Jing-Bao Nie for scholarly advice.

References

Advisory Committee on Human Radiation Experiments. 1995. *Final report*. Washington, DC: U.S. Government Printing Office.

Annas, G.J., and M.A. Grodin (eds.). 1992. *The Nazi doctors and the Nuremberg Code: Human rights in human experimentation*. New York: Oxford University Press.

Beecher, H.K. 1966. Ethics and clinical research. *New England Journal of Medicine* 274: 1354–1360.

Brody, H., S.E. Leonard, J.B. Nie, and P. Weindling. 2014. U.S. responses to Japanese wartime inhuman experimentation after World War II: National security and wartime exigency. *Cambridge Quarterly of Healthcare Ethics* 23(2): 220–30.

Fell, N. 1947a, April 22. Report by Norbert H. Fell. Interrogation of Masuda, Tomosada. In *Unit 731 and biological warfare: a CD-ROM collection,* S. Kondo, ed., Disc 2, pp. 1–2 (2003). Toyko: Kashiwa Shobo.

———. 1947b, June 24. Addendum to report by Norbert H. Fell In *Unit 731 and biological warfare: a CD-ROM collection,* S. Kondo, ed., Disc 2, p. 2 (2003). Toyko: Kashiwa Shobo.

Harris, S.H. 2002. *Factories of death: Japanese biological warfare, 1932–1945, and the American cover-up.* New York: Routledge.

Heller, J. 1972/2000. Syphilis victims in U.S. study went untreated for 40 years. Associated Press, July 26, 1972. In *Tuskegee's truths: Rethinking the Tuskegee Syphilis Study,* ed. S.M. Reverby, 116–118. Chapel Hill: University of North Carolina Press (Reprinted in 2000).

Hill, V. & Victor, J. 1947, December 12. Report by Edwin V. Hill and Joseph Victor. Introduction. In *Unit 731 and biological warfare: A CD-ROM collection,* ed. S. Kondo, Disc 2, p. 4 (2003). Toyko: Kashiwa Shobo.

Kevles, D.J. 1985. *In the name of eugenics: Genetics and the uses of human heredity.* Berkeley: University of California Press.

Lifton, R.J. 2000. *The Nazi doctors: Medical killing and the psychology of genocide.* New York: Basic Books.

Marrus, M.R. 1999. The Nuremberg Doctors' Trial in historical context. *Bulletin of the History of Medicine* 73: 106–123.

Miller, F.G., and H. Brody. 2003. A critique of clinical equipoise. Therapeutic misconception in the ethics of clinical trials. *The Hastings Center Report* 33(3): 19–28.

National Commission for the Protection of Human Subjects of Biomedical and Behavioral Research. 1979. *The Belmont Report. Ethical principles and guidelines for the protection of human subjects of research.* Washington, DC: U.S. Government Printing Office. http://www. hhs.gov/ohrp/humansubjects/guidance/belmont.html. Accessed 29 May 2013.

Nie, J.B. 2006. The United States cover-up of Japanese wartime medical atrocities: Complicity committed in the national interest and two proposals for contemporary action. *American Journal of Bioethics* 6(3): W21–W33.

Nie, J.B., N. Guo, M. Selden, and A. Kleinman (eds.). 2010. *Japan's wartime medical atrocities: Comparative inquiries in science, history, and ethics.* New York: Routledge.

Proctor, R.N. 1988. *Racial hygiene: Medicine under the Nazis.* Cambridge: Harvard University Press.

Rothman, D. 1991. *Strangers at the bedside: A history of how law and bioethics transformed medical decision making.* New York: Basic Books.

Shuster, E. 1997. Fifty years later: The significance of the Nuremberg Code. *New England Journal of Medicine* 337: 1436–1440.

———. 1998. The Nuremberg Code: Hippocratic ethics and human rights. *Lancet* 351: 974–977.

State-War-Navy Coordinating Committee (SWNCC). 1947a, August 1. Interrogation of certain Japanese by Russian prosecutor. In *Unit 731 and biological warfare: a CD-ROM collection,* ed. S. Kondo, Disc 3, Appendix A, p. 4 (2003). Toyko: Kashiwa Shobo.

———. 1947b, August 1. Interrogation of certain Japanese by Russian prosecutor. In *Unit 731 and biological warfare: a CD-ROM collection,* ed. S. Kondo Disc 3, p. 1 (2003). Toyko: Kashiwa Shobo.

Vollman, J., and R. Winau. 1996. Informed consent in human experimentation before the Nuremberg Code. *British Medical Journal* 313: 1445–1449.

Weindling, P.J. 2010. *John W. Thompson: Psychiatrist in the shadow of the Holocaust.* Rochester: University of Rochester Press.

———. 2012. Die Opfer von Menschenversuchen und gewaltsamer Forschung im Nationalsozialismus mit Fokus auf Geschlecht und Rasse. Ergebnisse eines Forschungsprojekts. In *Geschlecht und Rasse in der NS-Medizin,* eds. I. Eschebach, and A. Ley, pp. 81–100. Berlin: Metropol

Williams, P., and D. Wallace. 1989. *Unit 731: Japan's secret biological warfare in World War II.* New York: Free Press.

Chapter 14
In the Shadow of Nuremberg: Unlearned Lessons from the Medical Trial

Tom L. Beauchamp

The abnormalities in purportedly medical research in Nazi Germany were suffi- ciently extraordinary that they have often been judged a moral travesty of an exceptional nature with nothing to teach us about the ethics of commonplace clinical research. This assessment has for 67 years been a fixture of the American reaction to the Nuremberg Medical Trial and its offspring, the Nuremberg Code. The presumption that the findings of an American court sitting in judgment of Nazi physicians have no relevance to medicine and public policy has caused us to learn less from these events than we should have.

I will argue that the Nuremberg Medical Trial and some US biomedical research that occurred at the time of the trial offered a splendid opportunity to renovate and advance research ethics and to prevent tragedies of the sort that occurred in German medicine, but that this window of opportunity was lost because we lacked an appreciation of the promise of various of the principles in the Nuremberg Code. The failure to develop high-quality standards of research ethics and ethical peer review of protocols was costly both to medicine and to two generations of research subjects. It had profound consequences for those who became the victims of American abuses in clinical research, some of which began at the time of the German abuses and some of which continued for decades. These scandals could have been prevented had we attended to Nuremberg's lessons.

T.L. Beauchamp (✉)
Professor of Philosophy and Senior Research Scholar; Department of Philosophy and the
Kennedy Institute of Ethics, Georgetown University, Washington, DC 20057, USA
e-mail: beauchat@georgetown.edu

S. Rubenfeld and S. Benedict (eds.), *Human Subjects Research after the Holocaust*,   175
DOI 10.1007/978-3-319-05702-6_14, © Springer International Publishing Switzerland 2014

Human Subjects in the Yellow Fever Research of Sanarelli and Reed

I start with an analysis of Giuseppe Sanarelli's and Walter Reed's experiments on yellow fever at the turn of the twentieth century. My objective is to capture the then prevailing assumptions about the acceptable use of human subjects in medical research, assumptions that remained little changed during the 40-year period between this controversial research and the unparalleled extremes of the Nazi period (Lederer 1995).

In the design of Reed's study, one group of human subjects was intentionally exposed to the bite of a mosquito that was the vector of the disease. This research has been widely regarded as a scientific success—and even as monumentally important research. Reed has also been hailed as the first investigator to devise a consent form for healthy subjects of research.

The logical starting point for the examination of the ethics of Reed's research is the yellow-fever experimentation done immediately before him, in 1897, by Italian bacteriologist Giuseppe Sanarelli, director of the Institute for Experimental Hygiene at the University of Montevideo, Uruguay. He claimed to have discovered the bacillus that causes yellow fever, and he said that his discovery was supported by further work that intentionally and without consent injected into five patients a product that he said produced yellow fever in these patients. This work initiated some severe criticism in medicine (Lederer 1995, pp. 22–23, 49–50).[1] At a meeting in 1898, William Osler, Professor of Medicine at Johns Hopkins, where Reed himself trained under Professor of Pathology William Welch, rebuked Sanarelli's use of human subjects, calling it criminal to intentionally inject a potent poison into subjects without consent:

> The limits of human experimentation should be clearly defined in our minds ... To deliberately inject a poison of known high degree of virulency into a human being, unless you obtain that man's sanction, is not ridiculous, it is criminal. (Osler 1898, pp. 71–72)[2]

Osler's condemnations appeared two years prior to Reed's work, and Reed knew that he had been put on guard by these criticisms of Sanarelli. He knew that he should be careful not to act in a like manner, but Osler's rebuke noticeably pertained to research with a similar design to Reed's. It is perhaps for this reason that Reed frequently reported, questionably, that he received the "full consent" of his subjects, including those who died from the injections.[3] Nonetheless, when

---

[1] Some critical comments claimed that Sanarelli had not isolated the causal bacillus; other criticisms were moral in nature.

[2] Further on Osler's assessment of Sanarelli, see Royal Commission on Vivisection (1908, pp. 1502–1504, especially p. 1502).

[3] Sanarelli's work was discredited in 1902, but originally had been hailed in the medical community, including by Reed. In 1908 Osler appeared before the Royal Commission on Vivisection, an occasion on which he chose to discuss the subject of Reed's research on yellow fever. When asked by the Commission whether risky research on humans is morally permissible—a view Osler

confronting pointed criticisms by a Cuban physician, Reed defended himself, in part, by saying that he had been in a state of moral uncertainty and anguish over his use of human subjects. He offered two justifications: First, because medicine lacked an animal model, it was essential to use human subjects even if some subjects may die from an intervention; and second, the consent of subjects allays moral concern about deliberately infecting them with a deadly disease (Lederer 1995, p. 22).

Sanarelli's and Reed's studies show the ease with which vulnerable research subjects were enrolled, the risks of harm involved, and the failure, worldwide, to learn morally from this research how to correct physicians' presumptions and how to protect human subjects. Reed, in particular, engaged in what might be called, by today's standards, ethical shortcuts. The following are four worrisome parts of the conduct of his research.[4] First, he was a US army physician assigned to discover the cause of yellow fever in a context in which men were directly under his command. Many young men in the military at the time were conscripted, leaving it less than clear what it meant to voluntarily consent during their military service. Conflict of interest could have been a factor as well, because Reed was intensely motivated by the goal of achieving success in his research. Second, Reed's so-called consent form was not indicative of an informed consent form by today's standards; it was principally a disclosure statement, employment contract, and liability instrument. One of its provisions was that any volunteer for the research "understands perfectly well ... that he endangers his life" (Lederer 2008, p. 12). Forms at the time were regarded more as waivers and releases than as consents in the way we now understand "informed consent." It is not known whether Reed himself initiated the consent contract or the US Army required that he use it. (Sanarelli had had no consent or contract document.) Third, Reed used 15 poor Spanish immigrants as roughly 50 percent of his subjects. All were intentionally exposed to "loaded" mosquitoes, and at least six were infected, though none died. Calls for an investigation of these questionable uses of subjects were issued by Cuban newspapers and at the Pan-American Medical Congress held in 1901 in Havana (Lederer 2008, p. 13). Fourth, Reed did not himself participate as a subject in the yellow fever experiments, did not expose himself to the mosquitoes, and did not contract yellow fever; others under his command did.

Osler, in 1907, would praise the bravery and heroism of the soldiers under Reed who volunteered for his research, but in the same paper Osler presented criteria of

---

attributed to Reed—Osler answered as follows: "It is always immoral without a definite, specific statement from the individual himself, with a *full knowledge* [emphasis added] of the circumstances. Under these circumstances, any man, I think is at liberty to submit himself to experiments." When then asked if "voluntary consent ... entirely changes the question of morality," Osler replied "Entirely" (Cushing 1940, pp. 794–795). Joseph V. Brady and Albert R. Jonsen describe this testimony as reflecting the "usual and customary ethics of research on human subjects at the turn of the century" (1982, p. 4). However, this historical thesis about customary practice needs more supporting evidence than they supply.

[4] On Reed and his experimentation in general, see the sympathetic view in Bean (1977) and also Lederer (2008, pp. 9–16).

justified research at a demandingly high level that Reed's work almost certainly did not satisfy:

> The limits of justifiable experimentation upon our fellow creatures are well and clearly defined. The final test of every new procedure, medical or surgical, must be made on man, but never before it has been tried on animals. ... For man absolute safety and full consent are the conditions which make such tests allowable. We have no right to use patients entrusted to our care for the purpose of experimentation unless direct benefit to the individual is likely to follow. (Osler 1907, pp. 1–8)

I suspect that Osler was none too clear in his own mind which boundaries of justified research are "well and clearly defined." He unquestionably admired what Reed had done and was undoubtedly a person of high standards and strong intuitions about moral justification, but even he could not have been confident about how to morally assess Reed's research. The likely explanation is that there were at the time no "well and clearly defined" criteria that could confidently be applied to Reed.

Despite problems surrounding Reed's work, his procedures were at a reasonably high ethical level by the standards of the medical work at the time on healthy human subjects (e.g., Senate Bill 3424, 1900).[5] Reed's research has commonly been presented as morally warranted and courageous. Even the celebrated Henry Beecher praised Reed's experimentation, judging it justifiable despite the prospect of death that existed for volunteers. On the cusp of contradiction, Beecher despairingly observed that Reed's yellow-fever experiments would *not* pass requirements of ethical acceptability in the US military in the 1960s and, therefore, would not have been permitted (1970, p. 28, 53; see also Lederer 2008, p. 15). However, Beecher turned out to be wrong about whether the Army would have taken such a decision in the early 1960s. The decision to proceed with such experimentation had already been taken by the Army, and in the reverse direction from Beecher's prediction, as we will see.

During the period in which Reed's research occurred and was assessed by Osler and their contemporaries, Russian physician Vikentiy Smidovich, writing under the pseudonym V. Veresaeff, wrote an impassioned, carefully reasoned, and well documented critique of clinical and research practices in various countries. It was published in 1901–1902 in Russian, and shortly thereafter, published in English. This extraordinary book, *The Confessions of a Physician*, received some attention in educated circles in Russia, but its cultural impact was negligible, as far as I have been able to discover (Smidovich 1904; see also Katz, Capron, and Glass 1972, pp. 284–291),[6] just as Osler's biting criticisms of Sanarelli and others had only a

---

[5] See also the documentation of low standards in Lederer (1995, passim); her Appendix (pp. 143–146) contains the Congressional bill.

[6] Jay Katz et al. reprinted excerpts from "Veressayev's" book in his influential *Experimentation with Human Beings* (1972, pp. 284–291). Katz appended the following note: "Wherever possible the references in Dr. Veressayev's book were checked against the original sources; their accuracy was confirmed in every instance" (p. 284). Smidovich or Veressayev (or Veresaeff, as sometimes spelled) had a special interest in the moral unsoundness of venereal disease research. He examined

minor social, political, or professional impact. After these criticisms were published, public protests occurred in several nations about experiments that had gone badly and that had abused human subjects (Lederer 1995, Chapters 4–6; Brieger 1978). The US Congress and several state legislatures considered bills to control experimentation with humans, but none ever became law—not even in the District of Columbia, over which Congress had direct control (Lederer 1995, p. 72; 2008, p. 12). Nothing came of this bill or any other bill in the United States.

Summing up, the criticisms of research investigators presented by Osler, Smidovich, and others had little impact at the turn of the twentieth century. Medical research was uncontrolled by any form of meaningful professional self-regulation and government regulation. This absence of oversight mechanisms would persist through the 1930s and early 1940s. The door was, therefore, left open to utilitarian justifications of research for public benefit, and medical morals continued to be ill-equipped to confront the alliance of Nazi political ideology and medical research zealotry forged in Germany. Consequently, American medicine and government were poorly prepared to make the judgments Nuremberg demanded.

The Nuremberg Trials and the Nuremberg Code

In August 1945, it was determined by the US, British, Russian, and French governments that there would be a prosecution at Nuremberg by an International Military Tribunal ("Trials of War Criminals" 1948–1949, pp. ix–x). In January 1946, President Harry Truman approved supplementary Nuremberg trials to be conducted solely by the United States. The Nuremberg Medical Trial (*U.S. v. Brandt*) was the first. This trial commenced on December 9, 1946. The trial of each defendant and the moral principles they generated—The Nuremberg Code—became, symbolically, the most important watershed events of medical abuses in the more than 2,000 years of experimentation with human subjects.

The unprecedented murders, experimental injections, and assorted other cruelties were often administered by prominent physicians, several of whom occupied high positions in the Third Reich's medical hierarchy. The experiments were extensive and extreme in the physical harm and psychological suffering to which they knowingly exposed their many victims. Using subjects drawn from the populations of concentration camps, Nazi scientists explored the effects of ingesting poisons, intravenous injections of gasoline, immersion in ice water,

---

many published sources in countries around the world, including the United States. Using extensive documentation, he argued that the "zealots of science" had failed to distinguish between "humans and guinea pigs," and were callously disregarding "that consideration" of respect "due to the human being." He recounted instances of "classically shameless" and "criminal" experiments in which subjects were not informed of what was being done to them and, indeed, were actively deceived about the nature of the investigation.

infection with epidemic jaundice and spotted fever virus, killing people to obtain organs and brains for study, transmission of yellow fever, and the like.

In the opening statement of the prosecution, Telford Taylor, on behalf of the US government, declared as follows:

> The defendants in this case are charged with murders, tortures, and other atrocities committed in the name of medical science. . . . In many cases experiments were performed by unqualified persons; were conducted at random for no adequate scientific reason, and under revolting physical conditions. All of the experiments were conducted with unnecessary suffering and injury and but very little, if any, precautions were taken to protect or safeguard the human subjects from the possibilities of injury, disability or death. (Mitscherlich and Mielke 1949, p. xviii, xxv)

The extreme disregard of ethics and human welfare in the Nazis' exploitation and abuse of subjects is remarkable in light of the fact that in 1931 Germany had enacted strict regulations or guidelines (*Richtlinien*) to control both human experimentation and the use of innovative therapies in medicine—the strictest regulations in any nation. Issued by the Reich's Health Department, these regulations remained binding law throughout the Third Reich, but no evidence exists of serious attempts to enforce this law, in medicine or government, after the Nazis came to power in 1933. This law demanded that consent (first party or proxy, as appropriate) be given "in a clear and undebatable manner." Disclosure of appropriate information, bona fide consent, careful research design, and special protections for vulnerable subjects were all required in these guidelines.[7]

The Tribunals unambiguously condemned the sinister motivation behind the experiments, classifying them as "crimes against humanity" (Mitscherlich and Mielke 1949, p. xxxi).[8] The defendants were found to have corrupted the ethics of the medical profession and science and to have repeatedly and deliberately violated subjects' rights. During legal arguments, the accused physicians either defended themselves or were defended in their actions by lawyers using an aggressive attack on the thesis that voluntary participation by human subjects *generally* occurs in medical experimentation. Defendant Gerhard Rose stated:

> Aside from the self-experiments of doctors, which represent a very small minority of such experiments, the extent to which subjects are volunteers is often deceptive. At the very best they amount to self-deceit on the part of the physician who conducts the experiment, but very frequently to a deliberate misleading of the public. In the majority of such cases, if we ethically examine facts, we find an exploitation [by physicians] of the ignorance, the frivolity, the economic distress, or other emergency on the part of the experimental subjects. (as quoted in Katz et al. 1972, p. 304)[9]

---

[7] For a reprinting of the guidelines, with historical commentary, see Sass (1983).

[8] The quoted phrase is from Leo Alexander's "Statement," cited in Mitscherlich's and Mielke's book.

[9] See also the legal arguments by Robert Servatius in *United States v. Karl Brandt*, as cited in Katz, et al., p. 305. For the impact on the Nuremberg judges of Servatius's examination of the prosecution's medical expert witness, Werner Leibbrand, see Ulf Schmidt's account (2004, pp. 205–208). For a fuller, but edited, record, see *Trials of War Criminals Before the Nuremberg Military Tribunals under Control Council Law No. 10. Military Tribunal I* (1948–1949).

Nazi defendants argued that the Allies were in no position to judge them because the Allies too had engaged in similar research, including most notably American malaria experiments ("Trials of War Criminals" 1948–1949, p. 287)[10]—a reference to the malaria studies conducted at three prisons in Illinois, New Jersey, and Georgia[11] in the 1940s and stretching for years thereafter. Prisoners "volunteered" both to be bitten by the mosquitoes and then to take varying doses of investigative antimalarial drugs to determine safe dosage levels. Conduct of this research was a collaborative project of the US Army and Navy, the US State Department, the National Research Council, and the Public Health Service (Alving et al. 1948; Coatney et al. 1948; Laurence 1945).[12]

Had the information been released, the Nazi defense would undoubtedly have featured on its list of unjustified American experiments those just begun in 1946 by the US Public Health Service on sexually transmitted diseases in Guatemala. The study involved deliberate infection of subjects without consent. This inoculation study was hidden from public awareness for 65 years and has been judged "clearly and grievously" unethical by the President of the United States as well as the Presidential Commission for the Study of Bioethical Issues (Presidential Commission 2011).

The fact that the US government and some of its premier universities engaged in moral misconduct does not amount to a justification of the conduct of Nazi experimentation, as their defense counsel appeared to maintain. To the contrary, even if Walter Reed's work with soldiers and immigrants, the use of American prisoners in malaria research, and the inclusion of soldiers, psychiatric patients, and commercial sex workers in sexually transmitted disease research in Guatemala raise significant moral issues, as they do, the actions of Karl Brandt and associates in Germany nonetheless constitute serious moral wrongs on a different and incomparable scale of evil.

The History and Moral Basis of the Nuremberg Principles

The judges in the Nuremberg Medical Trial provided a catalog of "certain basic principles [that] must be observed in order to satisfy moral, ethical and legal concepts" in the conduct of human subjects research (Katz et al. 1972, p. 305). This list of ten principles was a component part of the justificatory argument in the

---

Servatius's various arguments, including his final plea (Servatius 1947) can be seen at Harvard Law School Library's useful online collection, Nuremberg Trials Project, at http://nuremberg.law. harvard.edu/php/search.php?DI=1&FieldFlag=1&PAuthors=278.

[10] See Closing Brief for Defendant Sievers in *Trials of War Criminals*. The claim is that "experiments in this field must unquestionably be permissible from an ethical point of view" by the precedent set in the use of hundreds of prisoners in the United States (p. 287).

[11] At the Stateville (or Illinois State) Penitentiary near Joliet, IL; the United States Penitentiary at Atlanta, GA; and the New Jersey Reformatory near Rahway, NJ.

[12] Ethical issues were not raised in these publications.

opinion of the judges and later was designated the Nuremberg Code. The most widely mentioned principle was then, and remains today, the first principle, which states, without qualification, that the primary consideration in research is the subject's voluntary consent, which is "absolutely essential" (*United States v. Karl Brandt* as cited in Katz et al. 1972, pp. 305–306).[13] This principle is demanding. It requires that consent have at least four characteristics: It must be voluntary, competent, informed, and comprehending. The rest of the Code sets general bounds within which an investigator may conduct research and delineates the conditions under which a subject has the ability to volunteer and to withdraw from the research after it has begun.

The judges apparently believed that their framework of ten principles did, and morally must, govern the conduct of research investigators. There are interesting *historical questions* about the origins (or source) of these principles and *moral questions* about whether they have the authoritative basis they were assumed by the judges to have. The answers to both questions bring forth unattractive features of how US authorities positioned themselves in presenting these principles in the context of a war crimes trial. The question is not whether the Nuremberg principles are morally valid principles. I do not question their validity or their applicability to the conduct of the Nuremberg defendants, but critical inquiry is in order about the moral basis on which Americans advanced these principles and the authority given to them.

The Nuremberg judgment makes it appear that the ten principles were authored by the judges, who stated immediately before listing the principles that, "All agree. ... that certain basic principles must be observed in order to satisfy moral, ethical, and legal concepts [the ten principles are then listed]" (*United States v. Karl Brandt* as cited in Katz et al. 1972, p. 305).[14] It is unclear to whom the "All agree" refers or what the moral force of this agreement is, but we know that this claim and the wording of the principles descended through the creativity and influence of Andrew Conway Ivy, who had been dispatched to Nuremberg by the American Medical Association (AMA) as an official consultant to the Nuremberg prosecutors. Ivy's appointment and its history are pertinent to the drafting of the Code, which needs to be set in this historical context.

---

[13] The text of Principle One reads as follows:

1. The voluntary consent of the human subject is absolutely essential. ... [T]he person involved ... should be so situated as to be able to exercise free power of choice, without the intervention of any element of force, fraud, deceit, duress, over-reaching, or other ulterior form of constraint or coercion; and should have sufficient knowledge and comprehension of the elements of the subject matter involved as to enable him to make an understanding and enlightened decision. This ... requires that ... there should be made known to him the nature, duration, and purpose of the experiment; the method and means by which it is to be conducted; all inconveniences and hazards reasonably to be expected; and the effects upon his health or person which may possibly come from his participation in the experiment. (Katz et al. 1972, pp. 305–306)

[14] Only later did the principles listed become known as the Nuremberg Code.

In May 1946, seven months prior to the opening statement of the prosecution by Brigadier General Telford Taylor ("Trials of War Criminal" 1947, p. 27), US Army Surgeon General Norman T. Kirk turned to the AMA for consultation, and its board of trustees then nominated Ivy, who had a history in his own research involving human subjects in the United States that, in topic, was similar to some of the experiments that occurred in the concentration camps, notably seawater desalination and the effects on humans of high altitudes in aviation. Ivy was judged well suited in credentials to understand the nature and objectives of some of the most gruesome medical science that occurred in Nazi Germany, and the AMA regarded him as a leading American investigator who could be counted on to protect the integrity and good name of genuine medical research.

Ivy travelled to Germany to meet with the Nuremberg prosecutors to afford them technical medical assistance. In Ivy's recounting, he discovered that the prosecutors needed more than *medical* knowledge: They "appeared somewhat confused regarding the *ethical and legal* aspects" of the German human experiments (Ivy as quoted in Advisory Committee on Human Radiation Experiments [ACHRE] 1996, p. 76; see also Ivy 1947, p. 133, 1949, p. 131).[15] The AMA trustees then asked him to write "a report as to the manner in which these [German] experiments [were] infringements of medical ethics." Ivy wrote a 22-page typescript given to the Nuremberg prosecution team. This submission included a statement of what he represented as "the [moral] rules" of human experimentation (ACHRE 1996, p. 76, 93).

These rules are the genesis of "The Nuremberg Code." The first eight of the ten principles were either directly or indirectly drawn from Ivy's rules, and much of the wording is verbatim from him. Some of what Ivy said in his document and statement of principles raises problems of accuracy, honesty, and justification. Without qualification, hesitation, or documentation, Ivy stated that these rules had been "well established by custom, social usage, and the ethics of medical conduct" (ACHRE 1996, p. 76). No empirical or moral argument was offered to support this claim. The judges at the Nuremberg Medical Trial nonetheless deferentially accepted Ivy's claim that the rules were the standard of practice and were generally acknowledged by research investigators (Katz et al. 1972, p. 306; see also ACHRE 1996, p. 77).

The historical origin of this claim of well-established rules is revealing. When Ivy's report of mid-September 1946 went up to the Judicial Council of the AMA for consideration, its chair, E. R. Cunniffe, claimed that the moral rules of human experimentation in Ivy's document were already inbuilt in the "Principles of Medical Ethics of the American Medical Association," a claim without foundation (ACHRE 1996, p. 77).[16] All of these claims about what was generally accepted in

---

[15] The Final Report of ACHRE, 1995, was subsequently published as *The Human Radiation Experiments* in 1996, which is the publication cited here.

[16] Far from making the rules prominent by a forthright announcement to the 126,000-member AMA, the rules were published in small print in a part of *Journal of the American Medical*

medicine and at the AMA—from Ivy's to Cunniffe's to the judges'—are venture-some speculations lacking in evidence. Nonetheless, the Judicial Council of the AMA used Ivy's representations to distill his rules, and on December 11, 1946—two days *after the opening prosecution* at Nuremberg—the AMA declared that these rules were the official policy of its House of Delegates (American Medical Association [AMA], p. 1090). Nothing like these rules had ever before been adopted or discussed by the AMA (AMA 1946; ACHRE 1996, p. 77).

In mid-June 1947, Ivy was called to the stand as a rebuttal witness in the Nuremberg Medical Trial and asked whether the AMA-approved rules were the "principles upon which all physicians and scientists guide themselves" in medical experimentation in the United States. Ivy asserted that they were. When asked by Judge Harold E. Sebring whether they were universal principles embraced across the civilized world, Ivy answered that they were identical worldwide, according to his information. These statements rested on unsupported presumptions lacking evidence. Ivy was then taken apart on the facts in cross-examination by German defense counsel Fritz Sauter, who forced Ivy to admit that before he had gone to the AMA, there had not been any printed form or codification of the AMA principles and that these principles were conceived in the form delivered only after the Nuremberg Medical Trial had begun and after Ivy had drafted them for the AMA. Ivy could say in response only that the rules were "understood as a matter of common practice," and therefore, were appropriate standards of practice (ACHRE 1996, p. 78).

Later discoveries of the actual practices in the United States in experimentation with human subjects in the 1940s and beyond raise doubts about the existence of such standards. Today, 67 years subsequent to Nuremberg, we have a fairly good picture of how much pretense went into Ivy's testimony about the culture of the civilized world. Fundamentally, a fallacy was at work: From the fact that the German medical experiments constituted egregiously uncivilized behavior on a scale never before witnessed, it does not follow that Ivy's rules had any credibility, stature, or general acknowledgment in biomedical research at the time.

Ivy needed an argument for his conclusions, and he did not have one. He needed to show that moral wrongdoing had occurred that violated either (1) basic moral principles (i.e., human rights whose violation constitutes crimes against humanity) or (2) guidelines in government policies or professional ethics of the period. Preferably both moral approaches would have been defended by evidence and argument. That is, Ivy could have addressed the question whether *universal rights* or *universal principles* were violated; and he could have addressed a second question regarding whether German investigators violated either well-articulated *rules of professional ethics* or *government policies*. Instead, he assumed that his rules delineated the standards of practice, and the judges accepted the claim.

Association unlikely to be read by the membership, buried with miscellaneous business items. For details on Ivy's rules and influence, and the surrounding events, see Weindling (2004, pp. 243–269, 278–283); a stern critique of Ivy's zeal is at pp. 279–280. See also Schmidt (2004, pp. 134–141).

The Damaging Consequences of American Misrepresentations

I turn now to some damaging *consequences* of these misrepresentations. The primary such consequence is that medicine and government in the United States came to assume, falsely, that there existed a world of well-entrenched medical standards of practice on which "all agree." Our eyes were averted by the judgment at Nuremberg from the need to articulate such standards and to give them a meaningful place in the oversight of research protocols and practices. Had we created a system of meaningful oversight, we might have kept history from being repeated in the research scandals that came to deeply scar American medicine. Instead, we came to the false belief that all was well in the ethics of research involving human subjects. Not for another quarter of a century did Americans come to appreciate how various research scandals had occurred in the United States at the time of the Nuremberg Medical Trial and how ill-designed and fragile our system was for protecting human subjects. This inattention in the post-Holocaust period is a reminder of what we ought to have learned from the Holocaust, but did not learn.

This conclusion takes me to the subject of whether the Nuremberg Code has had any practical influence or has been the basis of meaningful reform.

The Nuremberg Code's Failure to Influence Research Ethics

Insight into how Americans in the postwar years assessed the standing and relevance of the Code for US clinical research was illustrated by a written critique, presented on behalf of the Harvard Medical School by assistant dean Joseph Gardella in 1961, of a Nuremberg-inspired proposal governing the use of human volunteers in research by the Surgeon General of the Department of the Army (Gardella 1961). Gardella stated that, "The Nuremberg Code was conceived in reference to Nazi atrocities. . . . However suitable for the purposes for which it was conceived it is in our opinion not necessarily pertinent to or adequate for the conduct of medical research in the United States" (ACHRE 1996, p. 90).

Thirty years after this rejection of the Code, Jay Katz sarcastically described the way Americans for five decades had weighed the proper place of the Code: "It was a good code for barbarians but an unnecessary code for ordinary physician-scientists" (1992, p. 228; see also Macklin 1992, pp. 240–247). Katz was irritated by this embedded American belief because he thought it rendered the Code "buried soon after its birth" and was not the correct interpretation of how the Nuremberg judges intended the code to be understood and used (1996, p. 1662). Katz was right that the Nuremberg Code was screened from influence on the American frame of mind on the absurd grounds that it was not needed in the civilized world. Americans as well as citizens of other nations should have recognized that the Code captures necessary conditions of a civilized society and the grounds for judging uncivilized conduct. Ivy had had a competing worry: He was concerned that publicity generated by the Nuremberg Medical Trial would taint medical research generally, as if all experimentation were no better than Nazi experimentation (Hazelgrove 2002, pp. 112–113).[17]

Leonard Glantz has noted that the Code has never had a significant effect on US federal *regulations* (1992; see also Moreno 1996, pp. 351–356; 2001, Chapters 5–6; Lederer and Moreno 1996, pp. 229–235).[18] He could have added that the Code has had little effect on ethical oversight more generally. American Courts and research ethics, clinical ethics, and research medicine have essentially ignored the Code. So have German courts and research ethics. The judgments at Nuremberg were received in West Germany as victors' justice, ex post facto law, and one-sided legal reasoning (Burchard 2006).[19] East German parliamentary debates spoke of the "*victims* [emphasis added] of the Allied military tribunals" (Burchard 2006, p. 813).

The US government cannot claim, prior to roughly 1966, any abiding and serious interest in protecting research subjects against the kinds of abuses it had discovered, and then prosecuted, in Germany. The year 1966 initiated a critical wake-up decade in the development of moral awareness in the United States to problems of research abuses, not because of what had happened in Germany 20 years earlier but because of what was happening in the United States 20 years after Americans failed to learn from and apply the Nuremberg findings.[20]

The fact that the United States did not begin to develop a research ethics and body of regulatory guidelines until 1966 stands in sharp contrast to Ivy's claim that we already had well-delineated standards of practice in 1946. It is striking how little we apparently took to heart from the Holocaust and German experimentation in these two decades. Evelyn Shuster's assessment that Nuremberg served as a "blueprint for today's principles that ensure the rights of subjects" (1997, p. 1436) and that it has left a great "human rights legacy in research" is historically implausible (1997, p. 1440). Her statement grasps what should have happened while missing what failed to happen. The Nuremberg Code may be, symbolically, a vital research code, but it does not appear to have had a traceable or measurable effect anywhere, and certainly not in the United States or Germany.

Things could have been different. The Nuremberg Code could have served as a wake-up call and promising set of initial principles on the road to a sound research ethics, but in a disappointing moment of moral blindness, we missed this opportunity.

---

[17] Hazelgrove argued that, "the Nuremberg Code had virtually no impact on the research community, or the medical elite in Britain" (p. 132).

[18] Arguably an exception is that the US Department of Defense did accept the Nuremberg Code. But, having accepted the Code, Defense never thereafter seriously attempted to implement it, and public discussion of the Code by Army officials was prohibited because the Army's acceptance was itself classified material.

[19] This claim is defended and documented by Christoph Burchard.

[20] By 1965 a number of problems about human experimentation appeared on the horizon in the United States. In the 1965–1967 period, threats both to science and to human subjects came to the attention of the Director of the US National Institutes of Health (NIH). On the history of developments in these years, see Faden and Beauchamp 1986, pp. 205–215.

Postwar Research in the United States: The Disregard of Nuremberg

In 1966 Henry Beecher brought to public light what ought to have been recognized in 1946 after the Nuremberg Medical Trial. Beecher pointed to the persistently questionable conduct that abounded in *common research situations*—in effect contradicting what Ivy had asserted to occur in ordinary clinical research. Beecher started with a 1959 monograph entitled *Experimentation in Man*. In this little book, he announced that the atrocities disclosed at Nuremberg and the continuing advance into new areas of human biomedical research called for "a long, straight look at our current practices" (p. 3),[21] which he was convinced were "widespread" and "troubling" forms of medical exploitation that constituted significant abuses. His "Ethics and Clinical Research" (Beecher 1966a; see also Beecher 1966b) subsequently presented short case studies of contemporary biomedical research that contained serious or potentially serious ethical violations that "will do great harm to medicine unless soon corrected." The article centered on the "reasons why serious attention to the general problem is urgent" (1966a, p. 1354). It eventually became the most widely mentioned article ever published on failures of ethics in biomedical research.

Several of the experiments Beecher examined had been performed with a high ratio of risk to benefit, some involved vulnerable or disadvantaged subjects unaware of their participation in research, and consent was mentioned in only two of the 50 cases he collected. In various cases, a known effective and definitive treatment was withheld (in at least one case substituting placebos instead); in another case, a known inducer of potentially fatal aplastic anemia (chloramphenicol) was administered to patients without their knowledge; and in another case "artificial induction of hepatitis was carried out in an institution for mentally defective children" (Beecher 1966a, examples 1–3, 5, 16). Beecher argued that the ease with which he had collected samples from published articles in medical journals meant that if only a small percentage of the cases showed unethical behavior, medicine still would be faced with a situation in need of steep correction.

In looking back to 1966, it is sobering both that Beecher did not know, and could not have known, about the major abuses that had occurred and were occurring in the United States at the time he wrote the article (Beecher 1959).[22] That is, he did not

---

[21] See also Beecher's *Research and the Individual* (1970, p. xi), for a similar statement.

[22] It is also sobering that Beecher did not think the Nuremberg Code could have been constructively used to help counter the very abuses that he did identify. In commenting on the various codes that had been adopted, Beecher shied away from principles, rules, and regulatory approaches to the control of research: Though rules and regulations were occasionally needed, he thought they were in general "more likely to do harm than good" and would fail to "curb the unscrupulous." He speculated that the Nuremberg Code's unqualified insistence in its first principle on the consent of all subjects "would effectively cripple if not eliminate most research in the field of mental disease," as well as render impermissible all use of placebos (Beecher 1959, p. 50, 52, 58, and see also pp. 15–17, 43–44).

know the stories of the most significant abuses that had occurred over the two decades between Nuremberg and his article, namely, (1) the human radiation experiments (secret government experiments discussed below), (2) the Tuskegee syphilis experiments (Tuskegee Syphilis Study Ad Hoc Panel 1973; Benedek 1978; Jones 1981), and (3) the Guatemala sexually transmitted diseases experiments (mentioned above).[23] These scandals could all have been avoided had we taken the findings at Nuremberg seriously and created a research ethics and body of regulations that incorporated a system of oversight of protocols.

Tuskegee would become the most famous of the cases, and remains so today, but the human radiation experiments and the Guatemala experiments are especially interesting for purposes of this paper because they were undertaken precisely at the point Americans were condemning German physicians for their abuses. The US government failed to recognize that the principles American judges announced to the world as universal were being violated in US government-planned, secret research that had all of the dangers Beecher warned about. These violations and their embarrassing consequences for the United States were indirect results of ignoring Nuremberg's principles.

The reason the American scandals are compelling is that some were planned and underway in the United States on the day the Nuremberg Medical Trial began. The judgment at Nuremberg did nothing to stop them. No sooner had Americans reached for justice at Nuremberg than their representatives proceeded to undermine this judgment by injustices in medical research at home and abroad—the subject to which I now turn by taking a closer look at the human radiation experiments.

The Human Radiation Experiments

The radiation experiments started around 1944 with a goal of discovering the risks to workers of handling radioactive materials. It was not until 1995, a half-century later, that the Advisory Committee on Human Radiation Experiments presented a report to President William Clinton showing the vast scope of what American investigators had done in human-subjects radiation research in the 30-year period from 1944 through 1974 (ACHRE 1996).[24] The Advisory Committee showed a sea of wrongdoing in US government-sponsored research. This research included the injection of plutonium into uncomprehending hospitalized patients, US Air Force experiments on prisoners, feeding radiation-infused breakfasts to institutionalized children, total-body irradiation of cancer patients, experimentation in the course of

---

[23] Tuskegee first emerged in the public eye in 1972, but it was a Public Health Service study initiated in the early 1930s.

[24] See also Lederer and Moreno (1996) on ACHRE's importance for research ethics.

bomb testing, use of pregnant women and newborn infants, irradiating the testicles of prisoners, etc. (ACHRE 1996, "Case Studies"; see also Moreno 2001).

I cannot explore the wide array of experimental protocols that were involved, so I will concentrate on the plutonium experiments, which had their origins at the newly created Atomic Energy Commission (AEC). The patient-subjects in these radiation experiments were part of a secret research protocol initiated by the Manhattan Engineer District, a government program responsible for the production of atomic weapons. The purpose of the research was to determine the excretion rate of plutonium in humans so that the government could establish safety levels and standards for workers who handled this radioactive element. Performed largely in secret, some of this work was known publicly as early as 1951, and some was discussed in the 1970s and 1980s, but nothing generated public controversy until 1993, when it came to public light as a result of the work of a persistent investigative reporter in Albuquerque, New Mexico (Welsome 1993; ACHRE 1996, pp. xxi–xxii).[25] This journalistic outlet was reminiscent of the disclosure of the syphilis experiments in Tuskegee, which had attracted virtually no interest in the biomedical research community until it was exposed on the front page of the *New York Times* (Heller 1972).

The salient facts of the plutonium injections are as follows: From April 1945 to July 1947, 17 patients were injected with plutonium at three university hospitals—The University of California San Francisco (UCSF), the University of Chicago, and the University of Rochester (and one patient injected at Oak Ridge Hospital in Tennessee) (ACHRE 1996, pp. 144–156.). Three of the 18 (known as CAL-1, CAL-2, and CAL-3) were injected at UCSF. In a report completed 50 years after these secret experiments occurred, in February 1995, an ad hoc committee of the UCSF Medical School faculty confirmed both the UCSF involvement and that at least one of the three initial patients had been included in the secret government protocol. The report stated that the "injections of plutonium were not expected to be, nor were they, therapeutic or of medical benefit to the patients" (University of California at San Francisco Ad Hoc Committee [UCSF] 1995, p. 27; cf. ACHRE 1996, pp. 148–152, 154–155).[26]

It is decidedly unlikely that informed consents were given in this experimentation, though the AEC had stated that knowledgeable consent was required. The ad hoc committee found that written consent was rare, disclosure narrow, and the permission of patients not typically obtained even for nontherapeutic research during this period, as was "consistent with accepted medical research practices at

---

[25] Eileen Welsome's series of articles on the plutonium experiments earned her the Pulitzer Prize for National Reporting in 1994.

[26] The report was unpublished, but released to the public. Compare to ACHRE, Final Report, pp. 149–52, 154–155, especially pp. 150–151.

the time" (UCSF 1995, p. 33)[27]—another conclusion that contradicts Ivy's estimation. The committee found that the word "plutonium" was classified; it is, therefore, certain that it was not used in explanations to patient-subjects. The committee noted that, in a recorded oral history in 1979, Kenneth Scott, one of the three original UCSF investigators, said that he never told the first subject (CAL-1) what had been injected in him. Scott went on to say that the experiments involved deception, were "incautious," and were "morally wrong." A second nontherapeutic experiment (on CAL-2) was conducted on a four-year-old boy flown from Australia to UCSF for treatment of a rare form of bone cancer. Within days of his arrival, he was injected with plutonium (UCSF 1995, pp. 26–27, 34; ACHRE 1996, pp. 150, n. 86, p. 167).

This experimentation is a paradigm case of what the principles of the Nuremberg Code had prohibited in all biomedical research. The AEC had been aware of the potentially explosive public criticism that might occur were the secret facts publicly disclosed. AEC officials also knew about the judgments at Nuremberg, which worried them. Shields Warren, the chief of the AEC's Division of Biology and Medicine, objected in 1950 during a meeting with the Department of Defense (DOD) to discuss a proposal that "prisoner volunteers" be used in the radiation experiments. Warren said, "It's not very long since we got through trying Germans for doing exactly the same thing" (Moreno 1996, pp. 352–354). The AEC and the DOD apparently felt trapped at the time: They thought that they needed to do invasive nontherapeutic research on human subjects to protect their employees from plutonium risks and, thereby, to protect national security in the Cold War, but that the research—*military* medical research—could constitute, and ultimately did constitute, a violation of the strictures handed down at the same time in the US military's judgment at Nuremberg.

Conclusion

The murders and atrocities committed in the name of medical science and prosecuted at Nuremberg were on an unparalleled scale of evil. Nonetheless, this unmatched scale of horror does not excuse the many dubious actions of American research investigators and representatives at Nuremberg. Nor does it excuse our

---

[27] This committee found that it is not known and that we "cannot know" exactly what these research subjects were told or what they understood. In an appendix to the ad hoc committee's report, a lawyer and committee member, Elizabeth Zitrin, concluded that even if the experiments were consistent with accepted medical practices at the time, "it does not make them ethical. And they were not consistent with the highest standards of the time articulated by the government, the profession, or the public" (UCSF 1995, Letters, Appendix). The chairman of the Ad Hoc Committee, Roy Filly, responded that this comment held investigators to an unrealistically high research standard (UCSF 1995, Letters, Appendix). Filly's response was reported by Keay Davidson (1995) in *The San Francisco Examiner*.

thoroughgoing failure to improve research ethics and its oversight by introducing a peer review system. The later scandalous American abuses were preventable had we attended to Nuremberg's moral lessons.

Why the US military condemned German physicians at Nuremberg and almost simultaneously began experiments conducted by physicians that violated the strictures laid down in the Nuremberg Code remains today an unexplained moral incoherence. No less puzzling was Andrew Ivy's curious gilding of American medical ethics to look like the gold standard that it was not.

The conduct of American medical representatives at the time was, in my judgment, deeply disrespectful to the victims of the experiments and to the survivors of the Holocaust. They looked to the United States as a moral beacon and a force to prevent a recurrence of the shameful course of events in Germany, only to discover that the United States mismanaged the moral implications of the judgment at Nuremberg by letting the potential influence of its Code slip away.[28]

References

Advisory Committee on Human Radiation Experiments (ACHRE). 1996. *The human radiation experiments*. New York: Oxford University Press. Reprinted from *Final report*, 1995, Washington, DC: U.S. Government Printing Office.

Alving, A.S., C. Branch Jr., T.N. Pullman, C.M. Whorton, R. Jones Jr., and L. Eichelberger. 1948. Procedures used at Stateville penitentiary for the testing of potential antimalarial agents. *Journal of Clinical Investigation* 27: 2–5.

American Medical Association (AMA). 1946. Supplementary report of the Judicial Council. Proceedings of the House of Delegates (December 9–11, 1946). *Journal of the American Medical Association,* 132 (17):1075–1095 (December 28). doi: 10.1001/jama.1946. 02870520029014

Bean, W.B. 1977. Walter Reed and the ordeal of human experiments. *Bulletin of the History of Medicine* 51: 75–92.

Beecher, H.K. 1959. *Experimentation in man*. Springfield: Charles C. Thomas.

———.1966a. Ethics and clinical research. *New England Journal of Medicine* 274(24): 1354–1360.

———. 1966b. Some guiding principles for clinical investigation. *Journal of the American Medical Association* 195(13): 135–136.

———. 1970. *Research and the individual: Human studies*. Boston: Little, Brown and Co.

Benedek, T.G. 1978. The "Tuskegee study" of syphilis: Analysis of moral versus methodologic aspects. *Journal of Chronic Diseases* 31: 35–50.

---

[28] The years 1973–1974 may be the first true awakening in the United States to overzealousness in research involving human subjects, but the awakening is not traceable to Nuremberg. In the aftermath of Tuskegee and other controversies surfacing at this time, Senator Edward Kennedy held hearings on human experimentation in 1973 (*Quality of Health Care* 1973) The US Congress also appointed the National Commission for the Protection of Human Subjects of Biomedical and Behavioral Research in 1974. The first US regulations governing biomedical research also appeared in 1974 (Department of Health, Education and Welfare Rule 1974, pp. 18914–18920). None of these events was triggered by Nuremberg.

Brady, J.V., and A.R. Jonsen. 1982. The evolution of regulatory influences on research with human subjects. In *Human subjects research*, ed. R. Greenwald, M.K. Ryan, and J.E. Mulvihill, 3–5. New York: Plenum Press.

Brieger, G.H. 1978. Human experimentation: History. In *Encyclopedia of bioethics*, vol. 2, ed. W.T. Reich, 684–692. New York: Free Press.

Burchard, C. 2006. The Nuremberg Trial and its impact on Germany. *Journal of International Criminal Justice* 4(4): 800–829.

Coatney, G.R., W.C. Cooper, and D.S. Rube. 1948. Studies in human malaria. VI. The organization of a program for testing potential antimalarial drugs in prisoner volunteers. *American Journal of Hygiene* 47: 113–19.

Cushing, H. 1940. *The life of Sir William Osler*. London: Oxford University Press.

Davidson, K. 1995. Questions linger on 1940s UCSF plutonium shots. *The San Francisco Examiner*, February 23, A6.

Department of Health. 1974. Education and Welfare Rule, 45 CFR part 46.

Faden, R.R., and T.L. Beauchamp. 1986. *A history and theory of informed consent*. New York: Oxford University Press.

Gardella, J.W. 1961. [Memorandum]. From 'JWG" (Assistant Dean Gardella) to "GPB" (Harvard Medical School Dean Berry). Criticisms of "Principles, policies and rules of the Surgeon General, Department of the Army, relating to the use of human volunteers in medical research contracts awarded by the Army." In ACHRE document no. IND-072595-A, October.

Glantz, L.H. 1992. The influence of the Nuremberg Code on U.S. statutes and regulations. In *The Nazi doctors and the Nuremberg Code: Human rights in human experimentation*, ed. G.J. Annas and M.A. Grodin, 183–200. New York: Oxford University Press.

Hazelgrove, J. 2002. The old faith and the new science: The Nuremberg Code and human experimentation ethics in Britain, 1946–73. *Social History of Medicine* 15(1): 109–135.

Heller, J. 1972. Syphilis victims in U.S. study went untreated for 40 years. *New York Times,* July 26, 1, 8.

Ivy, A.C. 1947. Nazi war crimes of a medical nature. *Federation Bulletin* 33: 133.

———. 1949. Nazi war crimes of a medical nature. *Journal of the American Medical Association* 139(3): 131–135. doi: 10.1001/jama.1949.02900200001001.

Jones, J.H. 1981. *Bad blood: The Tuskegee syphilis experiment*. New York: Free Press. Reprinted *Bad blood: The Tuskegee syphilis experiment (New and expanded edition)*.(1993). New York: Simon & Schuster.

Katz, J. 1992. The consent principle of the Nuremberg Code: Its significance then and now. In *The Nazi doctors and the Nuremberg Code: Human rights in human experimentation*, ed. G.J. Annas and M.A. Grodin, 227–239. New York: Oxford University Press.

———. 1996. The Nuremberg Code and the Nuremberg Trial: A reappraisal. *Journal of the American Medical Association* 276(20): 1662–1666. doi: 10.1001/jama.1996. 03540200048030.

Katz, J., A.M. Capron, and E.S. Glass. 1972. *Experimentation with human beings*. New York: Russell Sage.

Laurence, W.L. 1945. New drugs to combat malaria are tested in prisons for army. *New York Times*, March 5, 1, 30.

Lederer, S.E. 1995. *Subjected to science: Human experimentation in America before the Second World War*. Baltimore: Johns Hopkins University Press.

———. 2008. Walter Reed and the yellow fever experiments. In *The Oxford textbook of clinical research ethics*, eds. E.J. Emanuel et al., 9–16. New York: Oxford University Press.

Lederer, S., and J.D. Moreno. 1996. Revising the history of cold war research ethics. *Kennedy Institute of Ethics Journal* 6(3): 223–237.

Macklin, R. 1992. Universality of the Nuremberg Code. In *The Nazi doctors and the Nuremberg Code: Human rights in human experimentation*, ed. G.J. Annas and M.A. Grodin, 240–257. New York: Oxford University Press.

Mitscherlich, A., and F. Mielke. 1949. *Doctors of infamy: The story of the Nazi medical crimes* (trans: Norden, H.). New York: Henry Schuman.

Moreno, J.D. 1996. Reassessing the influence of the Nuremberg Code on American medical ethics. *Journal of Contemporary Law and Policy* 13: 347–360.

———. 2001. *Undue risk: Secret experiments on humans.* New York: Routledge (previously published 1999, New York: W. H. Freeman).

Osler, W. 1898. Comments by Dr. Wm. Osler. In *Transactions of the Association of American Physicians 13* (thirteenth session). *The Bacillus Icteroides (Sanarelli) and the Bacillus X (Sternberg).* Symposium conducted at the meeting of the Association of American Physicians, Washington, DC. http://www26.us.archive.org/stream/transactionsass18physgoog/transactionsass 18physgoog_djvu.txt. Accessed 10 November 2012.

———. 1907. The evolution of the idea of experiment in medicine. *Transactions of the Congress of American Physicians and Surgeons,* Seventh Triennial Session, May 7–9, Washington, DC.

Presidential Commission for the Study of Bioethical Issues. 2011. "Ethically Impossible": STD research in Guatemala from 1946 to 1948. Washington, DC: U.S. Government Printing Office. http://bioethics.gov/sites/default/files/StudyGuide_EthicallyImpossible_508_Nov26.pdf. Accessed 8 December 2013.

Quality of Health Care—Human Experimentation, Parts 1–4: Hearings before the Subcommittee on Health of the Committee on Labor and Public Welfare, U.S. Senate, 93d Cong. 1 (1973).

Royal Commission on Vivisection. 1908. Fourth report. Evidence of Professor Osler (continued). *British Medical Journal* 2: 1502–1504.

Sass, H.-M. 1983. Reichsrundschreiben 1931: Pre-Nuremberg German regulations concerning new therapy and human experimentation. [reprint of guidelines]. *Journal of Medicine and Philosophy, 8,* 99–111.

Schmidt, U. 2004. *Justice at Nuremberg: Leo Alexander and the Nazi Doctors' trial.* Houndmills: Palgrave Macmillan.

Senate Bill for the Regulation of Scientific Experiments upon Human Beings in the District of Columbia, S. 3424, 56th Cong. 1900.

Servatius, R. 1947. Final plea for defendant Karl Brandt. *United States* v. *Karl Brandt.* Nuremberg Trials Project. A digital document collection. Harvard Law School Library. http://nuremberg. law.harvard.edu/php/search.php?DI=1&FieldFlag=1&PAuthors=278. Accessed 14 November 2012.

Shuster, E. 1997. Fifty years later: The significance of the Nuremberg Code. *New England Journal of Medicine* 20(337): 1436–1440.

Smidovich, V. [pseudonym V. Veresaeff]. 1904. *The memoirs of a physician* (trans. Linden, S.). London: Grant Richards. Reprinted 1916. New York: Alfred A. Knopf.

Trials of War Criminals Before the Nuremberg Military Tribunals under Control Council Law No. 10. Vols. 1 & 2. 1946–1949. Washington, DC: Government Printing Office.

Tuskegee Syphilis Study Ad Hoc Panel. 1973, April 28. *Final report of the Tuskegee Syphilis Study Ad Hoc Panel.* Washington, DC: U.S. Department of Health, Education, and Welfare, Public Health Service.

University of California at San Francisco Ad Hoc Fact Finding Committee (UCSF). 1995, February. *Report of the UCSF Ad Hoc Fact Finding Committee on World War II Human Radiation Experiments.* http://www2.gwu.edu/~nsarchiv/radiation/dir/mstreet/commeet/ meet12/brief12/tab_f/br12f1a.txt. Accessed 10 December 2013.

Weindling, P.J. 2004. *Nazi medicine and the Nuremberg trials: From medical war crimes to informed consent.* Basingstoke: Palgrave Macmillan.

Welsome, E. 1993. The plutonium experiment. *The Albuquerque Tribune,* November 15.

Chapter 15
The Ethics of Medical Experiments: Have We Learned the Lessons of Tuskegee and the Holocaust?

Patricia L. Starck and Doris S. Holeman

It has been 82 years since the Tuskegee experiments on Black males with syphilis began, and 42 years since their conclusion after the scandalous news became public. It has been nearly 70 years since the conclusion of the inhumane medical experiments conducted on Nazi prisoners during the Third Reich. And yet scientific misconduct, such as the Guatemala syphilis experiments, continues to be uncovered (Presidential Commission for the Study of Bioethical Issues 2011). The US Department of Health and Human Services maintains a list of current research integrity violations at its Web site, http://ori.hhs.gov/case_summary. The National Institutes of Health defines and describes types of violations, including falsification, fabrication, and plagiarism at its Web site, http://grants.nih.gov/grants/research_integrity/research_misconduct.htm.

This article briefly discusses the Tuskegee experiments and selected Nazi medical experiments and will grapple with two questions: How could supposedly caring and competent health care professionals turn on their patients and harm them? What external societal, scientific, and governmental pressures influenced health care professionals to abandon their ethics and sacred duties? We will examine the external pressures brought to bear in these two sets of experiments, and, with the hope of strengthening young professionals for future challenges, we will then discuss pressures that they may face. The reader is left to answer the question whether we health professionals have learned the lessons of these two significant human tragedies.

P.L. Starck, PhD, RN, FAAN (✉)
Dean and John P. McGovern Distinguished Professor; Huffington Foundation Chair for Nursing Education Leadership; Senior Vice President for Interprofessional Education, The University of Texas Health Science Center at Houston School of Nursing, 6901 Bertner Avenue, Houston, TX 77030, USA
e-mail: Patricia.L.Starck@uth.tmc.edu

D.S. Holeman, PhD, RN
Associate Dean and Director; School of Nursing and Allied Health, Tuskegee University, 209 Basil O'Connor Hall, Tuskegee, AL 36088, USA
e-mail: dholeman@mytu.tuskegee.edu

S. Rubenfeld and S. Benedict (eds.), *Human Subjects Research after the Holocaust*,   195
DOI 10.1007/978-3-319-05702-6_15, © Springer International Publishing Switzerland 2014

The Tuskegee Syphilis Experiment

The Tuskegee Syphilis Study was a 40-year experiment, beginning in 1932 and ending in 1972, on "Negro" males in Macon County, Alabama. Located east of Montgomery, Alabama, this part of the state is known as the "Black Belt," so called because of its rich dark soil. It was primarily populated by poor farmers, 90 percent of them Black. It is home to Tuskegee University, formerly Tuskegee Institute, the foremost Black college in the nation. In the 1930s, Macon County was severely depressed economically. Throughout the South, public facilities, including hospitals, were rigidly segregated by race. The county had few doctors and only two hospitals, and one, John A. Andrew Memorial Hospital, was on the grounds of Tuskegee Institute where it served as the major source of medical care for much of the Black community (Gray 1998; Jones 1993; Smolin 2013).

"The Tuskegee Study of Untreated Syphilis in the Male Negro," the original name of the study, was designed and conducted by the US Public Health Service (USPHS) (Bioethics Center, n.d.; see also "The Tuskegee Timeline," n.d.). This nontherapeutic experiment was designed to compile data on the spontaneous evolution of syphilis in Black males, which was considered, at the time, to differ significantly from the spontaneous evolution of syphilis in Whites. The researchers theorized that Whites experienced more neurological complications from syphilis, whereas Blacks were more susceptible to cardiovascular damages. The researchers believed that racial differences affected the course of the disease and, therefore, expected the Tuskegee study to provide a useful comparison to the Oslo Study, which was a study of syphilis in White people (Harrison 1956). There was, however, a fundamental difference in the two studies: The Oslo study was a retrospective study that utilized health records of untreated syphilis cases, whereas the Tuskegee study was a prospective study deliberately designed to withhold treatment (Brandt 1978; Harrison 1956). As Ogungbure stated, ". . . the United States government went to great lengths to ensure that the men in the 'Tuskegee Study' were denied treatment, even after penicillin had become the standard cure for syphilis in the mid-1940s" (2011, p. 78). As a result, many of the participants died painful deaths. Some died from advanced syphilitic lesions; others suffered adverse effects, ranging from paralysis of limbs caused by dangerous spinal tap procedures to extreme neurological damage. Wives were infected with syphilis, and many of the offspring were born with congenital syphilis (Jones 1993; Ogungbure 2011; Reverby 2000).

The basic design of the study called for the selection of syphilitic Black males between the ages of 25 and 60 years. While the number of human subjects varied in the many published reports of this experiment (Brandt 1978; Ogungbure 2011; Reverby 1999, 2000), most sources suggest that there were approximately 399 men involved and another 200 healthy, non-syphilitic men were controls. Initially designed to last six months, because it seemed to be both a simple and inexpensive project, the experiment lasted for 40 years. False assumptions that Blacks were illiterate, had no concept of time, and were promiscuous and lustful contributed to the length of the study (Jones 1993). Also, the study took longer than initially planned because of resistance the researchers encountered from the Black communities. Convinced that the call for men only over the age of 25 years meant that the program was really for military draft physicals, there was an initial low response. To allay these fears, the researchers had to test a large number of both men and women. Therefore, the 1932 survey took more time and cost much more money than had been predicted (Brandt 1978). The men were seen neither as patients nor as subjects, but as "less than human" (Brandt 1978, p. 27), and the researchers had "no further interest in these patients until they die" (Brandt 1978, p. 24). Because data from the experiment were to be collected at autopsy, the value of the study was in the death of the men and not their day-to-day lived experiences.

Testimony from Physicians and Nurses

It is difficult to provide a justifiable explanation for the participation of health professionals, but it is especially difficult to explain the participation of Black health professionals and educators in a study in a community that was 82 percent Black. The study was not a government secret hidden from health professionals—it was publicized widely in the Black and White medical communities without evoking any protest. The study name came from the most visible and highly recognized Black academic institution, Tuskegee Institute. Its campus hospital, John A. Andrews Memorial, and the Veterans Health Care Hospital loaned the USPHS their medical facilities to conduct the study. These factors contributed to the engagement and participation of Black health care professionals and educators. Eunice Rivers, a Black public health nurse and a graduate of the Tuskegee Institute nursing program, was a central figure for the duration of the study. Available published reports did not document or allude to the participation of other nurses (Brandt 1978; Jones 1993; Reverby 1999; Smith 1996).

There is a paucity of analytical information on the Tuskegee Syphilis Study published by nurses. Nevertheless, many social and medical historians (Gamble 1997; Jones 1993) tried to rationalize or explain the participation of health professionals and to summarize the lasting impact of the performance of this study on the Black community. The most conclusive statement that can be made as a result of the numerous interviews is that many health professionals did not understand all the ramifications of the study. As reported in the *Journal of Women's History*, a Tuskegee doctor praised the "educational advantages offered our interns and nurses as well as the additional standing it would give the hospital" (Smith 1996, p. 102). Other insightful statements reported by Smith, which served as explanations for the participation of Black health professionals and educators, were that the study "was seen as directing federal attention towards Black health problems . . ." and that ". . . costly treatment may be demonstrated as unnecessary for people with latent or third-stage syphilis" (p. 103). These statements echoed the justifications provided by the USPHS: "The educational institution would get credit for the research"

(Smith 1996, p. 103), and "Black educators and doctors envisioned future financial benefit for the educational facility from the federal government by participating in the study" (p. 102).

Similar statements to explain the participation of health professionals and educators were offered by other researchers (Gray 1998; Jones 1993). There were, however, objections to holding Black educators and physicians responsible for the conduct of the study, for they were considered to be "as much the victims of the study as the . . . men who were eventually enrolled in it" (Gray quoted in Brody, n.d., "Faces of Tuskegee"). In referring to the participation of Black health professionals and educators, Gray (1998) stated that "in the 1930s, during the great depression, Tuskegee Institute was an African American educational institution struggling for its very existence . . . Tuskegee institute was asked to provide facilities and services; they were never invited to review or set policy" (p. 86). Jones (1993) suggested that the participation of Black health professionals was due to deceit in communicating the purpose of the study.

Nurse Eunice Rivers' role was described as being one of a "passive obedient." She explained, "We were taught that we never diagnosed, we never prescribed; we followed doctor's instructions!" (Brunner, n.d.). It is believed that she participated because she could not have understood the full ramification of the study. Second, as a Black female nurse, she was in no position to challenge the authority of White male physicians (Jones 1993; Reverby 1999; Smith 1996).

Nazi Medical Experiments

Horrendous medical experiments on human beings could never have taken place without the participation of nurses and physicians during the Nazi era. Because students in both professions often take part in an oath ceremony pledging to do no harm, it is worthwhile to examine what caused these professionals to hurt and even kill their patients.

Hitler won election as Germany's chancellor based, in part, on his passion to return the nation to its pre-World War I glory. He co-opted physicians who had already shifted their focus from the individual patient to the German people, or *Volk*, as the patient. "Healing" the German race was synonymous with strengthening its "stock." The loss of many vital young men in World War I was alarming, and therefore, the Nazi government offered incentives to women to produce as many Aryan babies as they could. Nazi officials admonished doctors and nurses that just as soldiers do their duty, medical professionals should do theirs.

According to a court reporter at the Nuremberg Doctors' Trial (Spitz 2005), former prisoners testified about medical experiments in which they were the experimental subjects. The Nazis studied the effects of hypothermia, high altitude, drinking sea water, and deliberate application of mustard gas to improve the chances of survival of injured German soldiers and pilots. Prisoners had experimental surgeries, including transplantation of bone, muscles, and nerves, and were subjected to various experimental methods of sterilization. They were given poisons and were infected with hepatitis, malaria, and typhus, all in the name of science to aid and protect the *Volk*. While physicians designed and carried out the research, falsified causes of death, and ordered patients' deaths to obtain autopsy material for their experiments, nurses were complicit as well in terminating the lives of research subjects.

Testimony from Physicians and Nurses

Many people looking back on this history would like to conclude that these events can be explained by the aberrant behavior of a few madmen or inhumane individuals. Unfortunately, this was not the case. German medicine was considered among the best in the world. A much higher percentage of physicians than any other professional group, roughly 45 percent, joined the Nazi party by 1939 (Kater 2002), compared to 25 percent of public health and hospital nurses (Survey of German Nurses 1945). At the Nuremberg Doctors' Trial, the physicians did not express remorse. Their staunch positions were that they were doing the right thing for Germany; that they were doing their duty to the Fatherland; and that their experiments would not only advance science but help injured German soldiers (Spitz 2005). They justified killing research subjects to advance science.

Some physicians and nurses refused to participate, or if they participated, tried to find ways to help the patients. Lifton (1986) speculated that those who willingly or even reluctantly participated may have had certain psychological traits, such as feelings of omnipotence, sadism, paranoia, or schizoid numbing that facilitated this violation of their professional oaths.

Nurse scholars have analyzed the testimony of nurses, searching for explanations of the perpetrators' behavior (Benedict 2003, 2006; Benedict et al. 2007; Benedict and Georges 2006, 2009; Benedict and Kuhla 1999; Benedict et al. 2009; Georges and Benedict 2006; McFarland-Icke 1999). The nurses claimed they were either following doctors' orders, acted out of fear of reprisals, or were called upon to do their duty. They did not accept responsibility, stating that their orders must have been legal or surely some official would have intervened. Money may have motivated those who received a bonus for euthanasia duty. Some nurses justified the killings as a release from suffering. In retrospect, the nurses presented themselves as helpless victims, with no choice but to carry out their orders for human subjects research and mass murder.

Milgram (1963) conducted his landmark study regarding obedience to authority figures to discover if subjects would administer what they thought were increasingly painful electrical shocks to persons if directed to do so by authority figures dressed in lab coats. With directives from the authority figure, some 65 percent administered the highest level of shock in spite of some reluctance on their part and what appeared to be extreme suffering by the persons to whom they administered the shocks. Milgram repeated his experiments under varying conditions—changes in location, loudness of screams, family member participants—and still authority prevailed over personal preferences. However, Perry (2013), who analyzed Milgram's notes and audiotapes, believed that Milgram may have underreported inconsistencies and was guilty of overstating his research results in spite of their limitations. Indeed, she found that with some variables changed, more than 60 percent of subjects did not acquiesce to the authority figure. The question whether most humans will ignore their own values in the presence of an authoritative directive remains, but clearly contextual factors are important.

Zimbardo's (1972) experiment with volunteer Stanford University students randomly assigned to be either prisoners or guards in a simulated prison also produced surprising psychological results. It was funded by the US Office of Naval Research to explore conflict between military Navy/Marine guards and prisoners. The hypothesis was that the chief cause of bad behavior was inherent personality traits. Although the experiment was designed to run for two weeks, it was stopped after six days because of the extreme reactions of both prisoners and guards. The students internalized their roles beyond expectations: the guards used stringent psychological abuse, and prisoners used what resistance to authority they could. An interpretation of the results was that the behavior was more a reaction to the situation than actual individual personality traits.

Obeying an authority figure in a hierarchical medical system is a very common practice and often the most expedient course of action, whether within the medical hierarchy of student, resident, chief resident, attending physician, or within the nursing hierarchy. Patient care plans are called "orders" and must be obeyed by those to whom these orders are delegated.

There may be other rationalizations for unethical behavior. When told that others will take responsibility for the whole, and when only carrying out a small piece of the whole, humans tend to not make waves. If accused of not doing their duty or of failing the heroes of a cause, humans may reluctantly take action. Believing that the desired end justifies any means is another pathway down the slippery ethical slope.

Contextual Factors in Twenty-First Century Health Care

Many of the contextual factors that influenced science and policy in the World War II era and in the Tuskegee experiments have resurfaced with a similar face in the twenty-first century. Once again, people are turning to science in hopes of creating a more perfect world. There are increasing concerns about the cost of health care, about care at the end-of-life, and about personal responsibility for deteriorating health status. Additionally, there is concern that illness burden and health-related resources are unevenly distributed across the United States causing significant health disparities, which unfairly disadvantage certain racial or ethnic groups (Jones 2010). Some health promotion and disease prevention interventions are raising ethical and moral issues, such as calls for mandatory vaccination of young girls with the human papillomavirus (HPV) vaccine (Javitt et al. 2008).

Conclusion

The Tuskegee study was designed to establish the natural history of untreated syphilis in Blacks. It has unquestionably become one of the most well-known human subjects research abuses in the United States. The design of the study was not the problem, as no definitive treatments were known at the time it was started in 1932. The problems were denial of effective treatments once they became available, decades of secrecy, and unethical medical behavior for which there is no excuse.

It is indeed apparent that the legacy of the Tuskegee Syphilis Study moved us toward the ethical conduct of science, finally motivating the federal government to regulate human subjects research. It led to the 1979 Belmont Report (National Commission for the Protection of Human Subjects), which in turn led to the establishment of the Office for Human Research Protections as well as the Institutional Review Boards that review all human subjects research proposals (Greely 2010).

In spite of the many social changes and the protective policies and procedures that govern biomedical research, the Tuskegee study is the most frequently cited contemporary event to justify Blacks' suspicion of research. Freimuth et al. (2001) note that there are many documented obstacles preventing some members of the Black community from participating in medical research. El-Sadr and Capps (1992) highlight, in particular, the lack of knowledge about the various components of human subjects research. Kass et al. (1996) also found that trust in the respondent's physician was the critical factor determining whether a subject enrolled in a study. Thomas and Quinn (1991) indicated that the history of the Tuskegee syphilis study, with its failure to educate the participants and treat them adequately, helped to lay the foundation for Blacks' pervasive sense of distrust of public health authorities today. Barriers to Blacks' participation in biomedical research include ". . .common recollection and interpretation of relevant historical of [sic] biomedical events where minorities were abused or exposed to racial discrimination. . .[and the fact that] African American males continue to be less educated and more disenfranchised from the majority in society. . ." (Byrd et al. 2011, p. 480).

In the twenty-first century, the Tuskegee study provides a useful framework for considering the ethical issues in human subjects research. The legacy of the Tuskegee study, directly and practically, continues to create some of the attitudes, social conditions, and problems of trust and communication that give rise to ethical concerns among Blacks concerning medical research. The wounds continue to be agitated, thwarting the healing process as a whole.

Many professionals today believe that what happened during the Third Reich, when mentally and handicapped Germans, Jews, gays, communists, Blacks, Gypsies, and Jehovah's Witnesses were targeted, could not happen again. They also may believe that today's generation is somehow less susceptible to the contextual factors of Nazi Germany. No doubt this exact scenario will not be repeated, but will citizens of the future look back and puzzle over how we were blind to some equally egregious ethical errors and ask again, have we learned the lessons of Tuskegee and the Holocaust?

References

Benedict, S. 2003. Killing while caring: The nurses of Hadamar. *Issues in Mental Health Nursing* 24: 59–79.

———. 2006. Maria Stromberger: A nurse in the resistance in Auschwitz. *Nursing History Review* 14: 189–202.

Benedict, S., A. Caplan, and T. Page. 2007. Duty and "euthanasia": The nurses of Meseritz-Obrawalde. *Nursing Ethics* 14: 781–794.

Benedict, S., and J.M. Georges. 2006. Nurses and the sterilization experiments of Auschwitz: A postmodernist perspective. *Nursing Inquiry* 13: 277–288.

———. 2009. Nurses in the Nazi "euthanasia" program: A critical feminist analysis. *Advances in Nursing Science* 32: 63–74.

Benedict, S., and J. Kuhla. 1999. Nurses' participation in the euthanasia programs of Nazi Germany. *Western Journal of Nursing Research* 21: 246–263.

Benedict, S., L. Shields, and A.J. O'Donnell. 2009. Children's "euthanasia" in Nazi Germany. *Journal of Pediatric Nursing* 24: 506–516.

Bioethics Center, Tuskegee University. n.d. About the USPHS syphilis study. Retrieved from http://www.tuskegee.edu/about_us/centers_of_excellence/bioethics_center/about_the_usphs_syphilis_study.aspx. Accessed 20 Nov 2013.

Brandt, A.M. 1978. Racism and research: The case of the Tuskegee syphilis study. *Hasting Center Magazine* 8(6): 21–29.

Brody, H. n.d. Faces of Tuskegee. The syphilis men. Center for Ethics and Humanities in the Life Sciences at Michigan State University. https://www.msu.edu/course/hm/546/tuskegee.htm. Accessed 29 Nov 2013.

Brunner, B. n.d. The Tuskegee syphilis experiment. Retrieved from http://www.infoplease.com/spot/bhmtuskegee1.html. Accessed 1 Dec 2013.

Byrd, G.S., C.L. Edwards, V.A. Kelkar, R.G. Phillips, J.R. Byrd, D.S. Pim-Pong, T.D. Starks, et al. 2011. Recruiting intergenerational African American males for biomedical research studies: A major challenge. *Journal of the National Medical Association* 103(6): 480–487.

El-Sadr, W., and L. Capps. 1992. The challenge of minority recruitment in clinical trials for AIDS. *Journal of the American Medical Association* 267: 954–957.

Freimuth, V.S., S. Quinn, S.B. Thomas, G. Cole, E. Zook, and T. Duncan. 2001. African American views on research and the Tuskegee syphilis study. *Social Science and Medicine* 52: 797–808.

Gamble, V.N. 1997. Under the shadow of Tuskegee: African American and health care. *American Journal of Public Health* 87(11): 1773–1778.

Georges, J.M., and S. Benedict. 2006. An ethics of testimony: Prisoner nurses at Auschwitz. *Advances in Nursing Science* 29: 161–169.

Gray, F. 1998. *The Tuskegee syphilis study: The real story and beyond*. Montgomery, AL: New South Books.

Greely, H.T. 2010. From Nuremberg to the human genome: The rights of human research participants. In *Medicine after the Holocaust: From the master race to the human genome and beyond*, ed. S. Rubenfeld, 185–200. New York: Palgrave Macmillan.

Harrison, L.W. 1956. The Oslo study of untreated syphilis: Review and commentary. *The British Journal of Venereal Disease* 32(2): 70–78.

Javitt, G., D. Berkowitz, and L.O. Gostin. 2008. Assessing mandatory HPV vaccination: Who should call the shots? *Journal of Law, Medicine & Ethics* 36(2): 384–395. doi:10.1111/j.1748-720x.2008.00282.x.

Jones, J. 1993. *Bad Blood: The Tuskegee syphilis experiment*, New and expandedth ed. New York: Basic Books.

Jones, C.M. 2010. Why should we eliminate health disparities? *American Journal of Public Health* 100(1): 47–51.

Kass, N., J. Sugarman, R. Faden, and M. Schoch-Spana. 1996. Trust: The fragile foundation of contemporary biomedical research. *Hastings Center Report* 26: 25–29.

Kater, M. 2002. Criminal physicians in the Third Reich. In *Medicine and medical ethics in Nazi Germany*, ed. R. Nicosia and J. Huener, 77–92. New York: Berghahn Books.

Lifton, R.J. 1986. *Medical killing and the psychology of genocide: The Nazi doctors*. New York: Basic Books.

McFarland-Icke, B.R. 1999. *Nurses in Nazi Germany: Moral choice in history*. Princeton, NJ: Princeton University Press.

Milgram, S. 1963. Behavioral study of obedience. *Journal of Abnormal and Social Psychology* 67 (4): 371–378.

National Commission for the Protection of Human Subjects of Biomedical and Behavioral Research. 1979. The Belmont Report. U.S. Department of Health and Human Services. http://www.hhs.gov/ohrp/humansubjects/guidance/belmont.html. Accessed 30 Sept 2013.

National Institutes of Health Office of Extramural Research. 2013. Research integrity. http://grants.nih.gov/grants/research_integrity/research_misconduct.htm. Accessed 25 Nov 2013.

Ogungbure, A.A. 2011. Tuskegee syphilis study: Some ethical reflections. *Journal of the Philosophical Association of Kenya* 3(2): 75–92.

Perry, G. 2013. Behind the shock machine: The untold story of the notorious Milgram Psychology Experiments. [Book review]. *Science News* 184(6): 30. http://online.wsj.com/article/SB10001424127887323324904579040672110673420.html. Accessed 09 Sept 2013.

Presidential Commission for the Study of Bioethical Issues. 2011, September. "Ethically Impossible" STD research in Guatemala from 1946 to 1948. Retrieved from http://bioethics.gov/node/654. Accessed 9 Sept 2013.

Reverby, S.M. 1999. Rethinking the Tuskegee syphilis study: Nurse Rivers' silence and the meaning of treatment. *Nursing History Review* 7: 3–28.

———. 2000. *Tuskegee's truths: Rethinking the Tuskegee syphilis study*. Chapel Hill: University of North Carolina Press.

Smith, S.L. 1996. Neither victim nor villain: Nurse Eunice Rivers, The Tuskegee syphilis experiment, and public health work. *Journal of Women's History* 8(1): 95–113.

Smolin, D.M. 2013. The Tuskegee syphilis experiment, social change, and the future of bioethics. Highbeam Research. http://www.highbeam.com. Accessed 9 Mar 2013.

Spitz, V. 2005. *Doctors from hell: The horrific account of Nazi experiments on humans*. Boulder, CO: Sentient Publications.

Survey of German Nurses, Midwives, and Medical Social Workers. 1945, November 27. File of Geneva Panel, US Army Nurse Corps Historical Collection, Office of Medical history (DAG-MH), Office of the Surgeon General, US Army, Falls Church, VA.

The Tuskegee Timeline. n.d. U. S. Public Health Service syphilis study at Tuskegee. Centers for Disease Control and Prevention. http://www.cdc.gov/Tuskegee/timeline/htm. Accessed 20 Nov 2013.

Thomas, S.B., and S.C. Quinn. 1991. The Tuskegee syphilis study, 1932 to 1972: Implications for HIV education and AIDS risk education programs in the black community. *American Journal of Public Health* 81(11): 1498–1505.

U.S. Department of Health and Human Services, Office of Research Integrity. n.d. Case Summaries. http://ori.hhs.gov/case_summary. Accessed 20 Nov 2013.

Zimbardo, P. 1972. Pathology of imprisonment. *Society* 9(6). http://www.angelfire.com/or/sociologyshop/path.html. Accessed 14 Sept 2013.

Chapter 16
Human Subjects Research during and after the Holocaust: Typhus Vaccine Development and the Legacy of Gerhard Rose

Wendy A. Keitel

Introduction

The imprisonment of millions of men, women, and children by Nazi officials in crowded ghettos, concentration camps, and other prison camps during World War II provided Nazi doctors and other sympathizers with the opportunity to conduct inhumane and unethical experiments on victims without their consent. Among these were a number of experiments purportedly designed to develop vaccines for control of epidemic typhus, a disease whose impact was potentiated by the conditions imposed by the Nazis on their victims. Indeed, the twisted Nazi mentality rationalized both illegal experimentation and mass murder based on the premise that the victims were the perpetrators of epidemic typhus, a disease that contributed to the deaths of millions of prisoners, soldiers, and private citizens alike. Dr. Gerhard Rose, then director of the Department of Tropical Medicine of the prestigious Robert Koch Institute, participated in typhus infection and vaccination experiments conducted in the concentration camps and was found guilty of "murders, brutalities, cruelties, tortures, atrocities and other inhumane acts" on June 16, 1947, at the Nuremberg Doctors' Trial (McHaney et al. 1947, p. 24). The goals of this chapter are to provide a brief background about epidemic typhus and the Nazi typhus experiments in the concentration camps; to discuss the ethical issues surrounding the human subjects vaccine research in which Gerhard Rose participated; to describe the impact of these and other experiments on current vaccine development in particular and medical research in general; and to explore the legacies of the Nazi trials and the lessons learned from these events.

W.A. Keitel, MD, FIDSA (✉)
Kyle and Josephine Morrow Chair in Molecular Virology & Microbiology; Professor,
Departments of Molecular Virology & Microbiology and Medicine; Baylor College of
Medicine, One Baylor Plaza, Houston, TX 77030, USA
e-mail: wkeitel@bcm.edu

S. Rubenfeld and S. Benedict (eds.), *Human Subjects Research after the Holocaust*,
DOI 10.1007/978-3-319-05702-6_16, © Springer International Publishing Switzerland 2014

Epidemic Typhus and World War II

Historical Considerations

Epidemic typhus is an infectious disease caused by the bacterium *Rickettsia prowazekii*, an obligate intracellular pathogen. Epidemics of typhus have been recorded for over five centuries, particularly in the setting of natural and unnatural disasters such as war, during which poor sanitation, poverty, crowding, famine, and other deprivations prevail. While the role of human body lice in the transmission of typhus was established in 1909, the etiologic agent was identified experimentally during a series of investigations conducted between 1910 and 1922 (for a review see Walker and Raoult 2010). By the time of World War II, it was recognized that control of lice with insecticides could prevent outbreaks and that a killed vaccine prevented deaths in Allied soldiers (Walker and Raoult 2010). It was subsequently demonstrated that antibiotics including tetracycline and chloramphenicol were effective for treatment of established disease.

Typhus: The Disease

Typhus is spread from person to person by the human body louse, which becomes infected after taking a blood meal from an infected human. Rarely, sporadic human cases of typhus have been transmitted by the fleas of infected flying squirrels. The bacterium multiplies in the gut of the louse and is present in the feces of the louse. Fecal material then contaminates the clothing and skin of uninfected persons, who become infected after rubbing or scratching the infectious feces into skin or eyes, or by breathing in the fecal material. The clinical manifestations of typhus typically develop within a week or two of exposure. Common symptoms and signs include headache, fever, chills, muscle aches, and rash. If untreated, patients may develop more severe manifestations including delirium, stupor or coma, cough, hemorrhagic rash, and death. Death rates from untreated typhus were high in the setting of epidemics. A milder form of the disease, Brill-Zinsser disease, may reappear in persons who previously recovered from an initial case of typhus (see Walker and Raoult 2010).

Typhus During World War II

Detailed descriptions of the history of typhus and the occurrence of typhus during the Nazi era have been provided by Dr. Naomi Baumslag (2005). Epidemics of typhus were common in many settings, including ghettos, concentration camps, and among the military. Delousing and good hygiene were the most effective means for control; however, the Nazis persisted in using ineffective and/or murderous methods for their prisoners. Notably, the Jews and other "undesirables" were forced to live under conditions conducive to the spread of typhus and then were scapegoated as the source of disease. Afflicted persons and populations were then allowed to die or were intentionally murdered. High mortality in concentration camps was often the result of typhus and starvation. "Blocks set aside for typhus cases became waiting rooms for the gas chamber" (Baumslag 2005, p. 59); mass murder of inmates became a typical response to a typhus outbreak and a justification for genocide. Jewish doctors and other sympathizers resisted these and other evils perpetrated by their captors using a variety of means, including attempts to prevent the spread of typhus, provision of supportive care to affected individuals, concealment of typhus cases, and containment of outbreaks. Nevertheless, eventually over 1.5 million prisoners died of this preventable disease (Baumslag 2005, pp. xxv–introduction).

Typhus Experiments During the Holocaust

Human experiments involving typhus were primarily conducted at two concentration camps: Natzweiler in Natzweiler, France from 1941 to 1944, and Buchenwald near Weimar, Germany from 1937 to 1945. Natzweiler was primarily a camp for political prisoners and was the first camp liberated by Americans in November of 1944. Prisoners at Buchenwald originated from all over Europe and the Soviet Union, and it became a Soviet camp after the war until its destruction in October 1950.

Typhus experiments at Natzweiler between 1943 and 1944 were directed in part by Dr. Eugen Haagen, who was Chair for Hygiene and Bacteriology at Strasbourg University, captain of the military corps, and Consulting Hygienist to the Chief of the Medical Service of the Luftwaffe. The experiments were under the supervision of Dr. Joachim Mrugowsky. Haagen developed a vaccine that likely was protective against typhus following an artificial challenge, or intentional inoculation, although in doing so, he probably caused a typhus outbreak. Dr. Erwin Ding-Schuler, who committed suicide at the end of the war and was never brought to trial, primarily conducted the typhus experiments at Buchenwald between 1942 and 1945. His diary survived and became evidence for the role of doctors in the experiments (Spitz 2005). Large experiments were conducted at Buchenwald, and many of the subjects died.

Gerhard Rose: Who Was He and What Was His Role in the Typhus Experiments?

Gerhard Rose was born in November 1886 in Danzig. He joined the Army in 1914 and was wounded and taken as a prisoner of war in France, then in Switzerland between 1914 and 1918. Disabled as a result of his wounds, he left the army and enrolled in the University of Breslau and Kaiser Wilhelm Academy in Berlin. He received his medical degree in 1922 and became an assistant at several prestigious institutions in Breslau, Berlin, Basel and Heidelberg between 1922 and 1929. From 1929 to 1936, he served as the Director of the State Institute for Public Health Matters and Chair of the Parasitological Section for Tropical Medicine in China, where he focused on schistosomiasis and public health. He then joined the Reich's Service and Luftwaffe Reserves, and was appointed Director of Tropical Medicine at the Robert Koch Institute in Berlin, where he remained until his capture in 1945 (Miller and Collins, "Generalarzt der Reserve Prof. Dr. med. Gerhard Rose"). At the time of his arrest, he was preparing to leave Germany for the United States in order to continue his career in medical research (Epstein n.d.).

Although the historical record is sparse, evidence obtained from letters and other testimony at the Doctors' Trial confirms Rose's role in the typhus experiments conducted at Natzweiler and Buchenwald (Introduction to NMT Case I, 1946–47, Indictments). Upon request from Haagen, Rose assisted with the procurement of Natzweiler trial subjects from other camps. Furthermore, Rose proposed experiments involving deliberate infection of prisoners; procured vaccines for the experiments; provided financial support; and reviewed and commented on reports of the studies. Encouraging the testing of several vaccines, in one communication Rose indicated that "the test of various vaccines simultaneously gives a clearer idea of their values than the test of one vaccine alone" (McHaney et al. 1947, p. 11). Rose also supplied vaccines to Ding-Schuler and Mrugowsky at Buchenwald; he observed 145 inmates infected with virulent typhus (McHaney et al. 1947, p. 17); and he proposed experiments with a vaccine derived from mouse liver followed by challenge with material derived from so-called "passage persons," that is, people deliberately and actively infected with typhus. Letters clearly indicate that both vaccination and infection experiments were conducted in over 850 prisoners (McHaney et al. 1947, pp. 17–24).

In summary, Gerhard Rose provided vaccines as well as financial and expert "scientific" input into the typhus experiments in the camps, and for his participation in these and other illegal activities, such as malaria experiments, he was arrested and put on trial at Nuremburg.

The Doctors' Trial at Nuremberg and Beyond: The Fate of Gerhard Rose

The Doctors' Trial at Nuremberg began with arraignments of the defendants on November 21, 1946 (Spitz 2005). Telford Taylor served as the Chief of Counsel for War Crimes. Twenty doctors and three medical assistants were indicted: seven were sentenced to death, nine were imprisoned, and seven were found not guilty. Like the other defendants, Gerhard Rose was tried on four counts: conspiracy, war crimes, crimes against humanity, and membership in a criminal organization (Introduction to NMT Case I, 1946–1947, Summary). During his final statement, Rose commented on his role in medical experiments in the concentration camps:

> A subject of the personal charges against myself is my attitude toward experiments on human beings ordered by the state and carried out by other German scientists in the field of typhus and malaria. Works of that nature have nothing to do with politics or with ideology, but they serve the good of humanity, and the same problems and necessities can be seen independently of any political ideology everywhere, where the same dangers of epidemics have to be combated. (Spitz 2005, p. 261)

In a signed affidavit, he further commented that he had initially objected to the experiments (Rose 1947); however, he ultimately concluded that the potential benefits outweighed the deaths of the victims.

The verdict in Dr. Rose's case was pronounced on June 16, 1947, and read, in part, as follows: "[Rose was a] principal in, an accessory to, ordered, abetted, took a consenting part in, and was connected with plans and enterprises involving medical experimentation on human subjects without their consent . . ." (McHaney et al. 1947, p. 24). Rose was found guilty on all counts except membership in a criminal organization and was sentenced to life in prison. He was released in 1955, and subsequently, the Federal Disciplinary Chamber X of Dusseldorf acquitted him of the charges in 1963 (Miller and Collins n.d.).

Ethical Issues Related to Experimentation in Concentration Camps

What could possibly have justified the actions of Rose, other physicians, nurses, and medical personnel? Robert Lifton (2000) described various ways that doctors rationalized their participation, including the "technicizing of everything," whereby doctors and other participants could focus on the technical aspects of their activities and leave ethical concerns to others (p. 453). Arthur Caplan explores this issue in his thought-provoking essay entitled "The Stain of Silence: Nazi Ethics and Bio-ethics" (Caplan 2010). Potential moral justifications for experimentation in the camps described in Caplan's essay (pp. 88–89) can be summarized, as follows:

- Subjects volunteered; study participants could be freed.
- Experiments were conducted only on those already condemned to die.
- Participation offered the subjects expiation from any sins or crimes they had committed.
- Doctors were not responsible for values, only for the proper design of research.
- Experiments were done for the security of Germany during war, and casualties of the experiments represented a sacrifice of a few for the benefit of many.

Caplan dismisses the first four justifications but takes the last one seriously, noting that, "The last rationale is the one that appears to carry the most weight among all the moral defenses offered" (Caplan 2010, p. 89). Regardless of whether any of these potential rationalizations could justify what was done to the victims of these experiments, it is clear that major ethical principles were violated. Among these are respect for persons (autonomy, informed consent), beneficence, and justice (National Commission for the Protection of Human Subjects of Biomedical and Behavioral Research 1979). These studies were also illegal, as they violated a standing German law concerning human experimentation (Reich Minister 1931) and an earlier Prussian Directive (1900) that emphasized the need for consent to experiment on subjects and protection of vulnerable populations (Cohen 2010).

Human Subjects Research in the United States After the Holocaust: Defining and Refining Ethical Guidelines

Perhaps the greatest legacy of the Holocaust related to the area of human experimentation was the development of the first international code of research ethics—The Nuremberg Code (*Trials of War Criminals*, 1949, pp. 181–182). This document articulated ten principles of ethical conduct of human subjects research:

1. Subjects should provide voluntary consent.
2. The experiment should yield useful information.
3. The anticipated results should justify the experiment, based on animal studies and the natural history of the disease.
4. All unnecessary physical and mental suffering and injury should be avoided.
5. Investigators should not anticipate that any intervention or study procedure likely will result in death or disabling injury.
6. The potential risk should not exceed potential benefits.
7. Subjects should be protected against injury, disability, and death.
8. Experiments should be conducted by qualified persons.
9. Subjects have the right to terminate their participation at any time.
10. The scientist in charge should be prepared to terminate the experiment at any time if at any time there arises a risk of a subject's injury, disability, or death.

The Code was based on legal concepts, and was to be used for the practice of human experimentation wherever it was being conducted. Of interest is the fact that the American Medical Association voted on the requirements for the ethical conduct of medical experiments in 1946 in advance of the publication of the Nuremberg Code (Katz 1996, p. 1663; see also Chap. 14).

It is instructive to consider the conduct of experiments involving humans in the United States during the twentieth century, both before and after the horrors of the Holocaust were revealed. For example, J. P. Baker (2000) reviewed a number of clinical trials conducted during this period, including many that were done *after* the endorsement of the Nuremberg Code. Among these were experiments conducted among institutionalized children, foundlings, "crippled children," and "mentally retarded" children. Krugman and colleagues conducted a series of trials that began in 1956 and lasted 14 years that involved inoculation of "mentally retarded" children and adults with live viruses at Willowbrook State School in New York (Krugman and Giles 1970). Although consent was obtained from parents, they were told that their children could only be admitted to Willowbrook through the hepatitis unit. Children who were wards of the state or who were orphans were not included. In addition to these experiments, numerous trials in the United States were performed in prison populations until limitations on the use of prison inmates were imposed in the 1970s (Good Clinical Practice Q & A, 2007).

Special ethical issues are raised when vulnerable populations, such as those just described, are enrolled in research, including the attainment of voluntary informed consent and the provision of alternatives to participation. On the other hand, denying prisoners the opportunity to participate in therapeutic studies abrogates their rights to access potentially life-saving measures (Thomas 2010). The infamous Tuskegee experiments conducted by the US Public Health Service, in which researchers observed the natural history of syphilis in poor African-American men even *after* the introduction of successful antibiotic therapy, stands out as an example of unethical research. Other examples are detailed in Beecher's landmark article addressing the ethics of clinical research (Beecher 1966). Parenthetically, Beecher himself had conducted secret medical research for the US Central Intelligence Agency on military personnel involving psychotropic drugs without their informed consent. "In his decade of secret drug testing, Dr. Beecher sacrificed his subjects to the cause of national security," a common theme underlying the Nazi's experiments (McCoy 2007, p. 413).

Although it took several decades for American bioscientists to recognize and adopt the principles articulated in The Nuremberg Code, defining and refining guidelines for the ethical conduct of research is now an ongoing process in the United States and elsewhere. Table 16.1 provides an abbreviated timeline of the development of ethical guidelines and regulation of human subjects research in the United States after publication of The Nuremberg Code. Pivotal events and documents are briefly summarized, including the institutionalization of research oversight, articulation of basic ethical guidelines, and harmonization of such guidelines across many countries (Goldfarb, "Clinical Research Milestones").

Table 16.1 Abbreviated Timeline for Development of Ethical Guidelines after the Nuremberg Code

Year(s)	Milestone
1963	US Food and Drug Administration (FDA) requires that investigators certify informed consent
1964	World Medical Association issues Declaration of Helsinki
1966	Beecher publishes "Ethics and Clinical Research"
1966	FDA defines elements of consent
1967	FDA requires written consent
1972	National Institutes of Health (NIH) forms Office for Protection from Research Risks, which recommended independent review bodies
1974	National Research Act requires Institutional Review Boards (IRBs) for Department of Health and Human Services (DHHS) trials; forms National Commission for the Protection of Human Subjects
1975–1978; 1983	National Commission for the Protection of Human Subjects of Biomedical and Behavioral Research issues standards for conducting research on vulnerable populations (fetuses, prisoners, children, institutionalized mentally infirm subjects)
1979	National Commission for the Protection of Human Subjects of Biomedical and Behavioral Research issues Belmont Report
1981	FDA issues regulations regarding protection of subjects, consent, and IRBs
1991	Recommendations of the DHHS National Bioethics Commission become federal law (45 CFR 46), also known as "the Common Rule."
1996 (updated in 2002)	The Council for International Organizations and Medical Sciences updates international ethical guidelines, ICH Good Clinical Practice (GCP) Guidelines of 1996

Vaccine Development Now: Enduring Legacy of Nuremberg Code

The process of vaccine development is now highly regulated, and it may take decades for a candidate vaccine to be approved for human use (for a review see Keitel 2009). In many cases, the development of guidelines and the establishment of regulatory authority were spurred by a tragic accident or by a clear ethical breach. For example, the occurrence of tetanus among recipients of a diphtheria vaccine contaminated with tetanus spores in 1902 led to passage of the Biologics Control Act regulating vaccines and antitoxins in the very same year. In 1927, the Food and Drug Administration (FDA) was created from its predecessor, the US Bureau of Chemistry.

Explicit recognition of and requirement for training and adherence to basic ethical principles has grown in the 67 years since publication of the Nuremberg Code. Data and safety monitoring of vaccine clinical trials have greatly expanded, and requirements for documentation have grown exponentially. The assessment of sound scientific design has been emphasized as an important component of ethical review. A critical feature of this process is that the ethical considerations and the regulatory guidelines are inextricably intertwined. The complexity and redundancy of research oversight reflects the commitment to ethical conduct of vaccine clinical trials. According to the FDA, all trials except phase I must be registered with ClinicalTrials.gov, and investigators must declare potential conflicts of interest ("NIH—Institutional Conflicts of Interest," n.d.).

Lessons Learned: Holocaust Typhus Experiments and Beyond

So what did Dr. Rose and his colleagues learn from the typhus experiments conducted during the Holocaust? Because the experiments were scientifically unsound and statistically flawed, little of any scientific value was learned; because the experiments were unethical, these data were obtained at a terrible cost. Vaccines were poorly standardized and of uncertain safety, purity, and potency. Variability in methods of infection—man-to-man via intravenous inoculation of blood from "passage persons" vs. the natural route of louse fecal matter scratched into skin—rendered results difficult to interpret. The "subjects" were not representative of German troops, the primary target for prophylactic vaccination against typhus. The doctors often were not qualified, and results were falsified (Baumslag 2005, p. 155).

We must ask ourselves whether egregious ethical violations in vaccine research could happen again. Constant attention to the balance between individual and public health is necessary, and many safeguards are required to avoid unethical research. Despite extensive ethical examination, tensions persist between competing ethical principles. Changes in how vaccine trials are conducted have undoubtedly made vaccine research, in particular, and human subjects research, in general, safer and more ethically and scientifically sound. However, as Jason Liebowitz reminds us:

> The desire to have a successful medical career, the willingness to appease professional superiors, the ability to rationalise nearly any human research in the name of advancing knowledge—these explicit motives behind the actions of Rascher and countless Nazi physicians represent potential traps for modern medicine in the struggle to uphold beneficence and non-maleficence even when self-interest comes into play. (2011, p. 54)

The same could be said for Gerhard Rose.

From beginning to end, laws and regulations cannot be a substitute for ethical standards and behavior on the part of investigators.

We must also be attentive to the possibility that the development of onerous regulation of human subjects research may have unintended consequences. Extensive documentation requirements interfere with communication with the subjects. The length and complexity of consent forms reduces the likelihood that subjects actually read and/or understand the contents of the document. Delays in the availability of life-saving vaccines, drugs, and other interventions may occur as a result of the lengthy and redundant review processes, and acceptable research may fail to receive approval by overzealous and/or uninformed review committees.

"Protection" of vulnerable populations may lead to their exclusion from clinical trials that could benefit them. Ethics committees may undermine the autonomy of research subjects in an attempt to protect participants and/or their institution (Guillemin et al. 2012). Finally, rising costs, including time and financial resources, limit the amount of work that can be done.

References

Baker, J.P. 2000. Immunization and the American way: 4 Childhood vaccines. *American Journal of Public Health* 90(2): 199–207.

Beecher, H.K. 1966. Ethics and clinical research. *New England Journal of Medicine* 274: 1354–1360.

Baumslag, N. 2005. *Murderous medicine: Nazi doctors, human experimentation, and typhus.* Washington, DC: Baumslag.

Caplan, A.L. 2010. The stain of silence: Nazi ethics and bioethics. In *Medicine after the Holocaust: From the Master Race to the human genome and beyond*, ed. S. Rubenfeld, 83–92. New York: Palgrave Macmillan.

Cohen, M.M. 2010. Overview of German, Nazi, and Holocaust medicine. *American Journal of Medical Genetics, A* 152A: 687–707.

Epstein, H. n.d. The courtroom and the defendants: Gerhard Rose. Retrieved http://www.hedyepstein.com/history/courtroom/gerhardrose.htm. Accessed 23 Nov 2013.

Goldfarb, N. Ed. n.d. Clinical research milestones. First Clinical Research website. http://firstclinical.com/milestones/?page=2&date=All&anchor=word. Accessed 24 Nov 2013.

———. 2007, June. Good Clinical Practice Q & A: Focus on prisoners. *Journal of Clinical Research Best Practices, 3*(6), 1–2. (Reprinted from *Good clinical practice: A question & answer reference guide 2007*. Needham, MA: Barnett International, Inc.). First Clinical Research website. http://firstclinical.com/journal/2007/0706_GCP23.pdf. Accessed 24 Nov 2013.

Guillemin, M., L. Gillam, D. Rosenthal, and A. Bolitho. 2012. Human research ethics committees: Examining their roles and practices. *Journal of Empirical Research on Human Research Ethics* 7: 38–49.

Introduction to NMT Case 1: *United States of America v. Karl Brandt et al.* (1946–47). Harvard Law School Library. Nuremberg Trials Project, Digital Document Collection. http://nuremberg.law.harvard.edu/php/docs_swi.php?DI=1&text=medical. Accessed 23 Nov 2013.

Katz, J. 1996. The Nuremberg Code and the Nuremberg Trial: A reappraisal. *Journal of the American Medical Association* 276(20): 1662–1666.

Keitel, W.A. 2009. Clinical trials of vaccines for biodefense and emerging and neglected diseases. In *Vaccines for biodefense and emerging and neglected diseases*, ed. L. Stanberry and A. Barrett, 157–170. London: Elsevier.

Krugman, S., and J.P. Giles. 1970. Viral hepatitis: New light on an old disease. *Journal of the American Medical Association* 212(6): 1019–1029.

Liebowitz, J. 2011. Moral erosion: How can medical professionals safeguard against the slippery slope? *Medical Humanities* 37(1): 53–55.

Lifton, R.J. 2000. *The Nazi doctors: Medical killing and the psychology of genocide.* New York: Basic Books.

McCoy, A.W. 2007. Science in Dachau's shadow: Hebb, Beecher, and the development of CIA psychological torture and modern medical ethics. *Journal of the History of the Behavioral Sciences* 43(4): 401–417.

McHaney, J.M., Hardy, A.G., Horlik-Hochwald, A., and Johnson, E.J. 1947. Closing Brief for the United States of America against Gerhard Rose. MilitaryTribunal No. I Case No. 1. Harvard Law School Library Item No. 19. Nuremberg Trials Project, Digital Document Collection.

http://nuremberg.law.harvard.edu/php/pflip.php?caseid=HLSL_NMT01&docnum=19&num
pages=25&startpage=1&title=Closing+brief+for+the+United+States+of+America+against+
Gerhard+Rose.&color_setting=C. Accessed 23 Nov 2013.

Miller, M., and Collins, G. Eds. n.d. "Generalarzt der Reserve Prof. Dr. med. Gerhard Rose." Axis
Biographical Research, an apolitical military history site. http://www.geocities.com/~orion47/
WEHRMACHT/LUFTWAFFE/Sanitatsoffizier/ROSE_GERHARD.html. Accessed 23 Nov 2013.

National Commission for the Protection of Human Subjects of Biomedical and Behavioral
Research. 1979. U.S. Department of Health & Human Services. Ethical principles and guidelines for the protection of human subjects of research. The Belmont Report. http://www.hhs.
gov/ohrp/humansubjects/guidance/belmont.html. Accessed 23 Nov 2013.

NIH - Institutional conflicts of interest. n.d. First Clinical Research website. http://firstclinical.
com/regdocs/doc/?showpage=5&db=OTH_NIH_Institutional_Conflicts&matches=4,7,10,11,
12,14,15,17,20,22,24,25&search=conflict+of+interest&type=word&page=1. Accessed 25 Nov
2013.

Reich Minister of the Interior. 1931, February 28. German guidelines on human experimentation
1931 [circular]. http://artnscience.us/Med_Ethics/reichsrundschreiben_1931.pdf. Accessed
18 Dec 2013.

Rose, Gerhard. 1947, January. Document book in Mrugowsky. Documents Joachim Mrugowsky
No. 48. [Affidavit]. Harvard Law School Library. Nuremberg Trials Project, Digital Document
Collection. http://nuremberg.law.harvard.edu/php/pflip.php?caseid=HLSL_NMT01&docnum
=735&numpages=2&startpage=1&title=Affidavit.&color_setting=C

Spitz, V. 2005. *Doctors from hell: The horrific account of Nazi experiments on humans.* Boulder,
CO: Sentient Publications.

Thomas, D.L. 2010. Prisoner research: Looking back or looking forward? *Bioethics* 24(1): 23–26.
doi:10.1111/j.1467-8519.2009.01777.x.

*Trials of War Criminals before the Nuremberg Military Tribunals under Control Council Law
No. 10, Vol 2.* 1949. Nuremberg Code. 181–182, Washington, DC: U.S. Government Printing
Office. http://www.hhs.gov/ohrp/archive/nurcode.html

Walker, D., and D. Raoult. 2010. Rickettsia prowazekii (Epidemic or louse-borne typhus). In
Principles and practice of infectious diseases, 7th ed, ed. G. Mandell, K. Bennett, and R. Dolin,
2521–2524. Philadelphia, PA: Churchill Livingstone Elsevier.

Chapter 17
Ethics in Space Medicine: Holocaust Beginnings, the Present, and the Future

Neal R. Pellis

Beginnings

Many of us recall the beginning of the space program as the Union of Soviet Socialist Republics' launch of an orbiting satellite in 1957. However, the interest in the space part of aerospace medicine actually began in the 1930s with the investigation of the human response to high-altitude flight in military aircraft. At that time, aircraft were not pressurized as they are today. Subsequently, pilots and passengers were exposed to very low oxygen tension environments due to low barometric pressures. In addition to the low oxygen content, pilots and passengers encountered acceleration forces as planes went into diving mode or did high-speed banked turns. During these events, humans experienced substantial G forces that redistributed the fluid inside the body and subjected the body to a series of potentially devastating stresses. The early investigations in space medicine were devoted to these areas of science.

The events leading up to World War II spawned military-funded research, and Hitler's regime was at the forefront of the investigations in aerospace medicine. Research was conducted at a number of academic medical institutions in Germany as well as in other developed nations. As the Nazis came to power, physicians and scientists volunteered or were co-opted to perform experiments, most of which were on unwilling human subjects.

A central figure in early German aerospace research was the physician Hubertus Strughold, who later became known as "The Father of Space Medicine." The following narrative is a compilation taken from a series of accounts (Campbell et al. 2007; Campbell and Harsch 2013; Lagnado 2012). Dr. Strughold was born June 15, 1898, in Westtünnen-im-Hamm, Germany. He began a long academic career of studies in medicine and natural sciences at the Ludwig Maximilian

N.R. Pellis, PhD (✉)
Director, Division of Space Life Sciences; Universities Space Research Association, 3600 Bay Area Boulevard, Houston, TX 77058, USA
e-mail: pellis@dsls.usra.edu

S. Rubenfeld and S. Benedict (eds.), *Human Subjects Research after the Holocaust*, 217
DOI 10.1007/978-3-319-05702-6_17, © Springer International Publishing Switzerland 2014

University of Munich. Dr. Strughold then went to the Georg August University of Göttingen where, in 1922, he earned his doctorate in science. His medical degree came from the University of Münster in 1927. Afterward, he matriculated at the University of Würzburg where his work continued on aviation medicine research as professor of physiology.

In 1928, Dr. Strughold traveled to the United States as a Rockefeller Foundation Fellow and conducted research in aviation medicine and physiology at what is now Case Western University in Cleveland and at the University of Chicago. He returned to Germany in 1929 and resumed teaching at Würzburg Physiological Institute where he became a professor in 1933. By 1935, he had garnered the attention of many for his work in aeronautic medicine. He was appointed by Germany's Nazi government to direct the Research Institute for Aviation Medicine, which was operated by Hermann Göring's Reich Ministry of Aviation. While there, Strughold conducted extensive studies on high-altitude, high-speed flight, concentrating mainly on the medical effects experienced by crew and passengers in aircraft. In addition, he is credited with the development of a sophisticated altitude chamber concept and is known for his work on the concept of "time of useful consciousness" and the following recovery.

The funding for his work and the oversight of his activities during this time were the responsibilities of the German Armed Forces. With the onset of the military actions leading to World War II, Dr. Strughold's organization was absorbed by the German military and rechristened the Air Force Institute for Aviation Medicine. As a consequence, Dr. Strughold was commissioned as an officer in the Luftwaffe. In this capacity, he oversaw scientific investigators at various installations in Germany where research was conducted on prisoners. His alleged role in these experiments would be revisited in postwar investigations for nearly 30 years.

In 1942, Dr. Strughold participated in a medical conference in Nuremberg at which Sigmund Rascher, an SS physician, outlined various medical experiments he had conducted in conjunction with the Luftwaffe medical service. At the Dachau concentration camp, some prisoners were subjected to decreased barometric pressures to observe the outcome, while others were immersed in near freezing water for various periods of time to study their recovery (Campbell et al. 2007). Some of the personnel involved in the experiments were convicted, but no original documentation has identified Strughold as conducting or ordering these experiments (Campbell and Harsch 2013).

The experiments have many disturbing aspects: (1) They may have been unnecessary even if experimental animals rather than human subjects were used; (2) Researchers consistently made the error of having the "rescued individual" maintain an erect posture in the seated position while putatively going through recovery (the standard of care today is the supine position); and (3) Obviously, the prisoners endured horrific suffering and, in many cases, death.

In other experiments, subjects were forced to drink seawater and undergo invasive surgical procedures without anesthesia (Campbell and Harsch 2013).

Estimates place the number of inmates who died as a result of these experiments in the hundreds. Although this number may be small compared to other experiments during the Third Reich, the performance of these atrocious experiments by putative subscribers to the Hippocratic Oath remains an ethical enigma. Dr. Strughold claimed to Allied authorities that he had no knowledge of the Dachau atrocities (Campbell and Harsch 2013).

At the close of the war, Dr. Strughold became director of the Physiological Institute at Heidelberg University and resumed what might be considered a normal academic life. Just as other physicians and scientists of the Third Reich were recruited by the United States after WWII, he eventually was approached to become chief scientist of the Army Air Corps' Aeromedical Center in Germany. In 1947, Strughold was recruited by Operation Paperclip, a project that brought valuable physicians and scientists from the Third Reich to the United States. He was sent to Randolph Field in San Antonio, the home of a major Air Force base at the time. Strughold then conducted some of the first research into the medical challenges posed by space travel, coined the term "space medicine," and was appointed professor of Space Medicine at the US Air Force School of Aviation Medicine (SAM). This school was dedicated to research on the human response to factors associated with manned spaceflight, such as atmospheric control, weightlessness, and circadian rhythm disruptions. Dr. Strughold diligently pursued these investigations and rose to prominence in space medicine in the United States and worldwide. In 1956, he became a US citizen.

In the meantime, the Soviets had successfully launched the first orbiting satellite, and the United States quickly responded by building and launching spacecraft, thereby entering the race to be the first nation to put a human in space. The clinical staff of the National Aeronautics and Space Administration (NASA) recognized Dr. Strughold's accomplishments and consulted with him on the design of the pressurized suit and the onboard life support systems for early spaceflights. He was elected to a number of august academic medicine groups, the most prominent being the Space Medicine Association. In fact, he held prominent positions in the organization, and the organization itself instituted an annual honor known as the Strughold Award. He retired from the Air Force position in 1968 and consulted for some time thereafter (Campbell et al. 2007).

Despite his postwar accomplishments, Dr. Strughold was unable to divest himself of allegations that he participated in human subjects research atrocities during the Holocaust. Of the numerous investigations by the United States and by advocacy groups into Dr. Strughold's past, the three major governmental investigations share one thing in common—no conviction was obtained. The 1958 investigation by the Department of Justice exonerated him fully because of the lack of compelling evidence. In 1974, the inquiry by the Immigration and Naturalization Service was abandoned due again to a lack of evidence. In 1983, the Office of Special Investigations in the Department of Justice believed that there might be sufficient evidence for a conviction and initiated a third major investigation. Unfortunately, the long judicial process was abandoned when Dr. Strughold died in September 1986 prior to his appearance before the justices.

There were numerous attempts to prosecute Strughold, none of which yielded sufficient evidence to convict him. Yet this enigmatic question remains unanswered: How could the director of an institute claim no knowledge of the heinous research ostensibly conducted by his subordinates? Dr. Strughold was present at Nuremberg pretrial reviews where some of his subordinates were convicted, but he evaded implication. Colleagues in the United States and elsewhere defended him despite the investigations claiming a tangible association with specific Dachau scientists and their atrocities. Indeed, it may remain a controversy because only a few records of his associations with actual perpetrators have been unearthed despite the diligent efforts of many individuals and organizations (Campbell et al. 2007; Campbell and Harsch 2013).

The Space Medicine Association's annual award has been a problematic legacy of Dr. Strughold's research for the United States. Although under constant attack from scientists, physicians, survivors of the Nazi atrocities, and ethicists, the award was not formally retired until 2013 (Space Medicine Association 2013). The last recipient was in 2012, and all renowned museums featuring Dr. Strughold's accomplishments in military and space medicine have removed displays about him. Strughold's position in history is now as a physician and scientist who worked for the Third Reich and who may have supervised atrocities in human subjects research conducted at Dachau. No doubt there will be continued pursuit of the truth about this enigmatic character and his role in the Holocaust. As recently as November 30, 2012, the front page of the *Wall Street Journal* featured an article entitled, "A Scientist's Nazi Era Past Haunts Prestigious Space Prize" (Lagnado 2012).

The Present

While the past atrocious experiments on unwilling human subjects cause bewilderment, there remains the lingering question how it could have happened. Today, medical research is conducted under extensive scrutiny. Research sponsored publicly or privately undergoes rigorous review for ethical and scientific merit by independent review boards. The oversight continues throughout the experiment, and the results are reviewed by the researchers as well as by medical review boards and journal review boards for possible publication. Personal medical data is protected and released only with consent of the subject. NASA, being a federal agency funded by Congress and supervised by the executive branch of government, is subjected to strict adherence to all federal regulations on the use of human subjects. In addition, the nature of space exploration is such that the ethics considerations go well beyond the scientific experiments. Space missions have unique aspects such as a high cost, long duration and complexity of training, and a high national priority that may require sacrifices by humans to achieve the mission.

NASA has an institutional review board and an informed consent process. The reviewers scrutinize not only the experiment protocol but also the informed consent, all of which are then reviewed by the NASA ethicist who is assigned to the chief medical officer. Institutional review boards are made up of a diverse group of

Table 17.1 Medical and operational challenges to humans in space	Medical and operational challenges
	Visual impairment
	Exposure to ionizing radiation
	Bone density decrease
	Muscle atrophy
	Cardiovascular deconditioning
	Psychosocial impacts
	Circadian rhythm disruptions
	Vestibular dysfunction
	Hematological changes
	Immune dysfunction
	Delayed wound healing
	Gastrointestinal distress
	Orthostatic intolerance
	Fluid shifting
	Renal stones
	Nutrition

individuals including scientists, physicians, managers, ethicists, subjects, attorneys, and citizens from the community. In addition to a review of the involvement of human subjects, there is extensive peer review of the scientific merit, the medical applications, and the ethical and legal appropriateness of an experiment.

The expanse of the medical and operational challenges to individuals involved in space exploration is large (see Table 17.1). The participation of virtually every subdivision of human research and medicine is, therefore, required to mitigate the risks in space flight and exploration. Many investigators conduct ground-based operations seeking answers to these challenges as well as an opportunity to validate them in space. Once we understand the deleterious changes that occur in humans in microgravity, we formulate countermeasures in the forms of pharmaceuticals, training, operations, physical equipment, and selection criteria.

Most of our current space exploration activities take place in low Earth orbit for relatively short durations (approximately six months) on the International Space Station (ISS). The ISS is robustly equipped for astronauts to live and work while orbiting 250 miles above the surface of the Earth. There is constant communication, especially about health and mission status. In the event of an emergency, options are assessed, and, if needed, the crew could easily be returned to Earth in several hours to approximately a day (Stepaniak et al. 2001). Medical advice is available in a confidential conference venue. When required, the crew may be asked to perform an intervention. A physician is frequently on board and a health care team is available in Mission Control. The crew is trained to stabilize an impaired individual and prepare for a relatively short return journey to Earth. To date, the United States has not had to evacuate a crew member.

The Future

Space exploration beyond low Earth orbit poses increasingly formidable challenges to human physiology, the technology related to transportation, health, and habitation and, in particular, to our ethics. For example, where can we go, can we get back, and what are the prospects for thriving in the targeted destination? Increasingly, longer mission scenarios also raise a portfolio of ethical considerations, especially about a non-return mission. Will we, as a society, underwrite a mission for which we know there is no return? There may be a number of reasons for undertaking such a mission. For example, when faced with a substantial impending threat to civilization on Earth, we may choose to send a selected group of individuals on a colonizing mission to another location. What will be the selection criteria? Who will go? How many will go? Another possible example is a competitive one-way mission where the goal is to be the first country to claim a specific planetary environment. In these examples, mere authorization of the mission will not be enough—the decision will require extensive commitment from multiple agencies of the government and the public, along with consideration of the many difficult ethical questions.

There are some space exploration scenarios with the potential for return, such as colonization of the Moon, Mars, and some of the moons of Jupiter and Saturn. The most likely in the near future are Earth-to-Moon missions and colonization of the Moon. Afterward, there will be a greater incentive to go further into the solar system to the more distant planets and their moons. As the distance from Earth increases and the likelihood for return diminishes, ethical and other challenges become more formidable. First, the financial investment required for these missions is significant and will require the backing of the people remaining here on Earth. Political and philosophical discussions and debates will follow. Questions will arise regarding ownership, international participation, toxins or organisms returning with the explorers, and repatriation. There could be multigenerational missions committing generations yet unborn to a mission they have no knowledge of and have not personally affirmed. Knowledge of the destination is going to be a critical ethical consideration because we must have a convincing prospect for a meaningful life for the explorers. We will need detailed plans for managing health, logistics, and procurement of the energy requirements for life. Missions operating at great distances from Earth must be entirely autonomous, not only for resources, but also for governance and for accomplishing the mission.

For missions operating at even greater distances, the likelihood for return diminishes very rapidly as does the ability to access earthbound resources or rescue. Under these circumstances, the concept of informed consent and the ethical operation of an exploration party become more complex. The fidelity of embryogenesis in microgravity and hypogravity is unknown and, until it is known, participants on shorter missions should avoid incidental fertilization. Conversely, where the distances are so great that a multigenerational mission is inevitable, how do we ensure that the ensuing generations are capable and technologically prepared for the challenges to come?

Some of these ethical concerns are currently mitigated by more mundane factors. Energy determines what we can do no matter where we are. Modern propulsion systems do not enable travel outside the solar system—a trip to Alpha Centauri at 4.1 light years away would require 115,000 years. It is evident that going beyond the solar system requires the development of substantially greater propulsion capabilities. Another important energy consideration is the ability of space explorers to garner the energy required to sustain the exploration party at a predetermined destination; it is unlikely that we will transport our energy sources with us. It will be necessary to use in situ resources, first, for the generation of energy and, second, for other logistics that will be necessary to support the exploration party. As we get closer to solving the energy problems, we will have to develop a sophisticated ethical framework that accommodates both the explorers and the hosting environment.

Summary

Although some of the history of space medicine is inextricably linked to deplorable acts committed during the Third Reich, the preponderance of NASA's activities over the past 56 years has been exemplary and ethically sound, which in no way mitigates the questions surrounding the origin of space medicine. Unlike the space programs of most other countries, NASA is a *civilian* space agency, which may lessen the likelihood of future unethical behavior: Little if any of the non-military research and exploration is withheld from the public. Technologies that may compromise security or enable advanced military capacity to other countries are not as readily available to the public. Neither are the personal medical records nor technologies that are proprietor-protected as intellectual property. The United States has reason to hope that its space program will meet the remarkable ethical challenges as it explores the solar system and beyond.

Acknowledgments The author gratefully acknowledges the expert assistance of Ms. Carmella Mongiardo in the preparation of this manuscript. The author thanks Drs. Mauro Ferrari and Sheldon Rubenfeld for the opportunity to participate in the Human Subjects Research after the Holocaust Conference and the invitation to contribute to this book.

References

Campbell, M.R., and V.A. Harsch. 2013. The controversy of Hubertus Strughold during World War II. In *Life and work in the fields of space medicine*, 1st ed, ed. Hubertus Strughold, 96–118. Germany: Rethra Verlag.

Campbell, M.R., S.R. Mohler, V.A. Harsch, and D. Baisden. 2007. Hubertus Strughold: The "father of space medicine. *Aviation, Space, and Environmental Medicine* 78(7): 716–719.

Lagnado, L. 2012. A scientist's Nazi-era past haunts prestigious space prize. *Wall Street Journal*. http://online.wsj.com/news/articles/SB1000142405297020434940457810139387021883. Accessed 15 Jan 2013.

Space Medicine Association. 2013. Strughold award. http://spacemedicineassociation.org/strughold.htm. Accessed 22 Nov 2013.

Stepaniak, P., G.C. Hamilton, D. Stizza, R. Garrison, and D. Gerstner. 2001. Considerations for medical transport from the space station via an assured crew transport vehicle. NASA Technical Memorandum NASA/TM-2001-210198.

Chapter 18
Reproduction Then and Now: Learning from the Past

Tessa Chelouche

Over 100 years ago, on July 24, 1912, some 750 people from Great Britain, Europe, and the United States attended the First International Eugenics Conference in London. Prominent participants in the conference included Winston Churchill, First Lord of the British Admiralty, Sir William Osler, and Alexander Graham Bell. Leonard Darwin, son of Charles Darwin, presided over the conference, and in his opening remarks stated:

> Their [the conference attendants] first effort must be to establish such a moral code as would ensure that the welfare of the unborn should be kept in view in connection with all questions concerning both the marriage of the individual and the organization of the State. They should hope that the twentieth century would be known in future as the century when the eugenic ideal was accepted as part of the creed of civilization. (Darwin, p. 253)

A century elapsed, and in May 2012, at the annual meeting of the German Medical Association in Nuremberg, the German Medical Assembly made a long-awaited official declaration (Baader et al. 2012) in which they formally apologized for the atrocities that were committed by Nazi physicians:

> The crimes were not simply the acts of individual doctors, but rather took place with the substantial involvement of leading representatives of the medical association and medical specialist bodies, as well as with the considerable participation of eminent representatives of university medicine and renowned biomedical research facilities.
>
> These human rights violations perpetrated in the name of medicine under the Nazi regime continue to have repercussions to this day and raise questions concerning the way in which physicians perceive themselves, their professional behaviour, and medical ethics.
>
> We acknowledge the substantial responsibility of doctors for the medical crimes committed under the Nazi regime and regard these events as a warning for the present and the future. (as quoted in Livingston 2012, pp. 657–658)

Nazi doctors violated professional ethics and human rights, especially in the area of procreation and reproduction and, in particular, in the area of abortion. This

T. Chelouche, MD (✉)
Family Physician; Kupat Holim Clalit, Israel, Gan Yoshiya, Emek Hefer 38850, Israel

Rappaport Medical School, Technion, Haifa, Israel
e-mail: tessa.chelouche@gmail.com

S. Rubenfeld and S. Benedict (eds.), *Human Subjects Research after the Holocaust*,   225
DOI 10.1007/978-3-319-05702-6_18, © Springer International Publishing Switzerland 2014

chapter discusses abortion as a case study and examines how medical crimes during the Nazi period and the Holocaust have influenced reproductive human subjects research and policies in modern Israel and in Germany. The objective of this chapter is to demonstrate how historical events, in particular those of medicine during Nazi Germany, have influenced the current bioethical discourse and, as such, are still relevant to human subjects research today.

Eugenics and Reproductive Medical Practices in Nazi Germany

It is imperative to note that Nazi medicine was not just simply a part of the Nazi way of thinking. As the 2012 Nuremberg declaration of the German Medical Association stated: "In contrast to still widely accepted views, the initiative for the most serious human rights violations did not originate from the political authorities at the time, but rather from physicians themselves" (as quoted in Livingston 2012, p. 657).

The origins of the Nazi atrocities do not lie in the Nazi concentration camps; they can be traced to society's underlying social beliefs about human dignity (O'Mathúna 2006). The roots of Nazi reproduction and abortion policies culminating in the Holocaust took hold in the first decades of the twentieth century, particularly in the years following World War I. These policies evolved from Anglo-American eugenics and its German counterpart, racial hygiene, which presupposed a belief in the innate inequality of individuals and races. Eugenics was an internationally accepted movement that viewed individuals and groups in terms of their genetic "value" (Weiss 2004).

Eugenic principles became enmeshed with racial beliefs, which led to the enactment of involuntary sterilization laws in the United States, the first of their kind in the world. German eugenicists, who praised and promulgated the work of their fellow American eugenicists, formulated sterilization laws even before Hitler came to power. The eugenics discourse, as well as government economic priorities during the Great Depression, formed the crucial background to the passage in 1933 of the Nazi Sterilization Law, or the Law for the Prevention of Genetically Diseased Offspring (Bock 2004), which made sterilization compulsory for all people suffering from perceived inheritable diseases. Within three years, the Nazis sterilized approximately 200,000 people, almost ten times the number sterilized in the previous 30 years in America; and during the 12 years of the Nazi regime, approximately 400,000 people were forcibly sterilized (Kevles 2004). The Nazi's sterilization law was one of the first steps leading to the "euthanasia" of the mentally disabled, and eventually, to the Holocaust.

German physicians were also in the forefront of formulating, enacting, and implementing Nazi racial hygiene policies that promoted "positive eugenics" to encourage desirable births. One of the initial paradigms of the Nazi policy was the removal of women from the workplace so they could return to the home, where they were expected to have as many children as possible. Marriage loans were provided only after a medical examination of the couple for the existence of genetic diseases. Women received awards for large, healthy families. Strict restrictions were imposed on contraception for healthy German couples (Proctor 1988, p. 120). The Nazis were opposed to abortions as part of their social policy of rejuvenating the German nation. As early as 1931, even before Hitler came to power, they introduced legislation to imprison for racial treason anyone who attempted to curb artificially the natural fertility of the German people. Once Hitler became Chancellor in 1933, Nazi officials reaffirmed their opposition to any form of voluntary legalized abortion, which they perceived as a criminal offense and an act of sabotage against Germany's racial future. A doctor could be sentenced to prison for years if convicted of performing abortions. Nazi laws allowed abortion only in cases in which the mother's life was at risk, and Nazi physicians generally advised marriage as the solution to out-of-wedlock pregnancy (Proctor 1988, pp. 121–122; see also Grossmann 2002).

However, Nazi racial hygienists and policy makers encouraged abortion if it was in the interests of racial hygiene, and abortions were permitted for women deemed racially inferior. The key measure was the amendment to the Sterilization Law that allowed abortions for women slated to be sterilized (Proctor 1988, p. 123). In 1938, a Lüneberg court declared that abortion was legal for Jews (Proctor 1988, p. 123), and in 1943 Polish and other "Eastern" forced laborers in German factories were encouraged and even forced to obtain abortions. Abortion was not permitted for some of these women if they were deemed to be "of German or German related blood" or if they made a "good racial impression" (Proctor 1988, p. 123; see also Chelouche 2007), but in many thousands of cases, forced abortion was used to increase the slave labor force and to reduce the Slavic population (Hunt 2000).

Abortion in Germany and Israel

The differences in the legal, medical, and ethical aspects of abortion in Israel and in Germany today correspond to a large degree to the cultural, religious, political, and legal beliefs that are inherent in each society, and have been significantly influenced by both countries' historical past.

In Germany, abortion continues, to the present day, to be one of the issues that provokes many ethical conflicts in German society in general, and specifically among medical professionals. The 1995 German abortion law allows abortion only if the pregnancy endangers the mother's health physically or mentally, or if the pregnancy is the product of criminal action. This law does not include an embryopathic (congenital defect) clause (Hashiloni-Dolev 2007). One of the reasons proposed for the exclusion was that the embryopathic indication was a reminder of the Nazi eugenic practices, and thus unacceptable.

In Israel, the subject of abortion tends to remain outside the framework of public debate. It does, however, surface in the context of coalition negotiations between the larger, secular parties and the smaller, courted religious parties. In 1977, after many years of contentious debate, the Israeli Knesset legalized abortion under the following conditions: (1) the woman is below the age of marriage or above 40 years of age; (2) the pregnancy occurred as a result of rape, incest, or adultery; (3) the fetus is suspected of having a physical or mental defect; (4) the continued pregnancy might cause the woman physical or mental harm; or (5) the continued pregnancy might cause serious harm to the woman or her children because of difficult family or social problems. In 1979, under heavy pressure from the religious parties, the fifth clause was revoked. In Israel, the abortion discourse does not refer to the Nazi eugenic practices, even though there has been criticism by some Israeli scholars that Israel is in fact a pro-eugenic society because the Israeli abortion law enables abortions for embryopathic reasons even when the severity or probability of expression of the condition are mild or unlikely (Hashiloni-Dolev 2006, 2007; Rimon-Zarfaty and Jotkowitz 2012).

IVF, PGD, and Embryonic Stem Cell Research

Due to the fact that abortion is a procedure involving an embryo (or fetus), the issues surrounding the debate on abortion are inevitably linked to reproductive human subjects research and assisted reproductive technologies. Subsequently, the legal, religious, medical, and cultural views and policies on abortion have asserted a great influence on policies concerning related reproductive research and treatment issues, such as in vitro fertilization (IVF), preimplantation genetic diagnosis (PGD), and embryonic stem cell research.

IVF assists infertile couples in achieving pregnancy by removing eggs from a woman's ovary, fertilizing them with sperm in a laboratory, and then returning the fertilized eggs or embryos to the woman's uterus. In PGD, one or more cells are removed from the embryos created in vitro, tested for certain harmful, genetic conditions, and then one or more unaffected embryos are implanted in a woman's womb for pregnancy and birth. The affected embryos are discarded.

In 1998, a team of scientists from the United States and Israel launched a new paradigm for understanding and treating disease when they succeeded in the laboratory culture of human embryonic stem cells (Thompson et al. 1998). Embryonic stem cells arise in the inner cell mass that forms the blastocyst stage of preimplantation embryonic development. Embryonic stem cells are pluripotent: They are capable of forming all the tissues in the body, and as such could be used to develop into specialized tissues that might replace diseased tissues or organs. Such uses of human embryonic stem cells may prove useful in research in regenerative medicine; however, many of the clinical applications are still in the experimental stage and much research needs to be performed if they are to become a medical reality (Revel n.d.).

German Policies

The German Federal Constitution Court's decision on abortion addressed the termination of pregnancy and applied only to embryos implanted in the uterus—it did not bear on the status of embryos created in the laboratory by IVF. In the 1980s, as IVF became more widely used in Germany, controversy arose over the legal and moral status of the created embryos and the need for regulation of the IVF procedures that created them. The German public was accepting of a certain level of women's autonomy in their abortion policy but unwilling for their scientists to destroy or manipulate embryos in their research (Brown 2004). In the German public's mind, embryo research raised the specter of Nazi eugenics and dehumanizing experimentation (Brown 2004). These concerns led to the 1990 German Embryo Protection Act, which strictly regulated the creation and disposition of in vitro embryos. This legislation, passed many years prior to the development of embryo stem cell research, was, in effect, an infertility law. It was the result of a strong public movement to protect even early human embryos, which were perceived as having basic rights to life, dignity, and physical integrity as laid down in the German Basic Law (Faltus 2010). The Embryo Protection Act allows for the creation of embryos for IVF, but makes it a criminal offense to fertilize more eggs than can be transferred in one cycle; to transfer more than three embryos; to engage in egg donation and gestational surrogacy; and to create embryos for research or nonmedical sex selection (German Act 1990). The law applies only to embryos created in vitro, and considers every stage in the embryonic developmental process as a life that should be protected. The central provision is the prohibition of producing human embryos for purposes other than for reproductive medical treatments (Faltus 2010). The regulations for in vivo embryos are laid down within sections of the abortion law. Due to the restrictions on embryo research that reflected Germany's fear of reliving its past, medical research on embryonic stem cells, which was already being performed in many parts of the world, was completely halted in Germany (Brown 2004). Germany's concerns with their history are evident in a 2001 speech in Berlin by the German president, Johannes Rau. He addressed the controversial procedure and urged his fellow citizens to exercise great care and moral restraint to ensure that the coming genetic age brings progress "benefitting humanity." Rau told the nation:

> It does not take a believing Christian to understand that new forms of genetic manipulation and control would run contrary to the conditions of human freedom and human dignity. Eugenics, euthanasia and selection—these are words with terrible connotations in Germany... In the biogenetic future, Germans should never forget the sins of the past. (as quoted in Brown 2004, p. 38)

In effect, the Embryo Protection Act rendered PGD illegal in Germany for many decades because of its ban on the use of in vitro embryos for medical research. So the German situation can be viewed as contradictory in many aspects. On the one hand, there is wide acceptance of a quite liberal position on abortion, and on the other, the human rights of the embryo are vigorously protected by the Embryo Protection Act.

In 2000, the German Medical Association published a proposal of guidelines for PGD. This gave rise to a vigorous public and political debate, and once again the opponents of PGD stimulated fear of the recurrence of Nazi German eugenics (Brown 2004) using language that recalled the horrors of the Nazi biomedical programs. One statement said:

> During the period of German fascism, the search for the perfectly functioning human being quickly killed those who did not correspond to the bodily and mental ideal. This highly functional and achieving human being is today more than ever in demand. In this context the resurgence of eugenics which we can observe today is remarkable. (Brown 2004, p. 45)

In 2001, after many years of intense debate, the German parliament passed a new law allowing PGD in very limited circumstances. Under this new law, PGD can be used to screen embryos only if the parents have a high predisposition to a serious genetic illness. An ethics panel must review all applications for PGD, and couples are required to undergo counseling. The bill outlines only exceptions to the part of the 1990 Embryo Protection Act that bans PGD; otherwise, it remains intact ("Controversial Genetic Tests," n.d.). Embryonic stem cell research is also limited in Germany because of the same moral controversies over the status of preimplantation human embryos.

As described above, these issues were dealt with very strictly, and Germany, until 2001, was almost alone among the developed countries in imposing a near total prohibition on the use of embryos in research (Robertson 2004). However, in 2001 there was extensive public debate when the German Research Foundation sought to import embryonic stem cells for their research in Germany. There were no provisions in the Embryo Protection Act covering the importation of embryonic stem cells for research. German Chancellor Schroeder argued in favor of importation. He said that it was constitutionally impermissible to limit biogenetic research and that the ethics of healing warranted as much attention as the ethics of creation. He warned about the German cultural tendency to view these issues with "ideological blinders and taboos" (as quoted in Brown 2004, p. 47). President Rau, who was alive during the Nazi regime, responded:

> Those who begin to instrumentalize human life, to differentiate between worthy of life and unworthy of life, are on a runaway train...No one should forget what happened in academic and research fields during that period...An uncontrolled scientific community researched for the sake of its scientific aims without any moral scruples...This memory is a perpetual appeal: nothing must be given precedence over the dignity of the individual. (as quoted in Brown 2004, p. 47)

In 2002, despite opposition, the German Parliament amended the 1990 Embryo Protection Act by the Stem Cell Act, which regulates the import of, and research with, imported embryonic stem cells. This made it legally permissible for stem cells to be imported for research as long as the requirements in the Stem Cell Act were met: only those embryonic stem cells derived before May 1, 2007, may be imported into Germany for research; these stem cells must be derived in accordance with the laws of the country in which they were derived; they must be derived free of charge; and they must be obtained only from embryos that were created in infertility treatments but could not be implanted. Human embryonic stem cells that are created solely for research purposes in other countries are not permitted to be imported into Germany (Faltus 2010).

In the context of human subjects research after the Holocaust, we can see that the German medical profession continues to invoke its Nazi past, which has led to very strict laws and guidelines that limit reproductive medical research in Germany as compared to the rest of the world.

Research on human embryonic stem cells has stirred German society and politics (Wiedemann et al. 2004). Among the proponents of a graduated protection of life, further debates focus on the balance between the freedom of research and the potential of future therapeutics, on the one hand, and the protection of the embryo, on the other. In this context, two questions are of acute concern. How can the change in the status of embryos in different contexts, such as stem-cell research and induced abortion, be justified? The second question concerns the perceived moral double standard behind allowing the import of human embryo stem cells while prohibiting their production in Germany (Faltus 2010). Ironically, Israel, among other countries, provides Germany with human embryonic stem cells for research purposes (Wiedemann et al. 2004).

Today, freedom of research has an important role in German academic and scientific circles, and promises of potential treatments are highly relevant to the present German public debate. The German debate is "fundamentally harsh" because of the potent combination of the "undividable human dignity of the embryo argument" with the argument linking PGD and embryo research to the eugenic measures of the Nazi Germany (Krones and Richter 2004, p. 629). However, attitudes might be changing. In a recent study to determine German stem cell research experts' opinions on the future of this research in Germany, the majority expressed an increasing open-mindedness towards embryonic stem cell research, as well as a relaxation of legal restrictions (Wiedemann et al. 2004).

Israeli Policies

The situation is different in Israel. In effect, Israel is a country that is constantly trying to apply the ancient Jewish tradition to contemporary life (Jotkowitz and Glick 2009). While the majority of the population is secular, some religious values, particularly Jewish bioethical values, are incorporated into Israeli society. Consequently, Israeli bioethics "manifests a unique mix of [Jewish] orthodoxy and secularism, of communal paternalism and assertive individualism, of proscription and permissiveness, of religious norms and liberal ethical values" (Shapira in press).

Traditional Jewish law (Halakhah) draws a clear distinction between the preimplantation and the implanted embryo. An in vitro preimplantation embryo with potential for implantation may neither be destroyed nor used for scientific research unless it is done for the purpose of saving life. An in vitro preimplantation embryo without the potential for implantation may be subjected to research. The creation of an embryo for research purposes may only be allowed if the contemplated research might facilitate the saving of human life (Shapira in press).

The Israeli legal system balances secular and religious concerns, and there is cooperation between the factions on reproductive technology and stem cell research. The Jewish tradition generally favors these new treatment modalities and for this reason Israel has more IVF fertilization centers per capita than any other country (Jotkowitz and Glick 2009). IVF was introduced to Israel in 1981, and the first IVF pregnancy occurred in 1982. Israel broadly accepts IVF as morally and socially legitimate and has not questioned the procedure in any significant way (Birenbaum-Carmeli 2004). According to the 1987 Israeli IVF law, every woman, irrespective of her marital status, is entitled to government funding for IVF treatment cycles until the live births of two children (Levush 2012).

PGD has not yet been addressed by the Israeli legislature (Shapira in press). However, in 2001 the Bioethics Advisory Committee of the Israel National Academy of Science and Humanities did address PGD in a report on human research on embryonic stem cells and cloning. This committee acknowledged the usefulness of PGD, and rendered it an acceptable, even welcome, technique with specific and expanding applications for current clinical practice (Shapira in press). The Israeli approach is reflected by the Jewish stance, which places a higher value on the health and life of a pregnant mother than on the moral status of the embryo, up to the moment of birth. This is the ethical stance that has been the basis for authorizing IVF and PGD in Israel (Revel 2003).

The Israeli approach to human embryo stem cell research is also based on this Jewish stance. Research on embryo stem cells is permissible within the framework of IVF treatments, using supernumerary embryos that have been donated (not bought or sold) for the purpose of therapeutic research. The Israeli bioethics report states that there must be regulations to insure free and informed consent for the donation and respect of the donor's human dignity, autonomy, and liberty. The regulations should also protect the rights of parents who find embryo research unacceptable (Revel 2003).

From the legal point of view, Israel has a Law of Prohibition of Genetic Intervention that prohibits germ-line gene modification and reproductive cloning in humans (Prainsack 2006; see also Revel 2005b). The Law, adopted in 1999, and renewed by the Parliament in 2004 for five years, specifies that what is prohibited is the implantation of a cloned embryo in a uterus for reproductive purposes. The Law requires a committee to report on potential new therapeutic benefits or dangers in assisted reproductive and related technologies (Prainsack 2006; see also Revel 2005b).

Because of its reproductive policies Israeli has one of the highest birth and fertility rates in the world (Sperling 2010), and is one of the leading countries in the field of embryonic stem cell research.

Summary and Conclusion

As one enters the German Hygiene Museum in Dresden, one confronts a large life-size glass statue of a human man, The Transparent Human, which has been on display there since 1930. The museum was founded in 1912 by August Lingner, a firm believer in social education on hygiene and disease prevention, who stated: "Every person must acquire unconditional trust in the science that is recognized by the state, and become convinced that this science alone brings the certainty of a cure for diseases" (as quoted in S. Sperling 2013, p. 123). Models of the statue were sent to museums around the world, including the United States where they served to promulgate eugenic ideas. When the Nazis seized power, the Hygiene Museum became a vehicle of transmitting their ideology of racial hygiene and of the perfect Nordic ideal. The aims of science education and disease prevention were reformulated to encompass the Nazi eugenic aims (Kass 2010). The statue is a model of human perfection.

It is probably safe to say that the modern reproductive and genetic era is advancing with the aim of improving human well-being and not with the purpose of creating genetically superior human beings or perfect ones. New technologies and research bring with them the hope of enhancing human reproductive freedoms, not restricting or controlling them. These new technologies touch all aspects of human life, from beginning through middle to end. But at the same time, this area of human subjects research poses many fundamental ethical and moral dilemmas for modern researchers. Whatever opinions are held on these issues, and whatever semantics are used, it should be noted that physicians have been the most significant force behind these new reproductive technologies (Shapira 1987). Dr. Leo Alexander, the chief American medical consultant at the Nuremberg Doctors' Trial, published his thoughts on the Nazi medicine in the *New England Journal of Medicine* in 1949: "Whatever proportions these crimes finally assumed, it became evident to all who investigated them that they had started from small beginnings. The beginnings at first were merely a subtle shift in emphasis in the basic attitude of the physicians" (p. 45).

In 2001, Germany's president, Johannes Rau, entered the debate about genetic research, particularly PGD, and argued that it was not just Germans who should keep the excesses of the Nazi past in mind: "What is permitted and what is not permitted in fundamental ethical issues does not change from nation to nation, even if some nations stand as a special reminder of both man's capacity for evil and man's inherent dignity" (as quoted in Brown 2004).

This statement reflects the situation today. Both in Germany and Israel, there are marked differences of opinion on abortion and reproductive research, and each society has developed its own regulations of these controversial areas of treatment and research while simultaneously protecting the dignity, equality, and liberty of human beings. These developments are in line with the pluralistic approach that has been adopted by the UNESCO International Committee for Bioethics in its report on "The Use of Embryonic Stem cells in Therapeutic Research: Ethical Aspects of Human Embryonic Stem Cell Research," which states:

> Every society has the right and duty to debate and decide upon ethical issues with which it is confronted . . . The use of human embryos for deriving stem cells would appear to be one such issue . . . Human embryonic stem cell research and embryo research in general, is a matter in which each community, or state, will have to decide itself. (Smith and Revel 2001, p. 13)

The acceptance of pluralism in ethics guarantees coexistence and mutual understanding in international research on reproductive technologies. In addition, pluralism in ethics and human subjects research requires lucid definitions and boundaries, and should always encompass respect for human dignity and fundamental freedoms. Every society should be aware that "the Nazi doctors violated human dignity by discarding the very principles that are the essence of bioethics" (Revel n.d., Section E, para. 3).

The Human Genome Project hopes to map genetic disorders, eventually leading to their effective treatment, while avoiding eugenic practices, especially in PGD (K. L. Garver and B. Garver 1994). While some view modern human subjects research in reproduction as eugenics (Kass 2010), it should be emphasized that there is an essential distinction between the coercive eugenic state policies of Nazi Germany and present day human subjects research and reproductive treatments. Modern research and practice in assisted reproductive technologies and genetics "can remain free of eugenic danger as long as it is only the individual concerned (the mother) who makes a free and informed decision about procreation" (Revel 2005a, "Genetic Testing and Genetic Counseling," para. 2).

The Holocaust is a cautionary tale from the past that can teach medical professionals about human dignity and respect for human life in the future. With only cursory examination, Nazi reproductive medicine appears to have been unique, but there is a growing consensus that the lessons of Nazi medical crimes are universal and important for every generation. Simply mentioning the word "Nazi" in current debates on reproduction has become a universal warning of potential evil lurking in human subjects research. While Nazi eugenics is too often used as a superficial rhetorical tool on all sides of these debates, simplistic allusions to the Nazi past are not enough, and, in fact, they can be misleading. If the past is to be used in current debates on human subjects research, as I believe it should, then a true in-depth interdisciplinary dialogue needs to be initiated. Studying and teaching medicine and the Holocaust challenge us to search ourselves and our cultures for the means to prevent unethical behavior in medical treatment and human subjects research.

References

Alexander, L. 1949. Medical science under dictatorship. *New England Journal of Medicine* 241 (2): 39–47.

Baader, E., et al. 2012. In remembrance of the victims of Nazi medicine. *Israel Medical Association Journal* 14(9): 529–530.

Bioethics Advisory Committee. 2001. *The use of embryonic cells for therapeutic research.* [Report]. Israel Academy of Sciences and Humanities. http://bioethics.academy.ac.il/english/main-e.html. Accessed 13 Dec 2013.

Birenbaum-Carmeli, D. 2004. Cheaper than a newcomer: On the social production of IVF policy in Israel. *Sociology of Health & Sickness* 26(7): 897–924.

Bock, G. 2004. Nazi sterilization and reproductive policies. In *Deadly medicine: Creating the Master Race*, ed. D. Kuntz and S. Bachrach, 61–87. Washington, DC: U.S. Holocaust Memorial Museum.

Brown, E.B. 2004. The dilemmas of German bioethics. *The New Atlantis* 5(Spring): 37–53.

Chelouche, T. 2007. Doctors, pregnancy, childbirth and abortion during the Third Reich. *Israel Medical Association Journal* 9(3): 202–206.

Controversial genetic tests: German Parliament allows some embryo screening. n.d. Spiegel Online International. http://www.spiegel.de/international/germany/controversial-genetic-tests-german-parliament-allows-some-embryo-screening-a-773054.html. Accessed 13 Dec 2013.

Darwin, L. 1912, July 24. First International Eugenics Conference. [Opening remarks] *British Medical Journal*, 2(2692): 253–255.

Faltus, T. 2010. *The German stem cell law – A balancing act between the protection of human embryos and the freedom of research.* (World Stem Cell Report, 2010). Genetic Policy Institute.

Garver, K.L., and B. Garver. 1994. The human genome project and eugenic concerns. *The American Journal of Human Genetics* 54(1): 148–158.

German Act for Protection of Embryos (*Embryonenschutzgesetz*). 1990. *Federal Law Gazette*, Part 1, No. 69

Grossmann, A. 2002. The debate that will not end: The politics of abortion in Germany from Weimar to National Socialism and the postwar period. In *Medicine and modernity: Public health and medical care in nineteenth- and twentieth-century Germany*, ed. M. Berg and G. Cocks, 193–212. Cambridge: Cambridge University Press.

Hashiloni-Dolev, Y. 2006. Between mothers, fetuses and society: Reproductive genetics in the Israeli-Jewish context. *Nashim: A Journal of Jewish Women's Studies & Gender Issues* 12(1): 129–150. doi:10.1353/nsh.2006.0021.

————. 2007. *A life (un)worthy of living: Reproductive genetics in Israel and Germany.* Dordrecht, The Netherlands: Springer.

Hunt, J. 2000. Out of respect for life: Nazi abortion policy in the eastern occupied territories. In *Life and learning: Proceedings of the ninth university faculty for life conference*, ed. J.W. Koterski, 295–304. Washington, DC: University Faculty For Life.

Jotkowitz, A.B., and S. Glick. 2009. Navigating the chasm between religious and secular perspectives in modern bioethics. *Journal of Medical Ethics* 35(6): 357–360.

Kass, L.R. 2010. A more perfect human: The promise and the peril of modern science. In *Medicine after the Holocaust: From the Master Race to the human genome and beyond*, ed. S. Rubenfeld, 107–122. New York: Palgrave Macmillan.

Kevles, D.J. 2004. International eugenics. In *Deadly medicine: Creating the Master Race*, ed. D. Kuntz and S. Bachrach, 41–59. Washington, DC: U.S. Holocaust Memorial Museum.

Krones, T., and G. Richter. 2004. Preimplantation genetic diagnosis (PGD): European perspectives and the German situation. *Journal of Medicine and Philosophy: A Forum for Bioethics and Philosophy of Medicine* 29(5): 623–640.

Levush, R. 2012. Reproduction and abortion: Law and policy. *Law Library of Congress*, Israel. http://www.loc.gov/law/help/israel_2012-007460_IL_FINAL.pdf. Accessed 13 Dec 2013.

Livingston, E.H. 2012. German medical group: Apology for Nazi physicians' actions, warning for future. *Journal of the American Medical Association* 308(7): 657–658.

O'Mathúna, D.P. 2006. Human dignity in the Nazi era: Implications for contemporary bioethics. *BMC Medical Ethics* 7: 2. doi:10.1186/1472-6939-7-2.

Prainsack, B. 2006. 'Negotiating life': The regulation of human cloning and embryonic stem cell research in Israel. *Social Studies of Science* 36(2): 173–205. doi:10.1177/0306312706053348.

Proctor, R.N. 1988. *Racial hygiene: Medicine under the Nazis.* Cambridge, MA: Harvard University Press.

Revel, M. n.d. Why take cultural and religious diversity into account in bioethical debates? The case of human embryonic stem cells. [Online article]. http://98.131.138.124/articles/BrachaLeAvraham/revel.asp. Accessed 13 Dec 2013.

———. 2003. Human reproductive cloning, embryo stem cells and germline gene intervention: An Israeli perspective. Israel Academy of Sciences and Humanities. Bioethics Advisory Committee website. http://bioethics.academy.ac.il/english/articles/bioethics_revel.htm. Accessed 13 Dec 2013.

———. 2005a. Limits of genetic determinism in medicine and behavioural sciences. Israel National Academy of Sciences and Humanities. Bioethics Advisory Committee website. http://bioethics.academy.ac.il/english/articles/limits.htm. Accessed 13 Dec 2013.

———. 2005b. Research on human embryonic stem cells and cloning for stem cells. The Israel Academy of Sciences and Humanities. Bioethics Advisory Committee. Bioethics Advisory Committee website. http://bioethics.academy.ac.il/english/articles/ALLEA_HuEScellsJuly05.htm. Accessed 13 Dec 2013.

Rimon-Zarfaty, N., and A. Jotkowitz. 2012. The Israeli abortion committees' process of decision making: An ethical analysis. *Journal of Medical Ethics* 38(1): 26–30. doi:10.1136/jme.2009.032797.

Robertson, J.A. 2004. Reproductive technology in Germany and the United States: An essay in comparative law and bioethics. *Columbia Journal of Transnational Law* 43(1): 189–227.

Shapira, A. 1987. Reproductive technology: In Israel, law, religious orthodoxy, and reproductive technologies. *Hastings Center Report* 17(3): 12–14.

———. The status of the extra-corporeal embryo. Country report: Israel. http://bioethics.academy.ac.il/english/articles/The_Status_of_the_Extra-corporel_Embryo.htm. Accessed 13 Dec 2013.

Smith, A.M., and M. Revel. 2001, April 6. *The use of embryonic stem cells in therapeutic research: Ethical aspects of human embryonic stem cell research.* Report of the International Bioethics Committee on the Ethical Aspects of Human Embryonic Stem Cell Research. Paris: UNESCO. http://www.eubios.info/UNESCO/ibc2001sc.pdf. Accessed 13 Dec 2013.

Sperling, D. 2010. Commanding the "be fruitful and multiply" directive: Reproductive ethics, law, and policy in Israel. *Cambridge Quarterly of Healthcare Ethics* 19(3): 363–371. http://dx.doi.org/10.1017/S0963180110000149.

Sperling, S. 2013. *Reasons of conscience: The bioethics debate in Germany.* Chicago, IL: University of Chicago Press.

Thompson, J.A., J. Itskovitz-Eldor, S.S. Shapiro, M.A. Waknitz, J.J. Swiergiel, V.S. Marshall, and J.M. Jones. 1998. Embryonic stem cells derived from human blastocytes. *Science* 282(5391): 1145–1147. doi:10.1126/science.282.5391.1145.

Weiss, S.F. 2004. German eugenics, 1890–1933. In *Deadly medicine: Creating the Master Race,* ed. D. Kuntz and S. Bachrach, 15–39. Washington, DC: U.S. Holocaust Memorial Museum.

Wiedemann, P.M., J. Simon, S. Schicktanz, and C. Tannert. 2004. The future of stem-cell research in Germany. A Delphi study. *EMBO (European Molecular Biology Organization) Reports* 5 (10): 927–931. doi:10.1038/sj.embor.7400266.

Chapter 19
Promoting Clinical Research and Avoiding Bad Medicine: A Clinical Research Curriculum

Roy S. Weiner and Brian J. Weimer

In 1996, Congress and the National Institutes of Health (NIH) recognized the critical decline of investigator initiated clinical research in American academic medical centers (Atkinson et al. 2012). They observed, for example, that acceptance rates for papers in American medical journals had fallen to 15 percent while foreign submissions had tripled. In 1999, in response to signs that American dominance in therapeutic innovation was being threatened, the National Heart Lung and Blood Institute (NHLBI) and the National Center for Research Resources funded Clinical Research Curriculum Awards (K-30 Awards) at 35 academic institutions to support the development of curricula to prepare physicians and other clinically oriented scientists for careers in clinical research (National Institutes of Health 1999). The K-30 scholars would then compete for individual research training funds to build skills and perspective that would position them for sustainable clinical research careers. The hope was that current and future trainees would fill the pipeline to populate academia and industry and, thus, forestall and even reverse the outward migration of therapeutic innovation. Tulane University was among those institutions funded by NHLBI in 1999 (Clinical Research Curriculum Award, Grant Number 5K30 HL 04521-05) to develop an academic program to prepare aspiring clinical researchers for productive, effective, and successful careers.

The target populations of trainees were junior faculty and postdoctoral fellows from clinical and clinically oriented disciplines. They already had earned their MDs and/or PhDs and had various amounts of specialty and subspecialty training. As an indication of the commitment of both the trainees and the institution, they were

R.S. Weiner, MD (✉)
Associate Dean for Clinical Research and Training; Medicine – Hematology/Oncology, Tulane University, 1430 Tulane Avenue, SL68, New Orleans, LA 70112, USA
e-mail: rweiner@tulane.edu

B.J. Weimer, JD
Director of Compliance, Office of the Vice President for Research, 1440 Canal Street, Suite 2425 TW-5, New Orleans, LA 70112, USA
e-mail: bweimer1@tulane.edu

S. Rubenfeld and S. Benedict (eds.), *Human Subjects Research after the Holocaust*,   237
DOI 10.1007/978-3-319-05702-6_19, © Springer International Publishing Switzerland 2014

prepared to devote 75 percent of their time and effort to their training. These trainees were born, by and large, after 1960—almost a generation after the start of WWII but before the Declaration of Helsinki in 1964 (World Medical Association General Assembly [WMA] 1964). Thus, our scholars would have matured in parallel with the declaration but would not have had primary knowledge of the history that propelled the evolution of the declaration and today's concepts of medical research ethics. Among the challenges that the curriculum had to meet— in addition to teaching the quantitative tools of research—were the precepts of what questions are appropriate to explore, how to pursue answers to research questions, and how to involve human subjects in clinical research. The quantitative knowledge base that clinical researchers need is fairly well-defined, and learning the subject matter is demanding. The spectrum of skills that our scholars were expected to acquire included biostatistics, epidemiology, protocol design, molecular mechanisms of disease, personnel management, budgeting and finance, grant writing, and more.

The relationship of ethics and human subjects research includes a didactic core that has evolved over the past 15 years at least as much as the subject matter in molecular pathobiology. Initially, our curriculum was guided by the Institutional Review Board (IRB) requirements for protocols at Tulane University. Our lectures to the scholars familiarized them with the Nuremberg Code (US Department of Health & Human Services 1949), the Declaration of Helsinki (WMA 1964–2013), the Belmont Report (National Commission for the Protection of Human Subjects of Biomedical and Behavioral Research 1979), and the Common Rule (US Department of Health & Human Services 1991). We added the Clinical Trial Volunteer's Bill of Rights (Getz and Borfitz 2002) to our core lecture content. We also added a mandatory online Collaborative Institutions Training Initiative (CITI) Certification (2000) to the curriculum. Tulane University took legitimate pride in going beyond compliance in offering and promoting didactic course material covering the ethical conduct of human research.

Tulane University now offers a seminar series each fall on the Responsible Conduct of Clinical Research and encourages all faculty, staff, and students who conduct or who are interested in conducting human subjects research to participate. Additional lectures facilitating accessibility to instructional material are offered during the spring. The chair of Tulane University's Biomedical IRB offers seminars throughout the year for clinical research coordinators, clinical research nurses, and clinical research assistants. These staff members are, after all, the individuals who in many cases identify clinical research subjects and obtain informed consent from participants. The non-physician research staff often includes the individuals who communicate with human research participants most frequently, and many times on a more intimate level than the physician/principal investigator. Involving staff in formal education and embracing them as part of the vigilant research team is necessary and valuable.

Although the training program in clinical research is focused on the senior postdoctoral fellows and junior faculty members, it cannot be isolated from the entire academic medical environment. In fact, our scholars are encouraged to involve medical students, residents, and fellows in research promoting the model of tiered mentorship (Fowler and Muckert 2004). The Tulane School of Medicine introduces clinical research ethics in the first year of medical school through the course entitled Foundations in Medicine. The content of this course regarding past atrocities committed under the guise of medical research has been greatly influenced and expanded by the US Holocaust Memorial Museum's (USHMM's) exhibit, Deadly Medicine: Creating the Master Race, hosted at the National WWII Museum in New Orleans (discussed in more detail below). The Foundations of Medicine course now begins with the study of atrocities committed in Nazi Germany during World War II. The course also discusses in detail the US Public Health Service Syphilis Study at Tuskegee (Centers for Disease Control and Prevention [CDC]), eugenic sterilizations in Indiana and North Carolina (Kaelber 2012), and more recent atrocities such as the sterilizations that occurred in California State Prisons (Johnson 2013). As stated by Fr. Donald Owens, one of the instructors of the medical student course and medical school chaplain:

> The course is designed and intended to put at the forefront of our medical students' minds that eugenics is unfortunately alive and present today. Eugenics is not a relic of the past, and our medical students must be ever vigilant of its continued existence and its clear violation of medical ethics. (personal communication 2012)

Such rich didactic instruction for medical students was lacking in the past. Jacques L. Courseault, MD, a student at Tulane School of Medicine from August of 2005 through May of 2010, noted that the medical research atrocities that occurred during the Holocaust and the Tuskegee Syphilis Study were never mentioned during his years as a medical student. The current Foundations in Medicine curriculum is a clear improvement, bringing history into the present and casting it forward into the future.

Our neighboring medical school, and often our collaborator in clinical research, Louisiana State University, also offers training on atrocities committed under the guise of medical research. Two courses are provided (Introduction to Medical Ethics and Legal and Ethical Issues in Clinical Medicine), which cover topics ranging from historical abuses in Tuskegee and Guatemala to modern abuses, such as the unauthorized removal of over 200 blood samples from the Havasupai Indian tribe of Arizona by an Arizona State University researcher in 1990 (Harmon 2010). These blood samples were the basis of unauthorized research on the genetic basis of schizophrenia and consequences of inbreeding.

Training in animal use was likewise introduced and promoted as a natural extension of research ethics for preclinical and translational researchers. Training on the responsible use of animals in medical research is included in Tulane University's annual fall Responsible Conduct of Research Seminar Series. Online training in the responsible use of animals in research is offered through the Tulane University's subscription to the CITI training certificate.

The faculty of the Master of Science in Clinical Research (MSCR) degree program at Tulane University, while part of the institution's commitment to training in the ethics of medicine and medical research, realized its leadership role in clinical research training as part of the rebuilding after Hurricane Katrina. First, we evolved from a Master's in Public Health (MPH) in the School of Public Health and Tropical Medicine to a Master's of Science (MS) in Clinical Research in the Biomedical Sciences Division of the School of Medicine. This change gave us more freedom in the curriculum and brought us closer to both graduate programs in the biomedical sciences and to other physicians and scientists whose research involved living human subjects and patients' specimens. Our trainees shared concerns with a broader group of scholars—protection of identifiable data, risk–benefit considerations, combining the roles of physician and researcher, patient empowerment, children and other vulnerable populations, and assuring access to clinical research. Our courses were open to scholars who were not candidates for the MS degree. We introduced cases on the proper conduct of research in our course. These cases were, initially, focused on adulterated data, responsibility for and supervision of research personnel, the approach to the subject who elects to withdraw from a protocol, the ethical mandate to publish valid results irrespective of outcome, the physician's role as a researcher, and the researcher's role as a physician. Discussions were lively, evidence-based where possible, and well received. The MS program in Clinical Research was well populated and well funded, and its message was reinforced by the success of previous graduates on their way to sustainable careers in clinical research.

A dramatic and positive change occurred when Tulane University cosponsored the USHMM's exhibit, Deadly Medicine: Creating the Master Race at the National WWII Museum in New Orleans from July 25 through October 15, 2012. This exhibit uses official Nazi records, photographs, artifacts, and first-person narratives to chronicle and document Nazi Germany's campaign to cleanse Germany of people who were thought to compromise the nation's health (see Kuntz 2008). During this era, medical geneticists, psychiatrists, physicians, anthropologists, and other members of the scientific community formulated policies that defined racial health and how to achieve it. Variances from the racial ideal were attributed to hereditary disease, which, in turn, led to the Nazi policy of eliminating the gene pool that compromised the racial ideal. Many of the professional collaborators were clinicians and scientists of excellent repute and international recognition—but not all. Gerhard Rose, for instance, was a productive scientist with a responsible academic position studying tropical diseases. He joined the military in response to a very real threat to German soldiers from infectious diseases, and then infected unwilling and uninformed prisoners with tropical diseases to develop effective vaccines (see Chap. 16 in this book). Josef Mengele, with doctorates in both anthropology and medicine, carried out horrific experiments on young twins contributing nothing of scientific value (see Chap. 2 in this book). While Rose served six years of the life sentence he received at the Nuremberg Doctors' Trial, Mengele was never brought to justice and died of a stroke while swimming in Brazil in 1979 (USHMM, n.d.).

This climate of loyalty to government and military priorities spawned the national program of sterilization of those viewed as unfit, which was, in fact, an extension of a multinational infatuation with eugenics (Kaelber 2012) that preceded the Nazi era; the logic of the eugenic sterilization policy was ultimately extended to rationalize the annihilation of people who were perceived to dilute or contaminate the Master Race. The exhibit traced this policy of creating a Master Race from the global interest in eugenics in the early twentieth century to the Nazi's pursuit and development of the "science of race." Our MSCR scholars contrasted the researchers' often sophisticated, disciplined experimental design directed at serving critical needs of government with their undisciplined, unethical, opportunistic subject selection. The exhibit showed both.

The exhibit also gave us an opportunity to recast the ethics instruction for our trainees. Rather than basing the scholars' didactic experience entirely on learning and understanding rules and codes of conduct in existence today, we had an opportunity to see and learn about world leaders in science and medicine who, in the name of science and medical advancement, perpetrated atrocities. We had an opportunity to ask why. We also had an opportunity to consider how our American culture processed knowledge of and formulated responses to the medical atrocities of WWII. More importantly, we had an opportunity to reflect on how we might cope with the advances in technology and resist the unknown temptations that will challenge our ethical principles as physicians and scientists today and in the future.

Beginning in September of 2012, we seized upon these opportunities and incorporated as much material as possible into our curriculum, including:

(a) Darwin, Galton, American eugenics, and German racial hygiene;
(b) Beneficence, better good, nonmaleficence, human rights, empowerment, and respect;
(c) Justice, distributive justice, access;
(d) Managing ill-gotten gains in science;
(e) Hypothetical ethical challenges to come and potential coping strategies.

We arranged for docent tours of the exhibit for our scholars and held seminars afterward. The rich roster of visiting thinkers associated with the exhibit presented our scholars a wealth of thought-provoking lectures. Among the highlights of this series were a book discussion and viewing of *Sophie's Choice*; Arthur Caplan's lecture, "Justifying the Unthinkable: The 'Ethics' of Nazi Medical Experimentation," which he graciously gave his permission to record and archive for use in our curriculum going forward (Caplan 2012); Eva Kor's stirring firsthand account of "Lessons from Dr. Mengele's Lab"; Baruch Cohen's lecture, "The Ethics of Using Medical Data from Nazi Experiments"; and Laurie Zoloth's address, "The Thief of the Future: The Holocaust—Women, Reproductive Science, Eugenics, and the State," which illustrated how well-reputed doctors were captivated by the potential benefit of eugenics and carried out accelerated species evolution by sterilizing women. This one-time bonanza of visiting lecturers will form the basis of research papers and class discussion in the MSCR curriculum.

We have dedicated a portion of our curriculum library to books and other material about the Deadly Medicine exhibit and about the Holocaust. Tulane University School of Medicine is devoting sufficient time in its curriculum in clinical research to document how wrong science can go and how distinguished, well-trained scientists can go wrong. It is particularly instructive and poignant that Michael DeBakey, an illustrious cardiac surgeon and an alumnus and benefactor of Tulane University School of Medicine, was among several outstanding American physicians and surgeons who chose to travel to Germany in the 1930s to hone their clinical and research skills (see DeBakey 2008). Paradoxically, the physicians and scientists who mentored these pioneering Americans frequently deployed the very same skills in their government's attempt to create the Master Race.

One of the first beneficial alterations to our course as a result of the exhibit and lectures was the recasting of regulatory oversight of human research as a necessary shared responsibility of regulators and scientists. The flagrant violations of human rights, dignity, and person perpetrated by the Nazis were juxtaposed with contemporaneous violations within the United States and funded by the US government. We discussed the Tuskegee experiments, the first of which began in 1929 with funds donated by Julius Rosenwald to the United States Public Health Service (USPHS) to determine the prevalence of syphilis in rural blacks and the feasibility of treatment (Brandt 1978). The study demonstrated that black males in Macon County, Alabama had a very high prevalence of syphilis and that treatment could be successfully implemented. The funding, however, never materialized and the patients were not treated. The better-known USPHS experiment, formally entitled the "Tuskegee Study of Untreated Syphilis in the Negro Male," began in 1932 and, although the men were told they were being treated, they were not (CDC; see Chap. 15 in this book). Even when penicillin became the standard treatment for syphilis in 1947, the patients were still observed and not treated, though they believed they had been. In 1972, the public revelation of the deception resulted in the end of the observational study, and in 1997 President Clinton issued an apology from the White House (Stout 1997). Similarly, decades-old experiments on syphilis infection and epidemiology in Guatemala, also under USPHS auspices, prompted an apology on page one of the *New York Times* in 2010 (McNeil).

We also reviewed the examples of deception by American scientists that occurred two or more decades after WWII. Dr. Saul Krugman studied the transmission of hepatitis virus at Willowbrook in New York, a home for children, by infecting the children. He told their parents that the children were being treated and obtained the parents' consents for the studies (Education Development Center 2009). Dr. Stanley Milgram studied human nature by asking paid subjects to act as teachers and deliver electric shocks to pupils to motivate learning—the "teachers" did not know that the shocks were fake and that the "pupils" were only pretending to be shocked. With the encouragement of the "scientist" who took responsibility for the experiment, the test subjects delivered what they believed to be dangerously increasing shocks as a teaching tool (Tavris 2013). Dr. Laud Humphreys infiltrated gay "tearooms" and violated the anonymity of the subjects as he studied their homosexual behavior (Humphreys 1970).

Understanding and conforming to regulations seemed less of a burden to our clinical research scholars when they understood what has come to light after the Holocaust. Oversight began making sense; active participation in oversight activities was seen as a legitimate demand on their time as researchers; in fact, service on peer oversight committees was beginning to be seen as a legitimate obligation.

Our scholars understood that the Nuremburg Code was not so much a codification of existing western values in 1946, but really a document that helped shape our values over several decades. The research standards that permitted the unethical human experiments in Tuskegee and elsewhere were clearly in violation of the Nuremburg Code. However, the World Medical Association (WMA) did not convene until 1964 to develop the Declaration of Helsinki that evolved through 2013 (WMA 1964–2013). This long incubation period, the continuing evolution of regulations, and the awareness that progress can be made while preserving the integrity, the dignity, and the autonomy and empowerment of research subjects became a reality to our scholars when they saw documentation of flagrant abuse.

The precepts of the Belmont Report also took on new meaning. It was refreshing to observe how much discussion ensued over the principle of distributive justice and the challenges of providing equal access to clinical trials. The principles of beneficence and utility engendered lively discussions on how to pursue phase I clinical trials when toxicity is the end-point. The ongoing promulgation of regulations governing clinical trials became a subject of active learning rather than a necessary evil that stands between a bright researcher and his/her answer to an important question. The IRB became a living group of responsible peers and consumers who assured quality and provided the oversight that the events of history demanded.

The scholars raised thoughtful questions. Is it ever warranted to use deception in pursuit of research? They discussed Milgram's study of teacher stress over the use of pain to motivate learning. They pointed to the knowledge gained about hepatitis from Krugman's experiments in an environment in which hepatitis was ubiquitous. The scholars had lively and erudite discussions on the relative value of the principles articulated in the Belmont Report. Can one principle override another? How much can be compromised for the greater good? Why can't science self-regulate and avoid external oversight? Predictably, the discussions did not unearth pat answers, rather they helped the scholars appreciate the need to remain alert, vigilant, and productively skeptical about the nature of the research question being asked and the research methods employed to seek the answers.

Another value of embracing the Holocaust and scientific misconduct (by current standards) in the MSCR curriculum is the repeated press coverage of Nazi era research haunting modern society. On December 1, 2012, a scholar called my attention to a front-page article in the *Wall Street Journal* describing the controversy of naming the award for outstanding work in aviation medicine after Hubertus Strughold, who may have countenanced experiments in Nazi Germany that killed subjects in the process of defining tolerable levels of hypoxia (Lagnado 2012; see also Chap. 17 in this book). A review of Gina Perry's book, *Behind the Shock Machine* in the *Wall Street Journal* on September 7, 2013, entitled "The Experiments That Still Shock," described Stanley Milgram's experiments with "teachers" delivering electric shocks to motivate students (Tavris 2013). Current exposures in prestigious popular newspapers help sustain interest, vigilance, and concern.

It is sobering to realize that our current, elaborate structure designed to protect human subjects evolved, though slowly, from the atrocities that came to light during and after the Nuremburg Trials in 1946. The WMA's Declaration of Helsinki was slow to evolve. US federal regulations evolved slowly, too, and were even slower to acquire the force of law. Our American policies continued to mature following the exposure of the Tuskegee experiments in 1972, and these policies may well be a product of our communal shame, shame at what we tolerated in the way of assaults on human rights, human dignity, empowerment, and autonomy, all in the name of science.

How will we rise to meet the future challenges, the nature of which we cannot know? How will the medical advances we are implementing today challenge our ethics in the future? Will the electronic medical record and the interconnectivity of medical record systems open new avenues to violations of patient integrity? What will we do with knowledge gained a decade from now on tissue we ethically obtain from a patient today? Do we apprise the patient's progeny of a genetic risk? What is our threshold for breaking the identity code on an anonymous specimen? Who will oversee our scientific experiments? Will we commit to oversee the scientific experiments of others? Will we remember the shame we felt as a society when Nazi atrocities were performed in the name of science, when prisoners were drowned so that an effective life vest design could be discovered, when prisoners died so that airline cabins and spacecraft could be appropriately pressurized, and when syphilis and hepatitis were spread in the name of science? What is the value of teaching tomorrow's clinical researchers about yesterday's communal ethical lapses?

Ken Jacobson, the Deputy Director of the Anti-Defamation League said it well at the end of his remarks at a fundraiser in New Orleans, December 2, 2012: "What keeps me up at night? Worry that we will lose the communal shame for the atrocities committed in the name of science and public policy."

References

Atkinson, R.D., S. Ezell, V. Giddings, L.A. Stewart, and S.M. Andes. 2012. Leadership in decline: Assessing U.S. international competitiveness in biomedical research. *Reports*. Information Technology and Innovation Foundation. http://www.itif.org/publications/leadership-decline-assessing-us-international-competitiveness-biomedical-research. Accessed 1 Oct 2013.

Brandt, A.M. 1978. Racism and research: The case of the Tuskegee syphilis study. *The Hastings Report* 8(6): 21–29. http://dash.harvard.edu/bitstream/handle/1/3372911/Brandt_Racism.pdf?sequence=1. Accessed 9 Oct 2013.

Caplan, A. 2012. *Justifying the unthinkable: The ethics of Nazi medical experimentation. [CD of speech]. Archive of scholar resources*. New Orleans, LA: Tulane University Curriculum Teaching Facility.

Centers for Disease Control and Prevention (CDC). 1932–1972. U.S. Public Health Service Syphilis Study at Tuskegee. Tuskegee study http://www.cdc.gov/tuskegee/. Accessed 1 Oct 2013.

Collaborative Institutional Training Initiative at the University of Miami. 2000. CITI Program. https://www.citiprogram.org/index.cfm?pageID=22. Accessed 1 Oct 2013.

DeBakey, M.E. 2008, May 16. Interview by S. Rubenfeld [video recording]. Michael E. DeBakey Lecture Series. http://www.medicineaftertheholocaust.org/lecture-series.php?v=30#play

Education Development Center, Inc. 2009. Willowbrook hepatitis experiments. In *Exploring Bioethics* (pp. 1–4). http://science.education.nih.gov/supplements/nih9/bioethics/guide/pdf/Master_5-4.pdf. Accessed 30 Nov 2013.

Fowler, J., and Muckert, T. 2004. Tiered mentoring: Engaging students with peers and professionals. In *Seeking Educational Excellence*. Proceedings of the 13th Annual Teaching Learning Forum, 9–10 February 2004. Perth: Murdoch University. http://otl.curtin.edu.au/professional_development/conferences/tlf/tlf2004/fowler-j.html. Accessed 1 Oct 2013.

Getz, K., and D. Borfitz. 2002. *Informed consent (The clinical trial volunteer's bill of rights)*. Boston, MA: Thomson CenterWatch. http://www.ciscrp.org/patient/infocenter/BillofRights.pdf. Accessed 1 Oct 2013.

Harmon, A. 2010, April 21. Indian tribe wins fight to limit research of its DNA. *The New York Times*. http://www.nytimes.com/2010/04/22/us/22dna.html?emc=eta1&_r=0. Accessed 1 Oct 2013.

Humphreys, L. 1970. *Tearoom trade: Impersonal sex in public places*. London: Duckworth.

Johnson, C. G. (2013, July 7). Female inmates sterilized in California prisons without approval. The Center for Investigative Reporting. *The Sacramento Bee*. http://www.sacbee.com/2013/07/07/5549696/female-inmates-sterilized-in-california.html. Accessed 1 Oct 2013.

Kaelber, L. 2012. *"Eugenics" and Nazi "euthanasia" crimes*. A presentation at 2012 Social Science Association. http://www.uvm.edu/~lkaelber/gateway/ Accessed 1 Oct 2013.

Kuntz, D. 2008. *Deadly medicine: Creating the Master Race*. Washington, DC: United States Holocaust Memorial Museum.

Lagnado, L. 2012 December 1. A scientist's Nazi-era past haunts prestigious space prize. *The Wall Street Journal* A1, A12.

McNeil, D.G. 2010. U.S. apologizes for syphilis test in Guatemala. *The New York Times*. http://www.nytimes.com/2010/10/02/health/research/02infect.html?_r=0. Accessed 1 Oct 2013.

National Commission for the Protection of Human Subjects of Biomedical and Behavioral Research. 1979. *The Belmont report*. http://www.hhs.gov/ohrp/humansubjects/guidance/belmont.html. Accessed 1 Oct 2013.

National Institutes of Health. 1999. *Clinical research curriculum award*. Retrieved from the National Institutes of Health website: http://grants.nih.gov/grants/guide/rfa-files/RFA-OD-00-002.html. Accessed 1 Oct 2013.

Stout, D. 1997. Clinton to apologize for tests on blacks. *The New York Times*. http://www.nytimes.com/1997/04/09/us/clinton-to-apologize-for-tests-on-blacks.html?scr=pm. Accessed 1 Oct 2013.

Tavris, C. 2013. The experiments that still shock. *The Wall Street Journal* C5, C7.

U.S. Department of Health & Human Services. 1949. Nuremberg code. http://www.hhs.gov/ohrp/archive/nurcode.html. Accessed 1 Oct 2013.

———. 1991. Federal policy for the protection of human subjects ("Common Rule"). Retrieved from http://www.hhs.gov/ohrp/humansubjects/commonrule/index.html. Accessed 1 Oct 2013.

United States Holocaust Memorial Museum. n.d. Josef Mengele. Holocaust encyclopedia. http://www.ushmm.org/wlc/en/article.php?ModuleId=10007060. Accessed on 24 Nov 2013.

World Medical Association General Assembly. (1964–2013). WMA Declaration of Helsinki—Ethical principles for medical research involving human subjects. http://www.wma.net/en/30publications/10policies/b3/. Accessed 1 Oct 2013.

Chapter 20
The Psychophysiology of Attribution: Why Appreciative Respect Can Keep Us Safe

Linda Emanuel

This essay has the marks of the wandering, pondering journey that produced it. The invited topic was human subjects research after the Holocaust. It started a deeper inquiry. Then, deeper reflection came from listening to the perspectives of other contributors to this volume when we gathered together for the meeting to present our work. Rather than distill and separate the end result from its origins, this chapter tracks the journey as it occurred.

Therefore, the first part of the essay is about contemporary human subjects research and its challenges from the perspective that medical professionals are the guardians who, by virtue of their role, should protect humans rendered vulnerable by illness. The second part of the essay considers the mechanisms of violation that make protection necessary and asks how, in view of these mechanisms, protection can best be achieved. The third part proposes a cultural value that may offer the most protection.

While this may seem logical enough, the wandering and pondering came from a deeply personal perspective that, though it was never a secret, I chose to share publically for the first time. As a result, my private perspectives changed. The Center for Medicine after the Holocaust (CMATH), our host for the meeting, achieved its apparent goal with me: to engage professionals in personal reflection. The sharing was inspired by the unadorned authenticity of the personal journey described by Eva Kor of her survival of Dr. Josef Mengele's experiments on her and her twin sister in Auschwitz. I hope that my gratitude for the opening to share publically is matched by some validation of these disturbing new perspectives, and of the value of the proposed protections for human subjects in medical research.

L. Emanuel, MD, PhD (✉)
Buehler Professor of Medicine; Director, Buehler Center on Aging, Health & Society; Feinberg School of Medicine, Northwestern; Principal, Education & Implementation in Palliative and End-of-life Care; Principal, The Patient Safety Education & Implementation Program, 750 N Lake Shore Drive, Suite 601, Chicago, IL 60611, USA
e-mail: l-emanuel@northwestern.edu

S. Rubenfeld and S. Benedict (eds.), *Human Subjects Research after the Holocaust*, 247
DOI 10.1007/978-3-319-05702-6_20, © Springer International Publishing Switzerland 2014

Human Subjects Research Today: Clear-Eyed Oblivion?

The Role of Professionals in Society

Standards of civilization are often measured by how a society treats its vulnerable members, even more so perhaps by how it treats its vulnerable values. Societies can forget the importance of things like care for the poor and the old; provision of education; rescue and securing of safety; health care; and justice. Physicians, nurses, and those working in the many newer health care disciplines, are entrusted by society with protecting the ill and with protecting the values of health and health care.

Broadly speaking, society often gets divided into the public sector, the private sector, and the professional sector. The sections have different social mandates. While the public sector is charged with protecting the interests of citizens, the private sector fosters market transactions. All sectors have their value systems. For example, the public sector is accountable for public service while the private sector is accountable for honest, legal returns to the stakeholders in the metaphoric marketplace. Each sector has its own fiduciary relationship to society, and each one is necessary for a balanced, stable, decent civilization (Hughes et al. 2004). In an ideal world, activities will be adequately fostered and protected by one or two or even all three of these areas.

For instance, an international HIV/AIDS project involved collaboration between the government of Botswana, companies in the for-profit pharmaceutical industry, and nonprofit medical institutions. Strikingly, this project's leaders attributed its success to each sector's encouragement of the others to stick to their own values and to not allow a party from another sector, including their own, to make incursions on those values. They were able to rely on one another's strong values and, thereby, achieved a balance that worked (Hughes et al. 2004). Many, many successful human endeavors rely on partnerships that adhere to this formula. Others, if incursions on the values are made, tend to fail after causing varying degrees of suffering in participants or in those around them.

Medical Professionals and Civil Disobedience

Professionals who take an oath at the time of graduation from their training are particularly bound to the values of their group. The importance of this commitment in the professional sector is appreciated most powerfully when its adherents abrogate its values and when they fail to carefully protect those in their care. Medical professionals have daily opportunities to promote their values and to prevent ethical lapses in relationships with patients and other colleagues and in their activities on committees and elsewhere. They have both the capacity and an obligation to speak out against lapses in as vigorous a manner as the extent of the lapse warrants, including private and public dissent in some cases. In circumstances as egregious as the Holocaust, however, professionals are obliged to engage in acts of civil disobedience that will protect their values and their patients to the fullest extent possible (Wynia et al. 1999). While it is gratifying that a few medical students and professionals were engaged in this manner to some extent during the National Socialist period (see Chap. 12 in this book), it is a deeply distressing and important lesson to all that many more were not so engaged or engaged on the wrong side. How could this have happened?

Antidotes to the Commodification of People: Culture Outweighs Regulation

At the heart of Nazi human subjects research was their view of selected groups of people as subhuman. Nazi bioscientists saw their subjects as enemies and useful commodities for achieving their Aryan ends. Modern informed consent requirements provide modest safeguards against commodification, being honored in the breach as much as in the observance. Review of studies by an Institutional Review Board (IRB) provides another procedural safeguard against abuse of human subjects in medical research. These procedural steps also instantiate a cultural endorsement of respect for humans, which prevents the researchers from commodifying vulnerable people who may choose to participate in medical research because of the excessive altruism arising from their vulnerable circumstances such as a terminal illness.

The reality is that cultural endorsement of respect for humans is much more powerful than regulation. Culture and regulation support one another, but if the power of one is assessed against the other, culture wins. History has seen the overthrow of specific laws, clusters of regulation, or entire bodies of regulation and law many times. Without the support of some cultural bedrock, regulations do not survive. By contrast, culture creates its own norms even without regulations. The implication of this reality is that while informed consent and Institutional Review Board review may help to prevent the repetition of unethical Nazi human subjects research, these regulatory mechanisms will be inadequate if culture no longer supports them.

Having witnessed the impact of the rising tide of hatred in Germany on my Jewish father, who in 1938 was sent out of Berlin alone as a child in order to save his life, I can personally testify to the power of Nazi culture and its multigenerational impact. Rules and regulations break like crystal in one night if culture so moves, and the implication for medicine after the Holocaust is profound: Under the right circumstances well-intentioned medical professionals who claim to care deeply about the values of their profession may disregard human subjects protection.

It is not too difficult to imagine a well-intentioned effort to deliver better patient care that has an unintended adverse effect. One such example resulted from the tragic death of Libby Zion, which was attributed to inadequate staffing and her resident physicians' fatigue (Lerner 2006). Enough cultural pressure built after the incident that new limits on resident work hours were legislated. However, the unintended effect of shifting the burden of care to older attending physicians, who were likely to be less resilient under conditions of sleep deprivation, was not assessed. Patients did not know that such a shift was made, did not consent to the shift, and could have been outraged if they knew that their older, attending physicians might be either burned out or more prone to error than their resident physicians. Believing that patients are unable to make an informed decision about care at an institution in which there are errors waiting to happen is a position no one would right-mindedly defend. Nonetheless, quality improvement continues apace without informed consent, and changes in hospital procedures and protocols are often hidden under the blanket of proprietary data. An example like this implies that the foundations for egregious violations of human dignity could be laid today by esteemed medical professionals, much as they were laid by our Nazi medical colleagues—in clear-eyed oblivion with excellent intentions.

What Can Be Learned? Listening When the Paradigm Changes

A paradigm shift in the cultural setting of medicine is occurring. At times of change, traditional values can get lost and new values may replace them. How values are to be sustained must be a question of active consideration in the new era. The greatest challenge is not so much noticing that the paradigm has changed, or examining how old values remain relevant, though both are challenging enough. The biggest challenge is seeing the new paradigm's blind spots created by oblivion to or denial of potentially serious problems. To see the fuller implications of oblivion or denial in human subjects research, it is important to examine the psychological mechanisms allowing us to violate other human beings.

But first, consider the import of the collection of work in this volume. Gross violations of human beings in medical research are found in every discipline, in every time, and in every country that has been studied carefully enough. To learn what is possible, it seems that this startling reality must come to the foreground. If no group and no setting are exempt from such violations, the mechanisms must be inherent to human nature. Any notion of good and evil that identifies a clear and reliable separation and lodging of evil in a particular group, or time, or setting, is incompatible with reality. Any lesson that teaches only about one group, time, or setting is an incomplete lesson: The lesson must include an understanding of our common human nature.

Mechanisms of Violation

What Happens, Psychologically, That Ends in Atrocity?

Groups chosen as targets for atrocity and annihilation must first be widely seen as something other than the dominant party. The Nazi government selected people with disabilities for sterilization, as did the governments of the United States and other countries prior to the Third Reich. People with disabilities, including plentiful WWI German veterans with severe injuries, were widely seen first as "other," then as a burden, and finally as a disease (Kass 2010, pp. 111–112).

Psychologically, when identification with another stops or fails to occur, then the possibility of scapegoating arises. Challenges to the dominant party can be displaced onto the scapegoat and the burden on the dominant party is relieved—it is a short psychological step from blaming to excluding the scapegoat. The Jews in the Hebrew Bible sent a goat into the wilderness to expiate sin. A distorted, humanized version of the same psychological and social process underlies scapegoating a person or population. From the playground bully to the Führer, this mechanism may be essentially the same.

The critical element in human scapegoating is that the scapegoat is no longer empathically experienced as a fellow human being. He or she is blamed without a sense that it hurts the person or perhaps with a sense that hurt is deserved. The scapegoat's perspective is severed, the scapegoat becomes a commodified object, and the dominant party can then label the human scapegoat as an enemy or any other psychologically satisfying designation.

Add to this situation some significant real or imagined need, such as food, power, or money, and the psyche often switches on the will to kill. If some creature other than one of our own has something that stands between us and physical or psychological death, most humans will take whatever they need even if it requires killing the creature. The inference from this view of human nature is that there is a normal, healthy pathway or set of pathways deep in each of us that, under the right circumstances, enables us to kill without compunction. This mechanism is possibly the one that was in play in Nazi Germany.

In the following section, I turn to a personal experience of how quick and slippery is the transition from unity to disunity with differentiation into the "hunted" and the "hunter" (prey vs. predator).

A Personal Narrative and Commentary

My father was a Berliner Jew, born to an assimilated family that owned one of the largest publishing houses in Europe. The family was German first and Jewish second. Hitler's rise meant that almost all German Jews, assimilated or not, suddenly became only Jewish—the option to remain German died in front of them. (As an American, I'm acutely aware of the privilege I've had of assimilating without having to join a different religion.)

My father did not talk much about the Holocaust, but I did learn that someone we never talked about had died in a concentration camp. My father, his brother, and two cousins survived and married Presbyterians. One female cousin married a secular Jew—a survivor she helped bring out from Austria—and affiliated with the Church of England.

The journey to adulthood for my father was paved with rejections. He was a sensitive, intelligent, shy young German boy who spent part of his childhood in an English boarding school while the Germans were bombing England. And yet, even though he did not know much about it or feel that good about it, he was also Jewish.

The impact of this story on my generation was manifold. The first message I received was loud and clear from my father: Above all, truth matters. When playing Monopoly, my father insisted on telling his unmindful opponents if his piece was on their property. My father was honest to a fault, and this trait is my touchstone to this day. My confusion began when I realized that I alone in the nuclear family felt wholeheartedly Jewish, and that my father's truth about our Jewishness was the flip side of mine. As I went through an orthodox conversion to become outwardly what I was already inwardly, his greatest fear for me was that everyone would reject me.

Two of my cousins, though we grew up far from one another, made a similar journey back to Judaism, and we all raised our children in the Jewish tradition. Each of us knew nothing of the others' journey until each was well underway; we, and our children, became close later. My childhood nuclear family continued to love me and I them. But it felt like a Herculean feat to see and respect one another's realities. It took many years to reach the point where appreciative respect made agreement superfluous, and the journey of transformations was paved with mysterious pain. To reach the point of resolution and peace was a lifetime of work—more than a lifetime in my father's case. My father, may his memory be a blessing, died a Christian. Was it all worth it? I believe so, for the reason that I hope is well enough set out below.

Why is this relevant? First, because the fact that my father's truth was not my truth (about what it meant to be Jewish) created within me a sense of being unseen, unheard. My family could not appreciate what I saw and loved and how I lived as a Jew. My narrative truth was invisible to my beloved family members, which was the source of the mysterious pain along my journey of transformation. While I could not initially understand my feeling of being in their blind spot, I ultimately realized it was a feeling of dread.

As I was doing the research for this paper, I read about Holocaust deniers, which reactivated my feeling of dread—I was shocked when I experienced the same sensation from a much different source than my family. Pondering the experience, I realized that the common thread was the fear of being unseen in the blind spot and, therefore, at risk of being identified as the endangered "other."

A similar dread must be what an animal feels when it is in danger of becoming its predator's next meal. At that moment, just as the predator is programmed to kill an animal for food, the prey is programmed to feel dread. Similarly, when a society rationally and with a clear conscience denies the humanity of certain groups of people and sees them as subhuman, most members of those groups will experience dread as they realize they have become the society's prey. At that time, the meaning each party has constructed from shared facts are incompatible and of ultimate importance.

What Can Be Learned? The Case for a Culture of Appreciative Respect

Bioethics has long taught four fundamental principles: beneficence, nonmaleficence, justice, and autonomy (Beauchamp and Childress 2001). By themselves, these four principles do not prevent, or promote recovery from, atrocities. What is missing? Appreciative respect.

Appreciative respect is the validation of another's truth or point of view without necessarily agreeing with it, as demonstrated in my family's religious choices. Appreciative respect is mandatory in a multicultural democracy, and when present, it can engender both a good legal system and good regulations that may be the strongest available "vaccine" against atrocities. Without it, the need–dread dynamic may become operative and generate passions that cannot be contained.

What Can We Do for Our Times? American Culture Today

American culture today can be enormously affirming and tremendously worrying for those who are concerned that the Holocaust or anything like it might happen again. Affirmation comes from an awareness that Americans of every nationality, creed, race, and orientation are proud of who they are with a twofold (or manifold) identity. People teach and learn about one another's traditions in a culture of appreciative respect. At the same time, America is developing an alarming culture of narcissism (Lasch 1979). Narcissism is a state of oblivion to others, which is a necessary precondition for atrocities such as the Holocaust. Furthermore, in our times and those of our children, rapid and revolutionary changes in medical practice and research, such as nanomedicine and genomics, will challenge us and create unintended and unforeseen consequences. We must, therefore, create a culture of appreciative respect in our schools, houses of worship, homes, businesses, and in political discourse if we are to meet these challenges without repeating the mistakes of the past. Much work remains to be done to create such a culture and to establish procedures and regulations that protect appreciative respect. As difficult as this effort may seem, if appreciative respect is to stand between peace and killing, it is an essential effort.

References

Beauchamp, T., and J. Childress. 2001. *Principles of biomedical ethics*, 5th ed. New York: Oxford University Press.

Hughes, G., S. Rammohan, and L.L. Emanuel. 2004. Corporate citizenship: Managing relationships with professionals and government. *Organ Ethics* 1: 3–20. Reprinted in *AIDS Public Policy Journal* 18(3/4): 61–76.

Kass, L.R. 2010. A more perfect human: The promise and the peril of modern science. In *Medicine after the Holocaust: From the master race to the humane genome and beyond*, ed. S. Rubenfled, 107–122. New York: Palgrave Macmillian.

Lasch, C. 1979. *The culture of narcissism: American life in an age of diminishing expectations*. New York: W. W. Norton.

Lerner, B.H. (2006). A case that shook medicine. *The Washington Post*. Retrieved from http://www.washingtonpost.com/wp-dyn/content/article/2006/11/24/AR2006112400985.html. Accessed 24 Nov 2013.

Wynia, M.K., S.R. Latham, A.C. Kao, J.W. Berg, and L.L. Emanuel. 1999. Medical professionalism in society. *New England Journal of Medicine* 341(21): 1612–1616. doi:10.1056/NEJM199911183412112.

Chapter 21
Confronting Medicine during the Nazi Period: Autobiographical Reflections

Volker Roelcke

This essay is an unusual one for me, completely different from all others I have written as a physician and medical historian. For the last 20 years, medicine during the National Socialist period has been one of the main focuses of my professional work. I have done detailed research on eugenics and racial hygiene, on the emergence of the modern understanding of euthanasia and programs of systematic patient killing, as well as on human subjects research in concentrations camps and psychiatric institutions. In addition to my specific community of medical historians, I have addressed physicians, historians, medical students, and other audiences, and I've been invited to lecture internationally on the related issues. Although the topic was (and is) a strong personal concern for me, I have now decided to turn to other topics in medical history without, however, completely abandoning the questions related to medicine during the National Socialist period. During this transition, Shelly Rubenfeld asked me to explain why, having started my professional life as a physician, I chose to make medicine during the Nazi period the focus of my professional work, and why I am leaving the field.

To put it in a nutshell: Confronting medicine during the Nazi period has been a transformative experience for me not only as a physician but also as a citizen. This confrontation did not occur in a single, circumscribed event, but rather in a continuous process related both to my personal life as a German and to a number of experiences during my medical education and career as a physician. In spite of some doubts about if and how I could meet Shelly Rubenfeld's expectations, I accepted his challenge and responded with this brief biographical account highlighting some formative experiences and the associated changing images and evaluations of medicine that may explain why this particular period in the history of medicine was (and is) so important to me.

Medicine appealed to me as a highly interesting and rewarding field when I left high school at age 18. I remember explaining my choice of medical school in an

V. Roelcke, Prof. Dr. med (✉)
Institute of the History of Medicine, Giessen University, 35392 Giessen, Germany
e-mail: Volker.roelcke@histor.med.uni-giessen.de

S. Rubenfeld and S. Benedict (eds.), *Human Subjects Research after the Holocaust*, 255
DOI 10.1007/978-3-319-05702-6_21, © Springer International Publishing Switzerland 2014

application letter for a grant, listing three reasons for this decision: I could help suffering patients and do "good" work; I could systematically look into the universe of the human body and study the structures and processes constituting man as a living organism; and I would have legitimate and privileged access to the broadest imaginable range of people and social contexts that otherwise would be out of reach for a "normal" middle class person. When I made this choice in the second half of the 1970s, the medical profession had a thoroughly positive image, and what I saw, read, and heard of medicine suggested that there was an even brighter future to come, both for the field of medicine itself, and—with medicine's help—for humanity.

Because my father was a physician who specialized in laboratory medicine, I had very early experiences with the everyday life of a medical doctor. He had his own practice in Heidelberg, offering diagnostic services to medical practitioners and hospitals of all kinds in a wide region of southwest Germany, including university departments that required special tests. Before we reached our teens, my two brothers and I accompanied our father in the evenings or weekends to his practice. We were fascinated by the laboratories with their technical equipment and strange smells. Not only could we look at a drop of our own blood or an autumn leaf under the microscope, but we could also observe our father doing some laboratory work, like checking the growth of bacterial cultures after defined intervals or reviewing preliminary diagnoses of stained pathological tissue specimens prepared by his assistants. We also stood beside him when he signed outgoing reports in the evenings, eagerly listening to his comments on the "cases" or on his physician-colleagues. Long before leaving high school, I knew at least a dozen types of bacteria, their growth patterns in special culture plates, and their susceptibilities to antimicrobial drugs, and I was able to do a differential blood count.

For me, however, the most intriguing aspect of my father's profession was his contact with people. For example, a shepherd outside the town looked after a wether and some sheep that belonged to my father, and every now and then, he collected blood from the animals that he needed for certain specific investigations. This blood collection was always an occasion for coffee and a chat with the shepherd, and I loved it when I had a chance to accompany my father on these visits. He also had patients with chronic illnesses like diabetes who could come to his practice for regular blood or urine checks, but he preferred to visit them in their homes, including those in the poor areas of Heidelberg's old town center. I looked forward to school holidays because then I could come with him to the exotic worlds of people that otherwise neither I nor my friends would ever have access to. My father also talked to people on these visits, learning about their living conditions and debating public affairs and politics. The patients and their families appeared to like him, invited him for coffee, and occasionally gave him presents like a freshly baked loaf of bread or cake or a big basket of just harvested cherries or apples. What a wonderful job!

Soon after commencing medical training, however, I became more and more disillusioned. Apart from an obligatory short course in medical psychology, the first two years of medical school were filled with laboratory work and accompanying lectures on material that we could readily read about in textbooks. Our professors expected, apparently, that we students learn the natural sciences as *the* relevant grounding of any medical work. But where was the sick human being in this kind of medical education, the suffering patient with his perceptions and interpretations of his body and his symptoms, with his biography and social relations?

Two big exceptions to the focus on the natural sciences were the facultative seminars on medicine and literature and on ethics in medicine, organized by the medical historian Dietrich von Engelhardt. Using novels and other literature, he addressed a broad spectrum of views, perceptions, and interpretations of human life, body and mind, birth and death, and normality and deviance. The implications of potential attitudes and behaviors of both patients and physicians concerning these issues were probed and fully discussed.

After the experiences in my father's laboratory, and the recognition of the priorities in the medical curriculum, I decided early during my studies to "really" learn about science in medicine. Therefore, I began research for a doctoral dissertation in my third year of medical school. I did not want to work with animals for ethical reasons or with statistics because of a concern about reaching inadequate conclusions from amalgamated numbers representing some aspect of the reality of human existence or suffering. I chose an immunological topic: to search for specific antigens or antigen patterns to differentiate in the laboratory among the various clinical forms of acute leukemia, which involved testing blood samples from leukemic patients with a range of naturally occurring human monoclonal antibodies.

Developing laboratory methods was tricky and, at times, intellectually challenging. I also found the long hours in the laboratory tedious, boring, and sometimes annoying—the antigens and the instruments appeared to have a life of their own and did not behave as I wanted or expected. I realized that generating scientific knowledge in the laboratory was not as direct and rational as I had assumed; instead it often involved tinkering, improvisation, contingencies, and impactful social dynamics among the laboratory personnel. I began reading about the processes of bricolage analyzed in the laboratory fieldwork of the anthropologist Bruno Latour in the Jonas Salk Institute for Biological Studies (Latour and Woolgar 1979).

After these disillusioning experiences with experimental laboratory work, I searched for another kind of academic medicine, one that systematically takes account of the patient as a unique human being living in a specific network of social relations that influenced the way he or she encountered his or her suffering. I had read that general practice was a distinct and obligatory subject at British medical schools. Hoping that this discipline might offer a different approach to patients than in the body- and laboratory-centered medicine we were taught at Heidelberg Medical School, I decided to spend a year at Glasgow University Medical School.

Another reason for my decision to go abroad was personal: My father, who had been responsible for my positive image of medicine, increasingly gave me cause for questions and doubts. We had been taught in school about National Socialism, the persecution of the Jews, and the atrocities committed during World War II. I realized that my father, born in 1907, had been in his mid-20s when the Nazi party was elected to power in 1933 and that he had been an ambitious young doctor in his early 30s when World War II started. During the early 1940s, my father was the deputy director of Heidelberg University's Institute of Hygiene. He told us that in the absence of its director, Professor Ernst Rodenwaldt, *Generalarzt* and chief physician of the German army who served in North Africa under General Erwin Rommel, he was responsible for surveying the hygienic conditions in the Heidelberg region. My father also told us that Rodenwaldt had been a great scientist whom he much admired. Now I found out that Rodenwaldt had not only been a renowned bacteriologist and specialist in tropical medicine but also been one of the most prominent German racial anthropologists and a fervent supporter of racial separation and the sterilization laws (Eckart 1998). After the war, the American occupational officers dismissed Rodenwaldt and my father from their academic positions (Eckart and Gradmann 2006).

This dismissal raised irritating questions: What was my father's attitude toward the Nazi party and its politics? Had he been involved in the eugenically motivated Nazi "hereditary health policy" that included the identification and sterilization of those supposedly suffering from hereditary disorders? Had he been involved in anti-Semitic activities? What had he known at that time about the Holocaust, and what did he think about it after the war? And why had he been forced to end his promising academic career immediately after the war and begin a private practice?

I turned to my father for answers in the early and mid-1970s. My father evaded any clear responses, pretended not to remember, and occasionally became verbally aggressive—a behavior I had never seen before—and finally withdrew more and more, refusing any coherent reply. He developed increasing apathy, difficulty with concentration, forgetfulness, and mood changes, and we slowly realized that he had a progressive illness that psychiatrists ultimately diagnosed as dementia.

In the beginning, my father's evasive reactions to my and my brothers' questions made me furious. Later, when we realized that his behavior was part of a disease process, we were very frustrated but abstained from further direct questioning, and I chose to find other ways to explore my father's past. But on a deeper level, I experienced a profound unsettledness: The personal knowledge that I had of my father as a cultivated and intelligent physician, interested in other peoples' lives and well versed in history, music, and philosophy, completely contradicted the possibility that he had been involved in medical atrocities or, more generally, in ideological activities infringing on the rights and well-being of other people. How could I reconcile these two fundamentally different images of my father?

I gathered more disquieting information. After Rodenwaldt's dismissal by the American occupational forces in 1945, he had initially been classified in the de-Nazification process as "involved [in Nazi activities] to a minor degree" [*Minderbelasteter*]. Following an appeal in 1948, the judgment was changed to "not guilty," and he was reinstated as provisional director of the Institute of

Hygiene, retiring in 1951 with all the rights and pensions of a full professor.[1] On the occasion of his death in 1967, the university and the Heidelberg Academy of Sciences honored Rodenwaldt in spite of his activities in racial anthropology and racial hygiene. Until 1998, a large institute of military medicine and hygiene of the German army in Koblenz was named after him; it was eventually renamed because of his Nazi past. How, then, were the quite different postwar developments regarding my father to be understood?

After a 32-month military internment by the American occupational forces, my father had also been classified as "involved to a minor degree," but this judgment was not altered any more. Although Rodenwaldt supported my father's attempt to regain his former position at the university in 1948, it was not successful (Eckart and Gradmann 2006, p. 717). Did that mean that my father had done more "wrong" (whatever that might have been) than Rodenwaldt himself? This question caused me considerable irritation and led to the decision to look more thoroughly into what had happened.

For me, both Rodenwaldt's and my father's biographies illustrated that physicians who had successfully acted in congruence with Nazi health and population policies were not one-dimensional monsters, but were appreciated and honored by their patients and colleagues in the postwar period. This was a deeply disconcerting thought. Rodenwaldt and my father also had apparently practiced scientifically sound medicine (in their view) during and after the war. Did that not imply—at least to them—that the principles and rationalities of medicine during the Nazi period had not been fundamentally different from the postwar period? Was it possible that scientifically coherent medicine was compatible with inhumane, indeed, atrocious effects?

At that stage, however, I was not able to find answers to these questions; I needed more distance, spatially and emotionally, but I was determined to research these issues later. During my year at Glasgow University, my father died. Although other concerns, including the completion of my medical studies and the dissertation project, moved into the foreground, I was now gripped by an interest in the history of medicine. In the 1980s, triggered by the German television series *Holocaust* (1978–79) and the film *Shoah* by Claude Lanzman (1985), a rapidly growing public interest in the exact circumstances and the developments leading to Nazi atrocities and the Holocaust coincided with my personal motivation to learn more about the role of medicine during this period.

In Glasgow, I was fortunate to get to know another kind of medicine. Stimulated by the teaching of Hamish Barber and David Hannay in the department of general practice and primary care, I decided to compete in a student research project in general practice. I did two extensive electives, one with a Glasgow general practitioner working in poor areas with unemployment rates up to 20 percent and one with the only medical practitioner on the Isle of Tiree, the most remote island of the

---

[1] On the de-Nazification procedures at Heidelberg University, including longer passages on Rodenwaldt, see Remy (2002).

Inner Hebrides off the west coast of the Scottish mainland. The project involved the study of the impact of life events on the well-being of individuals and on the occurrence of minor psychiatric illness in the general population using semi-structured interviews with a representative sample of probands in the local communities. The fascinating experiences in this project strengthened my feelings about the impact of social and cultural factors on the development and course of diseases, on the prevention of diseases, on illness behavior, and on the care of patients.

Back in Heidelberg, while completing my medical degree and the dissertation, I made further irritating discoveries at my medical school. Rumors implied that Hans-Joachim Rauch, a long-retired professor of forensic psychiatry, had been involved in unethical research on psychiatric patients during the Nazi time, but apparently, he had never been prosecuted for these activities. Some of my fellow students and I, therefore, were surprised to see the university and the medical school organize festivities to celebrate this distinguished medical scholar—as they saw it—on his 75th and then, in 1989, his 80th birthday. Rauch had not only been a much appreciated academic teacher and researcher after the war but also been one of the most prominent expert witnesses in forensic psychiatry in southern Germany's courts. It was, however, also clear that in the beginning of his career during the war, he had been a resident and junior lecturer at the Heidelberg University Department of Psychiatry under the notorious Professor Carl Schneider. It was already known by the late 1940s that Schneider had played a key role in the program of patient killings both as an expert advisor for the initial central *Aktion* T4 "euthanasia" program and as the coordinator of human subjects research on euthanasia victims (e.g., Platen-Hallermund 1948; Mitscherlich and Mielke 1947/1949; Chap. 10). Schneider had been arrested by the Americans in the immediate postwar period and then committed suicide in prison. Now it dawned on me that after his death, Schneider may have been a welcomed scapegoat both in postwar narratives on Nazi psychiatry in general and in the Heidelberg psychiatry department in particular, diverting attention from broader contexts and colleagues.

Rauch maintained that he had had nothing to do with the atrocious activities of Schneider. But had Schneider really acted in complete isolation within his department? What kind of research had actually been done there? What had Rauch and other faculty in the psychiatry department and the Heidelberg medical school before and after 1945 known about these research activities and Schneider's role in the broader programs of patient killing? After all, my father had been a member of the faculty until 1945. Even if Rauch or my father had not been directly involved in any barbarities, had they tried after the war to find out exactly what had happened in their immediate institutional surroundings? Had they addressed the responsibilities for and implications of the atrocities committed in their medical school in any relevant way?

Although I was looking for indications of self-reflection in the postwar period by either Rauch or my father, I found none. Nor were there any indications that the Heidelberg Medical School had undertaken any systematic investigations. As a consequence, the positive image I originally had of my father as an engaged, cultivated, and patient-friendly physician and the positive image that existed of Rauch in the public sphere as a medical scientist and legal expert advisor were profoundly put into question for me. I also saw that the implications of such questions were much broader than the behavior and culpability of individual doctors: If physicians, university medical schools, and professional medical organizations in Germany in the postwar era had not confronted the origins and the full range of atrocities committed by their colleagues prior to 1945, then something was deeply flawed with the ability or willingness for self-reflection by the medical profession. This reluctance to confront the Nazi past concerning medicine was not necessarily limited to Germany. For example, in later visits in Britain or in Switzerland where one of my brothers worked as a medical resident, I observed a "shortcut" response when the topic came up, usually a quick explanation that only a few fringe and fanatic Nazi physicians had been actively involved and that the rest of the profession had suffered from outside pressure by the regime. I knew that the behavior of Rauch, Rodenwaldt, and my father did not fit into this simple picture, and I had the impression that nobody really wanted to look into the relevant history in more detail, perhaps from fear of the unflattering, broader implications for medicine.

After graduation from medical school and completion of the doctoral project, I began studies in cultural and medical anthropology. This decision was not only a result of my interest in the social and cultural contexts of medical thought and practice but also a chance to get some distance from medicine proper in order to rethink what I really wanted to do as a physician. After completing a master's degree, I looked for a clinical discipline in which the biographical and social contexts of an individual's illness would be taken into account in a systematic way and found psychosomatics and psychotherapy. I took a position as lecturer and resident in the Department of Psychosomatic Medicine and Psychotherapy of Heidelberg Medical School, a department founded in 1950 by Alexander Mitscherlich, the official observer delegated by the West German Chamber of Physicians to the Nuremberg Doctors' Trial. His great awareness of both the subjectivity of the patient and the political dimension of medicine was a guiding legacy for some of us working there, including Waltraud Kruschitz, a psychoanalyst and consultant in charge of the outpatient services of the department.

In September 1989, on the occasion of the 50th anniversary of the German attack on Poland and thus the beginning of WWII, Waltraud, Sophinette Becker (a Frankfurt psychologist and former member of the department), and I organized a public conference to discuss the conditions leading to the war and, in particular, the postwar implications for German society using a kind of socio-psychoanalytic perspective. The reactions both in the department itself and in the medical school were overwhelmingly critical. For example, the newly appointed professor and chair of the department was very uncomfortable with our "political" activities, and he explicitly distanced himself from any broader social application of psychoanalytic theory and the legacy of Mitscherlich. We were urged to declare that the conference was our private initiative and that, while the event took place in departmental rooms, the department had nothing else to do with it. I was telephoned at home by a number of mostly elderly professors, acquaintances of either my father or my half-brother (also a member of the faculty) whom I did not personally know, to warn me about inconsiderate and imprudent political activities.

All this again illustrated that the Nazi period and its aftermath were particularly sensitive but not quite clear dimensions of medicine in the late 1980s. I wondered to what degree these dimensions affected not only the recollections and self-images of individual physicians and institutions but also the content of medical knowledge and practices or the structure of medical institutions. Waltraud and I looked into the history of our department and the findings were eye-opening. It turned out that the founding of the department was closely related to the Nazi past of Heidelberg University and its medical school, its closure in 1945 and comparatively early reopening (in contrast to other universities) by the American military government. Mitscherlich, one of only few politically "untainted" members of the medical faculty and short-term minister in the government of the American-occupied zone, was needed by representatives of the university for its reopening. The university and medical school were, therefore, reluctantly prepared to yield to Mitscherlich's requests for a psychoanalytically inspired psychotherapy department separate from psychiatry, which he argued was both too biologically oriented and also discredited by its involvement with the patient killings. The new discipline would differ from the other somatic disciplines by systematically looking into the psychological and social dimension of organic diseases. With external support from the Rockefeller Foundation, the new Department of Psychosomatic Medicine and Psychotherapy, the first academic unit of its kind, created a therapeutic approach radically different from either psychiatry or internal medicine, and it became a model for the institutionalization of similar departments, also separate from psychiatry, at almost all medical schools in West Germany.[2]

While researching the history of my department, I joined an already existing group of (then) medical students who looked into the history of the Heidelberg psychiatry department during the Nazi period, particularly the psychiatric research in the context of patient killings and the roles of psychiatrists Schneider and Rauch. After overcoming obstacles to gaining access to various archives, another disquieting picture emerged. Using patient files and physicians' correspondence for our analysis, we demonstrated that the psychiatric research practice had not been the work of a few fanatic and irrational Nazi doctors: The research question posed by Schneider and his group, namely, the differential diagnosis between genetic and acquired forms of clinically identical psychiatric conditions, was completely rational and in tune with scientific debates of the time. What is more, judged by the scientific standards of the time, their methods were up-to-date, if not innovative (Roelcke et al. 1994). However, the subjectivity and the suffering of the patients was completely neglected in the first phase of the investigations on living probands, and the killing of the patients for immediate correlation of clinical with postmortem

---

[2] A considerably revised and extended English version of an original German publication (Roelcke 1991) on this department may be found in Roelcke (2004).

findings was an obligatory part of the research. We also documented that Rauch had played a crucial role in the research process, as had three other psychiatrists with remarkable postwar careers either in academic contexts or in the German army (Hohendorf et al. 1997; Roelcke et al. 1998).

We also discovered an institutional link between the deadly research at the Heidelberg psychiatric department and the elite German Psychiatric Research Institute in Munich, an institute of the prestigious Kaiser-Wilhelm-Society, which had been a model for the Institute of Psychiatry in London in the 1920s. Further investigations elucidated broader issues. The director of this Munich institute, Ernst Rüdin, who was the president of the German Association of Neurologists and Psychiatrists until 1945, had been a fervent eugenicist and expert advisor for the Nazi sterilization policy (this had been known before), but, at the same time, he was considered an internationally leading figure in the field of psychiatric genetics and epidemiology. For example, as late as the eve of WWII in 1939, he was invited to be a plenary speaker to the World Congress of Genetics in Edinburgh. What is more, scientists who today are considered founding fathers of psychiatric genetics in the United States (Franz Kallmann), Britain (Eliot Slater), and Scandinavia (Eric Essen-Möller) were all postdoctoral researchers at Rüdin's institute during the Nazi period.[3] After Kallmann emigrated from Germany in 1936 (due to his Jewish origins) and Slater and Essen-Möller returned to their native countries, they had slightly modified, but not given up the eugenic motivations for their medical genetic research.[4] Finally, it turned out that Rüdin himself had supported the deadly genetic research on children in Heidelberg from the budget of the German Research Institute, and had himself situated this research within the broader measures aimed at the killing of handicapped children (Roelcke et al. 1998; Roelcke 2000).

All this (and similar historical research by colleagues) implied that the dominant postwar images and narratives of the atrocities of Nazi medicine were quite inadequate. In fact, I found that the issues involved were not specific for medicine during the Nazi period but had much earlier origins and broader implications: Backward scientists and fervent Nazis were not the only ones propagating and practicing eugenics (or racial hygiene, the term commonly used in Germany and

---

[3] For the Munich institute as an international "Mecca" for young researchers in psychiatric genetics and epidemiology, see Roelcke (2006, 2013).

[4] For example, Kallmann (1938b) talked of the "danger of the development of new schizophrenic cases" arising from the "unions" of heterozygous carriers of the supposed genes: "From a eugenic point of view, it is particularly disastrous that these patients not only continue to crowd mental hospitals all over the world, but also afford, to society as a whole, an unceasing source of maladjusted cranks, asocial eccentrics and the lowest types of criminal offenders. Even the faithful believer in the predominance of individual liberty will admit that mankind would be much happier without those numerous adventurers, fanatics and pseudo-saviours of the world who are found again and again to come from the schizophrenic genotype" (p. 105). He concluded that "there should be legal power to intervene, in addition to the general eugenic program of the biological education of all adolescents, marriage counsel, obligatory health certificates for all couples applying for a marriage license, and the employment of birth control measures" (p. 113). See also Kallmann (1938a); Slater (1948); Roelcke (2013a).

Scandinavia).[5] Rather, up until the 1940s, many, if not most of scientifically up-to-date medical geneticists in Germany, Scandinavia, Britain, and the United States were motivated by eugenic/racial hygienic concerns. The rationale for their genetic endeavors was a scientifically grounded biological improvement of the population.

Similarly, the programs of systematic patient killing were not the result of an ideological craze invented by Nazi politicians and imposed on physicians. Rather, these programs had been initiated by leading psychiatrists and pediatricians, and they were a logical and radical extension of ideas propagated widely by German and American physicians beginning in the late nineteenth century. These physicians argued that certain groups of people, such as chronic psychiatric patients or severely handicapped newborns, were not fully human. Supposedly, due to their mental incapacities, they were not able to suffer, and they constituted a burden to their families and communities. Physicians further argued that medical experts were able to judge the "value" of such "life" and, after a thorough evaluation by a state committee, these experts should be mandated to recommend and implement the termination of lives of low value. These pre-existing ideas were put into practice in the "Third Reich": Physicians in close cooperation with Nazi institutions pursued a radical and efficient "deliverance" of the collective *Volkskörper*, or German people's body.

Furthermore, the cases of atrocious human subject research were not, at closer look, the activities of a few fanatic and sadistic doctors acting beyond any standards of science. Rather, the research questions pursued in most cases were up-to-date and, in part, related to military questions during wartime. Scientists of international reputation, including Nobel prize winners such as Adolf Butenandt and Richard Kuhn as well as representatives of prestigious research institutions, were directly or indirectly involved in this research.[6] As a matter of fact, rational and brilliant medical scientists searched for research settings with the least regulations or where research subjects had no rights. In these contexts, scientists could pursue the answers to their research questions without regard for the well-being or even the lives of their probands.[7] Research subjects were treated exactly like animals, as is illustrated in an exemplary way by the term *Versuchskaninchen* (experimental rabbits) used by physicians in the Ravensbrück concentration camp to describe

---

[5] In the understanding of the time, racial hygiene (or eugenics) was focused on scientifically based studies of and interventions in one's own population to protect or enhance the biological-genetic quality of that population/folk body/"race," whereas—in contrast—racial anthropology looked into the (supposedly biological) differences between distinct populations ("races") in order to protect a particular race by prohibiting the mixing with other races.

[6] For a short overview of the core findings and implications, as well as further reading, see Roelcke (2010). On Butenandt and Kuhn, as well as the involvement of many other distinguished scientists of the elite Kaiser-Wilhelm-Society, see Heim et al. (2010).

[7] The search for deregulated places for research and a value hierarchy that prioritizes the production of new knowledge over the well-being of research subjects may be found internationally throughout the twentieth and early twenty-first century. For recent examples, see Angell (1997) and Rothman (2000b).

their objectified prisoners/human subjects (Weindling 2004; see also Chaps. 4 and 5). This kind of research not only completely neglected the probands' well-being but also systematically excluded the impact of the subjectivity, the quality of social relations, and the biography of a patient on the origin, symptomatology, and course of the suffering individual's disorder, a tendency apparently inherent in a body-centered medicine that had created considerable discomfort for me as a medical student.

A further irritating point for me was the behavior of physicians towards the "new state," as the "Third Reich" was called in contemporary terminology. More than 50 percent of German physicians became members of the Nazi party or one of its affiliated organizations, such as the SS (*Schutzstaffel*) or SA (*Sturmabteilung*) (Forsbach 2006, pp. 39–40). This is a much higher percentage than found in other, comparable academic professions such as lawyers or teachers, and points to a particular affinity of the medical profession for the regime. Conversely, this percentage is a clear indication that physicians were not forced to join—more than 40 percent of physicians were not party members. In fact, historical research has documented the existence of a range of possible behaviors in all fields of medicine. For example, although the Law for the Prevention of Hereditarily Ill Offspring made the reporting of patients with potential genetic disorders to the regional health offices mandatory, an exemplary study illustrated that only a small percentage of physicians in private practice followed this law, in contrast to physicians in public service or academic medicine who were much more willing to comply. Physicians who did not comply with the law did not experience any negative consequences (Ley 2004). Similarly, as far as we know, individual physicians were never coerced to participate in human subjects research[8]—rather, physicians themselves provided the initiative for almost all such research. Taken together, these findings imply that physicians acted much less in response to direct outside pressure than broadly assumed, and much more out of a tendency to adapt to the expectations of those in power or in possession of career resources. Thus, there was some degree of leeway for individual physicians to choose how to act (Roelcke 2006). Again, this pattern of adaptive or opportunistic behavior toward those in power is not at all specific for the Nazi context, but may be ubiquitously found in many other medical contexts.

Armed with this knowledge, I realized that medicine during the Nazi period was a particularly radical manifestation of problematic potentials in modern medicine in general. Thus, looking at Nazi medicine, it appears to me, we are not seeing something peculiar and specific to National Socialism. We are instead presented with an opportunity to study in microscopic detail the central features and dynamics of such problematic thinking, practices, and attitudes that are less visible but always present in medicine.

This growing insight, developing over time from my confrontation with Nazi medicine, led me to devote 20 years of my professional life to explore related

---

[8] The only known exceptions were prisoner physicians who were forced to assist in atrocious research.

questions. I also sought to bring the findings and conclusions to the attention of colleagues and medical students and to encourage (together with other colleagues) medical institutions and professional organizations to confront their Nazi past (e.g., Kolb et al. 2012). I found that the decades of refusal by both individual physicians and medical organizations to confront this past, to acknowledge extreme forms of wrongdoing, and to investigate their origins, were an extreme expression of a broader tendency in the profession not to admit mistakes and not to investigate the genesis and mechanisms of malpractice as described, for example, by David Rothman (2000a).

In sum, for me, confronting medicine during the Nazi period became a process of deep and continuous irritation with powerful repercussions and implications. At the core of this process was an emerging awareness of the persistently problematic potentials present in modern medicine, potentials that may become manifest in adverse economic, social, or political contexts, such as severe recessions or war. I had assumed that a systematic study of medicine and the Holocaust might "vaccinate" contemporary medicine against the temptation to behave badly: Exposing medical students and professionals to the choices made by Nazi physicians might help develop mechanisms for the early detection of similar moral hazards and the avoidance of unethical decisions, and thereby safeguard the physical integrity and well-being of the weak and suffering.

Why then, one might ask, have I decided to turn away from these topics in my professional work? As I now see it, several factors contributed to this decision: First, the continuous occupation with medical atrocities, the responsible physicians, their motivations and rationalities, and the suffering of victims did not lead, as one might assume, to habituation and inurement toward the dehumanizing and often brutal facts. On the contrary, my feelings of shock and shame have increased over time, especially at historic sites. For example, every semester for the last eight years, I have gone with medical students to the psychiatric asylum and euthanasia memorial in Hadamar. My feelings about the "banalities of evil"[9] embedded in the everyday life of this small German town—the provisions for the arrival of the patients earmarked for euthanasia, their final walk to the gas chamber disguised as a shower room, and the crematorium—have intensified and become more terrifying as I detect additional details with each visit.

Simultaneously, while the emotions evoked by the atrocities have intensified, talking about the historical events and their implications has become more and more repetitive and even tedious. I realize that in spite of a continuously growing number of well-researched publications on the topic, many physicians, medical students, and the broader public still clings to the simple explanations that have dominated so many decades of postwar narratives about medicine and the Nazi period, narratives that place blame and responsibility on political forces and a few ideological Nazi physicians. Persistent adherence to these false narratives allows us to characterize Nazi medicine as a perversion of inherently "good" medicine, the

---

[9] Hannah Arendt coined this term in her 1963 book about the Eichmann trial in Jerusalem.

kind of medicine that physicians learn and practice today. In contrast, history shows the inherent ambiguity of medicine itself, and the moral frailty of physicians and medical organizations. Therefore, physicians, medical organizations, and all other individuals and groups involved need to continuously strive to practice medicine in a way that is first and foremost orientated to the well-being of the suffering individual. I realize that it is becoming tiring and even frustrating for me to argue against this naïve, idealistic view of medicine. I have come to the conclusion that it is time for a younger generation of medical historians to develop their own analytical approaches to and narratives of Nazi medicine as well as to create new ways of communicating their insights.

Finally, I find that the core concerns that have motivated my historical work can be articulated and illustrated in ways other than focusing on this specific period in the history of medicine. The Nazi period is a particularly extreme case study in medicine that illustrates core concerns, but it is certainly not the only example suited to these purposes. The history of medical professionalism, the emergence and trajectory of the concept of the "animal model" to study human illness—a model that brackets out the psychological and social dimension of suffering[10]—and the history of the changing dynamics in the physician–patient relationship are but a few topics that have the potential to generate similar considerations and conclusions.

Acknowledgments I am very grateful to Berit Mohr and Shelly Rubenfeld for constructive criticism on earlier versions of this article.

References

Angell, M. 1997. The ethics of clinical research in the Third World. *New England Journal of Medicine* 337: 847–849.

Arendt, H. 1963. *Eichmann in Jerusalem: A report on the banality of evil*. New York: Viking.

Eckart, W.U. 1998. Generalarzt Ernst Rodenwaldt. In *Hitlers militärische Elite*, vol. 1, ed. G. Ueberschär, 210–222. Darmstadt: Primus Verlag.

Eckart, W.U., and Ch. Gradmann. 2006. Hygiene. In *Die Universität Heidelberg im Nationalsozialismus*, ed. W.U. Eckart, V. Sellin, and E. Wolgast, 697–718. Heidelberg: Springer Medizin Verlag.

Forsbach, R. 2006. *Die Medizinische Fakultät der Universität Bonn im "Dritten Reich"*. Munich: Oldenbourg.

Heim, S., C. Sachse, and M. Walker (eds.). 2010. *The Kaiser-Wilhelm-Society under National Socialism*. Cambridge: Cambridge University Press.

Hohendorf, G., V. Roelcke, and M. Rotzoll. 1997. Von der Ethik des wissenschaftlichen Zugriffs auf den Menschen: Die Verknüpfung von psychiatrischer Forschung und "Euthanasie" im Nationalsozialismus und einige Implikationen für die aktuelle medizinische Ethik. *Beiträge zur nationalsozialistischen Gesundheits- und Sozialpolitik* 13: 81–106.

---

[10] On the origins and implications of the concept of the "animal model" to study human disease and to gain knowledge of potential interventions, see Roelcke (2013b).

Kallmann, F. 1938a. Eugenic birth control in schizophrenic families. *Journal of Contraception* 3: 195–199.

———. 1938b. Heredity, reproduction, and eugenic procedure in the field of schizophrenia. *Eugenical News* 23: 105–113.

Kolb, S., P. Weindling, V. Roelcke, and H. Seithe. 2012. Apologising for Nazi medicine: A constructive starting point. *The Lancet* 380: 722–723.

Latour, B., and S. Woolgar. 1979. *Laboratory life. The social construction of scientific facts.* London: Sage Publications.

Ley, A. 2004. *Zwangssterilisation und Ärzteschaft: Hintergründe und Ziele ärztlichen Handelns 1934-1945.* Frankfurt: Campus.

Mitscherlich, A., and Mielke F. eds. 1947. *Das Diktat der Menschenverachtung,* Heidelberg: Lambert Schneider. English edition: Mitscherlich, A., and Mielke, F. eds. 1949. *Doctors of infamy: The story of the Nazi medical crimes* (trans: Norden, H.). New York: Schuman.

Platen-Hallermund, A. 1948. *Die Tötung Geisteskranker in Deutschland.* Frankfurt: Verlag der Frankfurter Hefte.

Remy, S. 2002. *The Heidelberg myth. The Nazification and de-Nazification of a German university.* Cambridge, MA: Harvard University Press.

Roelcke, V. 1991. Die Zähmung der Psychoanalyse durch öffentliche Institutionen: Zur Gründungsgeschichte der Heidelberger Psychosomatischen Klinik. *Psychoanalyse im Widerspruch* 6: 13–26.

———. 2000. Psychiatrische Wissenschaft im Kontext nationalsozialistischer Politik und "Euthanasie": Zur Rolle von Ernst Rüdin und der Deutschen Forschungsanstalt/Kaiser-Wilhelm-Institut für Psychiatrie. In *Die Kaiser-Wilhelm-Gesellschaft im Nationalsozialismus,* ed. D. Kaufmann, 112–150. Göttingen: Wallstein.

———. 2004. Psychotherapy between medicine, psychoanalysis, and politics: Concepts, practices, and institutions in Germany, c. 1945–1992. *Medical History* 48: 473–492.

———. 2006. Funding the scientific foundations of race policies: Ernst Rüdin and the impact of career resources on psychiatric genetics, ca. 1910-1945. In *Man, medicine, and the state: The human body as an object of government sponsored medical research in the 20th century,* ed. W.U. Eckart, 73–89. Stuttgart: Franz Steiner.

———. 2010. Medicine during the Nazi period: Historical facts and some implications for teaching medical ethics and professionalism. In *Medicine after the Holocaust. From the Master Race to the human genome and beyond,* ed. S. Rubenfeld, 17–28. New York: Palgrave Macmillan.

———. 2013a. Eugenic concerns, scientific practices: International relations and national adaptations in the establishment of psychiatric genetics in Germany, Britain, the US and Scandinavia, 1910-1960. In *Baltic eugenics: Bio-politics, race and nation in interwar Estonia, Latvia and Lithuania 1918-1940,* ed. B. Felder and P.J. Weindling, 301–333. Amsterdam: Rodopi.

———. 2013b. Repräsentation – Reduktion – Standardisierung: Zur Formierung des "Tiermodells" menschlicher Krankheit in der experimentellen Medizin des 19. Jahrhunderts. In *Tier – Experiment – Literatur, 1880-2010,* ed. R. Borgards and N. Pethes, 15–36. Würzburg: Königshausen & Neumann.

Roelcke, V., G. Hohendorf, and M. Rotzoll. 1994. Psychiatric research and "euthanasia": The case of the psychiatric department at the University of Heidelberg, 1941-1945. *History of Psychiatry* 5: 517–532.

———. 1998. Erbpsychologische Forschung im Kontext der "Euthanasie": Neue Dokumente und Aspekte zu Carl Schneider, Julius Deussen und Ernst Rüdin. *Fortschritte der Neurologie und Psychiatrie* 66: 331–336.

Rothman, D. 2000a. Medical professionalism – Focusing on the real issues. *New England Journal of Medicine* 342: 1284–1286.

———. 2000b. The shame of medical research. *New York Review of Books* 47(19): 60–64.

Slater, E. 1948. Genetics, medicine, and practical "eugenics". *Eugenics Review* 40: 62–69.

Weindling, P.J. 2004. *Nazi medicine and the Nuremberg trials: From medical war crimes to informed consent.* Basingstoke: Palgrave Macmillan.

Chapter 22
Teaching the Holocaust to Medical Students: A Reflection on Pedagogy and Medical Ethics

Joseph J. Fins

Every year that I teach medical ethics to second year medical students at Weill Cornell Medical College in New York, I worry that they won't believe what I have to tell them about the Holocaust and Nazi medicine. To today's medical students, that war, *the War*, is what their grandparents fought. For a generation that is often ahistorical and so forward-looking that they ignore the past, the Second World War and the Holocaust are foreign and distant. It all reflects a time that they have a hard time imagining. All of it happened 70 years ago. It is a lifetime away and almost inconceivable when memories recede into the past and the politics of the Holocaust are conflated with current atrocities and global politics. One oppression is contrasted with the other as if the comparison cancels out evil, instead of making us more attentive to past and present transgressions.

These are large pedagogical challenges. So as a teacher of medical ethics and the Holocaust, I seek to remain relevant. As I work through an introductory lecture concerning the evolution of medical ethics in the twentieth century, my perspective is distinctly parochial. I tell the larger history of medical ethics through the prism of local faculty and events as seen from our medical center, New York Presbyterian

J.J. Fins, MD, MACP (✉)
Chief, Division of Medical Ethics; The E. William Davis, Jr., M.D. Professor of Medical Ethics; Professor of Medicine; Professor of Public Health; Professor of Medicine in Psychiatry; Weill Cornell Medical College, New York, NY 10021, USA

Director of Medical Ethics & Attending Physician; New York Presbyterian Hospital-Weill Cornell Center, New York, NY 10021, USA

Adjunct Faculty & Senior Attending Physician; The Rockefeller University & The Rockefeller Hospital, New York, NY 10021, USA

Co-Director, CASBI, Consortium for the Advanced Study of Brain Injury, Weill Cornell and Rockefeller University, New York, NY 10021, USA

New York Presbyterian-Weill Cornell Medical Center, 435 East 70th Street, Suite 4-J, New York, NY 10021, USA
e-mail: jjfins@med.cornell.edu

S. Rubenfeld and S. Benedict (eds.), *Human Subjects Research after the Holocaust*,
DOI 10.1007/978-3-319-05702-6_22, © Springer International Publishing Switzerland 2014

Weill Cornell. Using vintage photographs from our archives, I try to make a link between our students and their predecessors.

Fortunately, there is a lot of historical material, from the rise of the doctrine of informed consent to the Karen Ann Quinlan case and the origins of the vegetative state and right-to-die movement (Fins in press). In between is the more generic story of the medical college at war in Europe and the Holocaust's impact on medical ethics. It was that part of the lecture that I worried had become increasingly attenuated over the years, as the Holocaust faded deeper into the past.

But this year's class would be different because I had participated in the 2012 Center for Medicine after the Holocaust (CMATH) conference in Houston. How that happened and how CMATH influenced my teaching and my own thinking is what this essay seeks to explain. It is a story of transformation and transcendence.

Teaching Medical Ethics

Because of the centrality of informed consent and the rise of patient autonomy to modern medical ethics (Fins 2001), I start the course with a discussion of the landmark informed consent case, *Schloendorff versus the Society of the New York Hospital*. The case adjudicated a lawsuit related to surgery done at The New York Hospital (our predecessor institution) in 1908 (*Schloendorff v New York Hospital*, 1914).

As the case transcripts tell it, Mary Schloendorff came to New York in 1906 after the earthquake in San Francisco. She was a guitar teacher and came to medical attention for the evaluation of stomach pains. Along the way, she was to have a pelvic examination under anesthesia—"an ether exam." Before the advent of imaging techniques, this was how clinicians ruled out the presence of a pelvic mass.

From what can be discerned from the trial court transcript, Mrs. Schloendorff agreed to the ether exam. But when she went to the operating room for the evaluation, there was confusion about the scope of the assessment as reported in a conversation between her and the clinician charged with providing anesthesia. She asked him what was to be done. She recounted, ". . . and he said he was going to give me gas." When she pressed him, asking if they were going to operate on her, he responded, "I don't know; I am simply detailed to give gas; that is all; I don't know" (*Schloendorff v New York Hospital*, 1914).[1]

She became alarmed and asked, "I want to see somebody—show me somebody, I want to tell them that I am not to be operated on." And then he replied, in a most unreassuring fashion, "Did you come for operation?" She answered that she had not and that, "I came for ether examination." He responded, "then you will only receive ether examination" (*Schloendorff v New York Hospital*, 1914).

---

[1] All quotations from the trial court are verbatim from the transcript. Usage errors have not been corrected to preserve the narrator's historical authenticity.

But Mrs. Schloendorff was not satisfied by his response and was terribly scared. She tried to escape the operating room and get away, but to no avail. Her testimony tells of her being forcibly treated, against her will:

> But I was frightened and tried to get up; I tried to get off the litter and get away. And he pushed me—I could only raise my hand; all this time I was so frightened and so nervous. He had some apparatus there with a rubber tube and mouthpiece, and he took his hand and pushed against my forehead and pushed me back, and put the mouth-piece to my mouth and said, "Take a deep breath." I was frightened at the gas and tried to get up, but took a deep breath, I guess, and did not know any more. (*Schloendorff v New York Hospital*, 1914)

Without her permission, Mrs. Schloendorff underwent a hysterectomy for the treatment of fibroids. The surgery was complicated by emboli to her fingers, which was a problem because Mrs. Schloendorff was a music teacher and a guitar player (Lombardo 2005). Her surgeon was Lewis Atterbury Stimson, surgeon-in-chief at The New York Hospital and father of Henry L. Stimson, who would become Secretary of War under Presidents Taft, Franklin Roosevelt, and Truman (Hodgson 1990).

I share this case with our students, not only for its historic provenance but also for its ability to teach. After I read Mrs. Schloendorff's testimony of her encounter with the anesthetist, I ask rhetorically, "Does anyone have a problem with their encounter?" Students quickly point out that there was no informed consent for the surgery, and I gently remind them that at that juncture, early in the century, patients were not routinely included in discussions about what was in their best medical interest (Katz 1986).

Other comments point to the patient being forcibly treated against her will; even if she had been given all the information, she could not leave. I take such comments to introduce the notion of bodily sovereignty and privacy, concepts that will inform later legal thinking about informed consent.

I also respond by introducing the concept of voluntariness, the ability to "walk with one's feet," that one should never feel captive to a treatment or a clinical trial. It is the ability to leave voluntarily, by one's own free will, and terminate a situation that is not desired. Voluntariness is critical to informed consent because without it, all the disclosure and dialogue that are part of the informed consent process would be for naught. While choices would exist in theory, they would not exist in practice. I note that the only time the doctrine of informed consent is tested is when a patient *refuses* a doctor's recommendation. Later in the session, I return to voluntariness when speaking about the Nuremberg Code.

But first I share the progress of the Schloendorff case on appeal, when it went to the highest court in New York State, the Court of Appeals in 1914. There, Judge Benjamin Cardozo—who would later be appointed to the US Supreme Court—wrote a truly landmark decision on informed consent:

> Every human being of adult years in sound mind has a right to determine what shall be done with his own body; and a surgeon who performs an operation without the patient's consent commits an assault for which he is liable in damages ... This is true except in cases of emergency when the patient is unconscious and when it is necessary to operate before consent can be obtained. (*Schloendorff v New York Hospital*, 1914)

Cardozo's considerable rhetorical skills (Posner 1990) gave birth to a brief articulation of informed consent that would stand the test of time, even though decades would pass before legal thinking and clinical practice would incorporate his jurisprudence (Fins 2006b). A year after he died, the Harvard, Yale, and Columbia law reviews all joined forces to pay tribute to his life and work. Interestingly, the Schloendorff case was remembered for its ruling on the "charitable exemption" immunizing not-for-profit institutions against malpractice, not the landmark statement on informed consent buried in his decisions (Stone et al. 1939). Of course, today the former is forgotten and the later is recalled whenever informed consent is discussed. It would take a war and its aftermath to effect that change.

Cornell Medical College at War

That change began with the mobilization of young Cornell medical students for the medical corps during the war. I share archival photos of their predecessors training for combat: the first and second year students being instructed by enlisted men, learning medical evacuation and how to work as medics; the more senior students being instructed with large field surgical chest sets in preparation for combat surgery for which, as novices, they are surely ill-prepared.

The look of these photos is crisp and clear. The young medical students are dressed in reassuring white coats. Their mission is sound and right. These boys will be the men who Tom Brokaw celebrates as "the greatest generation" (1998).

These images are then contrasted with a photo taken from the Office of the Chief Counsel for War Crimes at Nuremberg depicting a freezing experiment at Dachau conducted by two physicians, Professor Doctor E. Holzloehner of the Medical School of the University of Kiel and Dr. Sigmund Rascher, both wearing the uniform of the Third Reich. Despite his lack of qualifications as an investigator, Rascher "took charge of the project" with the help of part-time consultants, including Holzloehner (Berger 1990). More egregiously, Rascher continued the studies alone after Holzloehner and another consultant thought the studies concluded. With the backing of Reichsfürher Henrich Himmler, they were seeking to understand the effects of hypothermia, the limits of the body's endurance in freezing water in order to understand what might happen if a Nazi pilot was shot down over the North Atlantic. They also did equally gruesome experiments on the effects of rewarming (Berger 1990).

In the image, the two doctors sit next to a tub of water filled with large blocks of ice in which a political prisoner is floating suspended by a "life-preserver." His hands seem tied in the grainy image and he is blindfolded. It is impossible to tell if he is unconscious, shivering, or still alive.

The historical literature suggests that 360–400 experiments involving submersion in temperatures between 2° and 12°C were performed on 280–300 individuals, suggesting that some underwent the ordeal more than once. Later testimony from assistants indicated that between 80 and 90 victims died, although there were only two known survivors (Berger 1990). Some died in the unspeakable cruel process of "rewarming" when victims were put into a tub of boiling water (Pacholegg 1947). That one fact is a moral indictment of the whole enterprise.

But what of the science? With respect to its quality, experts have concluded that the studies were uncontrolled with respect to important variables like baseline temperature and other vital signs and demographics of the subjects. These omissions, coupled with the principal investigator's inadequate training and history of duplicity make the resultant data highly suspect (Berger 1990).

As the image of Holsloehner and Rascher is projected, I speak of what these men did, deliberately speaking of "our" profession to illustrate that they were like us, healers and helpers, doctors and professors who committed these atrocities. Their justification? If Nazi soldiers could put themselves at risk of being shipwrecked or shot down at sea, should not prisoners also contribute to the war effort? I then tell my students that similar high-altitude studies were also done, by Rascher and others, at Dachau to mimic hypoxia and decompression disasters that might befall a pilot, in which subjects were brought to and over the brink of suffocation (US Holocaust Memorial Museum n.d., "Nazi Medical Experiments").

I then tell the class of the complicity of German doctors and that physicians had the highest percentage membership rate of any profession in the Nazi Party (Barondess 1996; Taylor 2011). Indeed, the number of doctors who committed crimes was so large, and the actions so heinous, that the Doctors' Trial was the first of the 12 Nuremberg war trials conducted by the United States after the trial of the major Nazi war criminals conducted by the International Military Tribunal. Simply put, the Holocaust as we now know it could not have occurred without the complicity of the German medical establishment and scores of rank and file physicians (Barondess 1996).

I then recount briefly the "utilitarian" role that the profession played early in the war in euthanizing those who were infirm and the "mentally defectives" in order to clear hospital beds for wounded German soldiers who were injured on the Western Front and how these early efforts of state-sanctioned physician killing grew into the more ambitious and systematic efforts to eliminate European Jewry in the concentration camps.

Over the years, as I tell students of these atrocities (Annas and Grodin 1995), my sense as a teacher is that they have either heard it all before and, thus, have defended themselves against the hurt and threat to their own emerging professional selves, or that they simply cannot comprehend that something so bad could actually happen, much less happen again. Either way, they are highly defended psychologically, and these defenses can be very hard to breach.

The concept of "evil doctors" is a hard concept for young medical students (Lifton 1986), who are new to the profession and full of idealism. To accept the fact that were it not for their newly chosen profession, the Holocaust would not have happened is almost too much to bear and nearly impossible to believe. Their enthusiasm for healing and helping is so far removed from the transfiguration of the profession under National Socialism that the unspeakable acts committed by their medical brethren seem distant and far-removed from them, their experiences, and their future careers.

But of course, none of that is true. The potential of good people going far astray is a cancer that lies dormant in all of us. The "banality of evil" was just that (Arendt 1963), normal Germans transformed into monsters by day and returning home as loving mothers and fathers by night. It is a potentiality that must be raised precisely because it is a human potentiality that we all share. This history must be recognized to prevent its recurrence, or at least to mitigate the threat.

We would not teach the Holocaust if we did not think that knowledge of the past might immunize us from some future evil. So I do not want my medical students to stray when I teach the Holocaust. I do not want them to view the history as irrelevant to them because their time and place is different or because the acts of cruelty that were performed tests the credulity of the rational mind. No, they must not stray. They must remain engaged. *This* is just too important.

So as I project a picture of the dark paneled courtroom, I tell them that Nuremberg was so well-suited to prosecution of Nazi war crimes because the early racial anthropology laws were promulgated there (Shirer 1960). I add that much of the laws' eugenic provenance was from work done on Long Island at the Cold Spring Harbor Laboratory early in the century (Allen 1986). And to locate Nuremberg in their visual memories, I remind them that it was the city where Hitler had his massive torch-lit rallies, where *der Volk* (the people) were stirred into a hypnotic frenzy (Persico 1995).

And then, when I return to eugenics, I step out from behind the speaker's podium to the center of class and tell my students that what I am telling them is real. The images of Nuremberg rallies were not the creation of Hollywood but of a highly civilized culture gone mad. And as I try with all my being to make them feel the sorrow and grief I feel in my heart, I surprise them and comment on the parquet floor in front of the class. As I step out on to it I tell them that it reminds me of a dance floor—and I tell them of a wedding I attended in the mid-1990s.

A Wedding Dance

It is a departure from the expected script and it catches their attention. The non sequitur causes them to look up and wonder what I am talking about as I describe the wedding.

There was a woman wearing a chiffon dress—a light blue, robin's egg blue dress of the sort that could only be worn at a wedding. She was dancing on a parquet floor much like the floor beneath my feet. But unlike all the others out there celebrating, she was in a wheelchair, being swirled around by her partner. As my wife and I moved closer on the dance floor, I asked her, "Who is that lady?"

My wife, Amy, a dear friend of the bride's mother told me that she was the aunt of the bride. As I got closer, intrigued by her spirit and enjoyment of the moment, I looked at her forearm and saw blue-green numbers etched on the dorsal surface perpendicular to the radius and ulnar.

I had never seen the tattooed serial numbers of a Holocaust survivor before and the image seized me. Fifty odd years removed from a horror I would never completely know, this woman was enjoying life and celebrating a Jewish wedding in America. She was free. She had survived and her family had not been completely snuffed out.

And then I asked Amy why she was in a wheelchair? Oh, Amy told me, she was a Mengele twin and had multiple sclerosis (MS). Her brother—Amy pointed him out across the dance floor—was the *control*. I looked over at him and saw a healthy, good-looking man dancing with ease and enjoying the evening. He was not in a wheelchair.

To some extent, it was reassuring and satisfying that she was experiencing joy. I hoped it compensated for whatever pains she had experienced and that at least that night, the night of the marriage of her niece, she could forget what had happened decades before. That was my hope as I, with sober tears in my eyes, danced with my wife.

As my vision cleared and I looked over at that proud woman in the wheelchair at her niece's wedding, the doctor in me began to think about her illness and its etiology. We still do not know what causes MS, but I wondered if all the toxic experimentation and exposures had somehow prompted the inflammatory process that is the hallmark of the disease. We will never know if it was causative, but then again, the etiology of MS was not the point of my lecture to the medical students. It was about a whole different sort of inflammatory process.

Houston and Eva

Since 1998, I have taught this class, and to the best of my recollection, I have always spoken of the Mengele twins I had seen at that wedding, and this year was no exception. But as I returned from the Center for Medicine after the Holocaust conference at Houston Methodist Hospital in December 2012, my plans for the class had changed.

I was honored to have been asked to speak at the conference about my work exploring the ethics and neuroscience of patients with disorders of consciousness, those in the vegetative and minimally conscious state (Fins 2005, 2006a; Fins et al. 2007; Fins 2008). I spoke of what society owed patients and families who were in these liminal states of consciousness (Fins 2013), arguing that consciousness was a civil right that can neither be ignored nor abridged (Fins 2010; in press). I also spoke of the complicated question of informed consent, in part a legacy of the Nuremberg Code, which made research on patients who were decisionally incapacitated so ethically fraught (Fins 2000; Fins and Miller 2000). I argued that we needed to be cautious not to equate respect for persons with informed consent, maintaining that consent is only possible in those who can provide it and that efforts to restore functional communication in those who were silenced by brain injury was a laudable intervention that should be sustainable by surrogate authorizations (Schiff et al. 2007; Giacino et al. 2012).

That was what I intended to do when I arrived in Houston, but what I did most was *listen*. Each lecture was more compelling than the next. But the talk that most captivated me was the one given by Eva Kor, a survivor who told her story as a prisoner at Auschwitz (CANDLES n.d.). It was her presence, that actual living link that made her testimony so incredibly powerful. For the rest of my sentient life, I will remember her description of her arrival at the camp with her family. It was horrific and the human pathos overwhelming.

As her family disembarked from the train that transported them to Auschwitz, Eva quickly realized that her father and two older sisters had vanished. They were never to be seen again. Then Eva, her mother, and sister walked towards a concrete slab for an initial inspection by an SS Officer. I do not exactly recall how Eva described it, but the feeling was that there was no place on earth, no place so small, where evil loomed so large. On that slab, a soldier asked her mother, "Are they twins?"

Eva's protective mother, even then wanting what would be best for her children, asked, "Would it be good that they are twins?"

He answered, "Yes."

"Well, then they are twins."

And with that Eva and her twin were grabbed by the SS Officer, one in each hand and scuttled off. Eva, in a heart-wrenching moment, tells of her turning to look at her mother one last time. She was leaning forward with her arms extended, reaching out to her two little girls, like a sculpture for all eternity. That would be the last image Eva would have of her mother whose life was taken at Auschwitz. Her mother's admission would lead Eva and her sister Miriam straight to Dr. Mengele's lab and helped assure their survival because they were set aside from the gas chambers to be used for "experimentation."

As I listened to Mrs. Kor talk, I could only imagine a child of her age fending for herself, seeking comfort from her sister, and making it through to the end. She survived although she nearly died of sepsis. She made it through by stealing potatoes from the kitchen in order to maintain a minimal calorie count. An intuitive child, Eva learned quickly that the conventional sanctions for stealing potatoes did not apply to her because she was a Mengele twin. Paradoxically, she was under his protection.

It was a gripping talk and as she finished, I wondered, might she know the woman I had seen at the wedding? It was curious because until that moment, I knew her only as the aunt of the bride. I did not know her name. She was an image, a symbol that represented an historical moment, a pedagogical device, an historic relic. Although I used her story to condemn her exploitation at the hands of Dr. Mengele, I, too, had objectified her. Although I had used her story for the laudable purpose of educating students about the Holocaust, that itself was an insufficient excuse for not knowing *who* she was.

I only realized this omission as I walked across the front of the conference room in Houston toward the other corner where Eva Kor was standing, surrounded by other members of the audience. Finally, it was my turn to say something. I asked her if she knew the sister-in-law of my wife's friend who had been a Mengele twin and had multiple sclerosis. And before I could confess that I did not know her name, she volunteered, "Oh, Irene. . ."

I was relieved and amazed: Eva *actually knew* Irene. She had lived with her at Auschwitz . . . It was an historical validation of their story. Here was someone who had experienced what Irene had experienced, seen what she had seen, *survived* what she had survived. They bore witness for each other, a sorority of survivors whose presence calls for remembrance.

The recognition of, "Oh, Irene. . ." also spoke to the sad fact that only a select—and ever smaller—group of people can ever know what these poor souls had experienced. Each had become precious to the other as a reservoir of a shared reality that the rest of us cannot know.

But there was something more. Eva's "Oh, Irene. . ." had given this woman a name, had de-objectified her and returned her to personhood from an anonymous Holocaust icon. And at least in my moral imagination, I realized that my early descriptions of *Irene*, though pedagogically dramatic and useful in bringing the Holocaust into my classroom, failed to bring her personally into that space. Even though I had told her story, I had only told part of it. And even though I had done so in the service of memory and remembrance of things past, it was a moral omission because there was a personal story still hidden in the past.

Perhaps it is a stretch but my oversight, in not knowing their names, or not viewing them as persons, made me recall what Hannah Arendt (2003) had profoundly noted in her last book, *Responsibility and Judgment*. Writing of the rise of Nazism, she observed that, "In brief, what disturbed us was the behavior not of our enemies but of our friends" who failed to engage in "personal judgment." It was the loss of that critical function, its "almost universal breakdown," which led to the Nazi's early success and all that followed (all quoted phrases Arendt 2003, p. 24).

In my lectures on the Holocaust and invocation of the evils of Dr. Mengele and the image of that woman in the wheelchair, I too had been a "friend" on the side of right. But my "omission," to borrow Arendt's word, was *disturbing* because I had engaged, albeit in a minimal fashion, in the very objectification or dehumanization of a person that could lead to evil. It was a sobering and chilling realization and may appear to be a bit of hyperbole to you, the reader, but I am glad to have made it. It is only by such insights that we regain the requisite personal judgment necessary to counter the forces of dehumanization that can lead us astray, even when we think we are pursuing righteousness.

After Eva's "Oh, Irene. . ." I hope my susceptibility has been tempered. I would hope that my failure to name, and to know by name, would be decreased as would any "innocent" propensity towards depersonalization of any Holocaust victim. Thanks to my experience in Houston, I have learned to share these stories with the respect that they, representative of persons, each deserve. The Center for Medicine after the Holocaust (CMATH) conference affirmed that each memory is precious. It had *personalized* the Holocaust in a way that I had never experienced before.

A Different Holocaust Narrative

The personalization of the Holocaust, which occurred so powerfully in Houston, was vitally important to me. It provided another perspective, from which the good fortune of my family had shielded me. In contrast to many Jewish-American families, mine had been spared the pain of a direct loss from the Holocaust. Unlike Daniel Mendelsohn, author of *The Lost: A Search for Six of Six Million* (2007), I did not grow up in the shadow of loss or with the burden of carrying the legacy of a victim.

As a teenager, the Holocaust had been more of an idea that defied understanding than the extension of a family's grief. When I thought of my family and the Holocaust, I thought of my father, Herman Fins. His experience did not make me feel the usual sting of guilt or victimization experienced by some children of survivors.

Instead, his experiences prompted a counter-story quite different than the usual Holocaust narrative of victimization. It was its opposite: one of empowerment, borrowing a nosology offered by Rabbi Michael Berenbaum in his brilliant volume, *After Tragedy and Triumph: Essays in Modern Jewish Thought and the American Jewish Experience* (1990). Berenbaum, the former director of the Research Institute of the US Holocaust Memorial Museum, considered the ways postwar American Jewry has incorporated the Holocaust into its consciousness. He described a polarity of victimization against empowerment: The former occasioned by the terror of the Shoah and the later engendered by the defeat of Hitler and the birth of Israel as a state (Berenbaum 1990).

My father's experience was distinctly in the lineage of Berenbaum's empowerment saga. His was a soldier's story, or more precisely, that of a combat medic with the US Army, 38th Signal Battalion. He had landed at Normandy a couple weeks after D-Day and tended the wounded as his mates fought their way through France and Germany. Although he mostly shielded my sister and me from the horrors of the war, he did tell us of two notable experiences related to the Holocaust.

The first was when his unit crossed into Germany and liberated a factory of Jewish women prisoners who were assembling fuses for torpedoes and bombs. With the intercession of Army Chaplain Rabbi Lieutenant Joseph Shalom Shubow of Boston, the women were freed and placed in a nearby German town that had been appropriated for this purpose (H. E. Fins, personal communication, March 17, 2013). Later, near the end of the war, my dad celebrated Passover with a seder in Josef Goebbels's castle, led by Rabbi Shubow. Evidently the young lieutenant convinced Ninth Army General John P. Anderson to allow the use of the castle for this purpose (Jewish Virtual Library n.d.).

An *Associated Press* account, dated March 30, 1945, carefully handed down to me by my father, notes that Corporal Herman Fins served as a waiter for the ecumenical feast. The newspaper account highlighted the paradox of Jewish soldiers celebrating Passover in the *Schloss* (castle) of Hitler's propaganda minister. In it the Jewish soldiers are portrayed as the victors (Jewish Virtual Library n.d.). Theirs was not a tale of victimization but rather of liberation:

> Shubow delivered his sermon standing in front of a long oaken speakers' table. At the opposite end of the room was a large oil painting of Adolf Hitler, and beside it was a cardboard portrait of a dirty-faced American doughboy wearing the 29th Division shoulder patch. The 29th took Muenchen-Gladbach less than a month ago. ("G.I.s Observe Passover," 1945)

The account continued with an excerpt from Rabbi Shubow's sermon, invoking themes from Exodus, "Thank God, even as our ancestors were liberated we can see all of Europe being liberated soon from Nazism" ("G.I.s Observe Passover," 1945). My father's recollection of the seder, drawn from an interview conducted by my son, Harry Fins, for a Hebrew school project, captured the triumphant spirit of the Jewish soldiers who helped liberate Europe. Speaking of that Passover seder in 1945, my father recalled:

> It was jubilation because we were doing something good and genuine. Doing it in Hitler's #2 or #3 man's house, the man most hated by the Jews... And we were celebrating a Jewish holiday with a Jewish Rabbi and 800 soldiers ... Teaching a lesson. The Nazis were beaten and it was the end of the War. (H. Fins 2012)

This became my understanding of the Holocaust as I was growing up. It was more a parable than the tragedy that it was. Although I was horrified by Hebrew school lessons, a viewing of *Night and Fog*, in fifth grade, and a reading of Elie Wiesel's *Night* (2006), soon thereafter, my views of the Holocaust grew out of my dad's experience. His story, epitomized by the doughboy Yank next to a defeated Hitler, was that right overcame might and good prevailed. May it always be so, although we should never let the outcome obscure the intervening tragedy.

CMATH brought me to these realizations. It brought me closer to the lived experience of not only those who fought the evils of National Socialism but those who experienced its oppression. As one who was lucky enough to have been born far from the Shoah, I will long be grateful to CMATH for giving me the privilege of hearing Eva Kor and helping me apprehend a human tragedy, so large that it can only be understood one life at a time.

Back in Class

In January 2013, I again returned to the podium to teach medical ethics to students at Weill Cornell Medical College. Most of my slides were the same, and yet the lecture was different. I approached the task with a sense of duty and urgency that I always thought I had possessed but whose absence was only apparent in retrospect.

This year, when I stepped out onto the parquet floor to discuss that woman in a wheelchair swirled about at her niece's wedding, I told my students who she was. After Houston, I had read that Irene Hizme (Guttman) was born in Teplics Sanov, Czechoslovakia in 1937 (Hizme and Slotkin 1995). When the Germans invaded Bohemia and Moravia in 1939, she was living with her family in Prague. Her father was arrested there and deported to Auschwitz where he was killed in 1941. Irene and Rene were deported with their mother to the Theresienstadt Ghetto—where Rabbi Leo Baeck continued his ministry (Friedlander 1991)—and then to Auschwitz, when she was just a four-year-old child. There, the children were separated from each other and became part of the "experimentation" done by Josef Mengele. They were reunited in the United States in 1950 (Hizme and Slotkin 1995).

But my focus was on Eva, especially when the dance floor beneath my feet became a concrete slab at Auschwitz where she and her sister Miriam had been separated from their mother. Unlike past years, this year my discussion of the Holocaust, of the Mengele experiments had become personal, less abstract. I was able to share with my students a *personal* connection with the Holocaust and Mengele's studies of inheritance and rightly honor the personhood of Eva, Irene, and her brother Rene.

I had seen Mengele twins with my own eyes. Mine was now an extension of their personal testimony. It was to show them that the Holocaust was not an abstraction about six million Jews and millions of *others*, it was about this one lady on the dance floor and of course, her brother—people who had names. And it was about a family ripped apart, of two little girls standing with their mother for one last, precious moment on a concrete slab in Auschwitz. It was about what the Belmont Report would later describe as "respect for persons" (National Commission for the Protection of Human Subjects 1979), respect for the individual, so cruelly distorted by an errant ideology and an evil perversion of science.

This year, I felt that my time out on the parquet floor was different. As I left that space and walked slowly back to the podium, the room was sad, silent, and reflective. As I advanced the slide to discuss the Nuremberg Code, the students understood why that document mattered (*Trials of War Criminals*, 1949, pp. 181–182; Shuster 1997).

Acknowledgements The author thanks Shelly Rubenfeld and the Center for Medicine after the Holocaust for their collegiality and the reflective space they provide. He also gratefully acknowledges the editorial insights of Rabbi Amy B. Ehrlich and partial support from the Clinical and Translational Science Center (UL1)-Cooperative Agreement (CTSC) 1UL1 RR024996 to Weill Cornell Medical College and its Ethics Core.

References

Allen, G.E. 1986. The eugenics record office at Cold Spring Harbor, 1910-1940: An essay in institutional history. *Osiris* 2: 225–264.

Annas, G.J., and M.A. Grodin. 1995. *The Nazi doctors and the Nuremberg Code: Human rights in human experimentation*. New York: Oxford University Press.

Arendt, H. 1963. *Eichmann in Jerusalem: A report on the banality of evil.* New York: Viking Press.
———. 2003. *Responsibility and judgment.* With introduction by J. Kohn ed. New York: Schocken Books.
Barondess, J.A. 1996. Medicine against society. Lessons from the Third Reich. *Journal of the American Medical Association* 276(20): 1657–1661.
Berenbaum, M. 1990. *After tragedy and triumph: Essays in modern Jewish thought and the American Jewish experience.* New York: Cambridge University Press.
Berger, R.L. 1990. Nazi science – The Dachau hypothermia experiments. *New England Journal of Medicine* 322: 1435–1440.
Brokaw, T. 1998. *The greatest generation.* New York: Random House.
CANDLES Holocaust Museum. n.d. Eva Kor Biography. http://www.candlesholocaustmuseum. org/index.php?sid=26. Accessed 14 Mar 2013.
Fins, H. 2012. *My grandfather's version of the Jewish holidays as a soldier in WWII.* Unpublished manuscript.
Fins, J.J. 2000. A proposed ethical framework for interventional cognitive neuroscience: A consideration of deep brain stimulation in impaired consciousness. *Neurological Research* 22: 273–278.
———. 2001. Truth telling and reciprocity in the doctor-patient relationship: A North American perspective. In *Topics in palliative care*, vol. 5, ed. E. Bruera and R.K. Portenoy, 81–94. New York: Oxford University Press.
———. 2005. Rethinking disorders of consciousness: New research and its implications. *The Hastings Center Report* 35(2): 22–24.
———. 2006a. Affirming the right to care, preserving the right to die: Disorders of consciousness and neuroethics after Schiavo. *Supportive & Palliative Care* 4(2): 169–178.
———. 2006b. *A palliative ethic of care: Clinical wisdom at life's end.* Sudbury, MA: Jones and Bartlett.
———. 2008. Neuroethics & neuroimaging: Moving towards transparency. *American Journal of Bioethics* 8(9): 46–52.
———. 2010. Minds apart: Severe brain injury, citizenship, and civil rights. In *Law and neuroscience, current legal issues*, vol. 13, ed. M. Freeman, 367–384. New York: Oxford University Press.
———. 2013. Disorders of consciousness and disordered care: Families, caregivers and narratives of necessity. *Archives of Physical Medicine and Rehabilitation.* 94(10): 1934–1939.
———. in press. *Rights come to mind: Brain injury, ethics & the struggle for consciousness.* New York: Cambridge University Press.
Fins, J.J., and F.G. Miller. 2000. Enrolling decisionally incapacitated subjects in neuropsychiatric research. *CNS Spectrums* 5(10): 32–42.
Fins, J.J., N.D. Schiff, and K.M. Foley. 2007. Late recovery from the minimally conscious state: Ethical and policy implications. *Neurology* 68: 304–307.
Friedlander, A.H. 1991. *Leo Baeck: Teacher of Theresienstadt.* Woodstock, NY: Overlook Press.
Giacino, J., J.J. Fins, A. Machado, and N.D. Schiff. 2012. Thalamic deep brain stimulation to promote recovery from chronic post-traumatic minimally conscious state: Challenges and opportunities. *Neuromodulation* 15(4): 339–349. doi:10.1111/j.1525-1403.2012.00458.x.
Hizme, I., and Slotkin, R. 1995. Irene Hizme and Rene Slotkin describe deportation to Auschwitz [interview]. Personal histories. United States Holocaust Memorial Museum. http://www.ushmm.org/museum/exhibit/online/phistories/viewmedia/phi_fset.php?MediaId=1148. Accessed 13 Mar 2013.
Hodgson, G. 1990. *The colonel: The life and wars of Henry Stimson, 1867–1950.* New York: Knopf.
Jewish G.I.s observe Passover in Goebbels's Rhineland home. 1945, March 30. *Associated Press.* Dateline: Muenchen-Gladbach, Germany.

Jewish Virtual Library. n.d. Joseph Shalom Shubow. http://www.jewishvirtuallibrary.org/jsource/
    judaica/ejud_0002_0018_0_18421.html. Accessed 19 Dec 2013.
Katz, J. 1986. *The silent world of doctor and patient*. New York: The Free Press.
Lifton, R.J. 1986. *The Nazi doctors*. New York: Basic Books.
Lombardo, P.A. 2005. Phantom tumors and hysterical women: Revising our view of the
    Schloendorff case. *Journal of Law, Medicine & Ethics* 33(4): 791–801.
Mendelsohn, D. 2007. *The lost: A search for six of six million*. New York: Harper Perennial.
National Commission for the Protection of Human Subjects of Biomedical and Behavioral
    Research. 1979. *Ethical principles and guidelines for the protection of human subjects of
    research*. Washington, DC: U.S. Government Printing Office.
Pacholegg, A. 1947. Testimony of Anton Pacholegg. In *Nazi conspiracy and aggression*, Interna-
    tional Military Tribunal. Suppl A, 414–422. Washington, DC: Government Printing Office.
Persico, J.E. 1995. *Nuremberg: Infamy on trial*. New York: Penguin.
Posner, R.A. 1990. *Cardozo: A study in reputation*. Chicago, IL: University of Chicago Press.
Schiff, N.D., J.T. Giacino, K. Kalmar, J.D. Victor, K. Baker, M. Gerber, and A.R. Rezai. 2007.
    Behavioral improvements with thalamic stimulation after severe traumatic brain injury. *Nature*
    448(7153): 600–603.
Schloendorff v New York Hospital, 211 N.Y. 125, 105 N.E. 92. 1914.
Shirer, W.L. 1960. *The rise and fall of the Third Reich*. New York: Simon and Schuster.
Shuster, E. 1997. Fifty years later: The significance of the Nuremberg Code. *New England Journal
    of Medicine* 337: 1436–1440. doi:10.1056/NEJM199711133372006.
Stone, H.F, H.V. Maughan, E. Hand, and L. Hand. 1939. Mr. Justice Cardozo. *Harvard Law
    Review, 52*(3): 353–363. Also printed in *Columbia Law Review, 39*(1) and *Yale Law Review,
    48*(3).
Taylor, F. 2011. *Exorcising Hitler: The occupation and deNazification of Germany*. New York:
    Bloomsbury Press.
*Trials of war criminals before the Nuerenberg Military Tribunals under Control Council Law
    No. 10*. Vol. 2. 1949. Washington, DC: Government Printing Office.
United States Holocaust Memorial Museum. n.d. Nazi medical experiments. *Holocaust Encyclo-
    pedia*. http://www.ushmm.org/wlc/en/article.php?ModuleId=10005168. Accessed 10 March
    2013.
Wiesel, E. 2006. *Night*. Trans: Weisel, M. New York: Hill and Wang.

Chapter 23
No Exceptions, No Excuses: A Testimonial

Mauro Ferrari

I am neither a medical ethicist nor a historian. I have no direct personal or family relationship with the Shoah. I am not Jewish. If anything, because of my Italian origins, I am, in a broad sense, on the wrong side of this conversation. Though I have had substantial exposure to historical renderings of its tragic events, I simply have no title whatsoever to write about the Holocaust, or medical ethics in its aftermath. I was simply an occasional bystander, who happened to be caught in the winds of cyclone Sheldon "Shelly" Rubenfeld.

Early in my tenure at Houston Methodist, I was introduced to Shelly by then President Ron Girotto, who reported about the extraordinary personal impact he had experienced through his participation and support of Shelly's Center for Medicine after the Holocaust (CMATH) and, in particular, its Houston conference and trips to the medical sites of the Shoah. As he was stepping down from his leadership role, Ron expressed his enthusiastic appreciation for CMATH's mission and recommended with great emotion that I continue supporting it.

It has been my great privilege to do so ever since, and a true gift for my soul. The first event we organized together was the 2012 conference in Houston, which provided impetus for this volume. In this testimonial chapter, and despite the inadequacies of my knowledge base and my obviously non-native English writing skills, I will endeavor to provide a measure of the greatness of the gifts I have received and the lessons I have learned through my involvement with Shelly and CMATH's events. The fundamental lessons I have derived in my conscience are summarized in five statements, which form the titles of the sections comprising this chapter. In short: No exceptions. No excuses.

M. Ferrari, PhD (✉)
Ernest Cockrell Jr. Distinguished Endowed Chair; President and CEO, Houston Methodist Research Institute; Director, Institute for Academic Medicine at Houston Methodist; Executive Vice President, Houston Methodist, 6670 Bertner, M.S. R2–216, Houston, TX 77030, USA
e-mail: mferrari@houstonmethodist.org

S. Rubenfeld and S. Benedict (eds.), *Human Subjects Research after the Holocaust*, 283
DOI 10.1007/978-3-319-05702-6_23, © Springer International Publishing Switzerland 2014

Any Discussion of Medical Ethics that Ignores the Lessons from the Holocaust Cannot Have Any Real Depth or Societal Significance

Indeed, that is a strong statement. I first studied medical ethics as a middle-aged medical student in 2002. My framework has remained what I learned then from the monograph of Tom Beauchamp and James Childress, and it is with respect to their four axes of reference that I will articulate my supporting perspective. *Beneficence* and the *Holocaust* is a strident contrast of two words that appear impossible to place in the same sentence. Yet, the perspective of the Final Solution was not articulated as "us vs. them," at least not for the vast majority of the time that lead to its tragic execution. It evolved—a malignant and intentional choice of words—from a perspective of beneficence, from a vision of a better world with ever-improving genetic traits, with a corresponding enhancement of overall health for mankind. It crept into societal consciousness, sneaking its way in with suggestions of an ideal futuristic society, sung by mermaids who also whispered of the ethical imperative to avoid the suffering of those who were imperfectly born in their bodies or their mind. The killing machine was so perfectly tuned to implement the Final Solution because it had been used for years to murder thousands of mentally and physically disabled children in the Third Reich's world-leading hospitals. These were mostly blue-eyed, blond-haired German children. Their extermination was not based on racial differentials; it was based on differences in ability and the implication that these differences were not compatible with somebody's notion of a life worth living. Are these differences different than those based on race? Are they more justifiable? Absolutely and obviously not—the tragic perversion of ethics is not in which differences are sufficient for excommunication from mankind, with all privileges of dignity pertaining thereto. It is the notion that there may be differences that warrant the exclusion of some from mankind—an exclusion that unavoidably exposes them to death and suffering by lowering or eliminating the ethical prohibition against bringing death and suffering to those who have been selected. No differences ever justify such exclusion, regardless of any consideration of "greater good" or beneficence for a class of citizens of humankind, no matter how large. This is why the purely utilitarian perspective has no citizenship in medical ethics; it must be bounded and balanced by considerations of *nonmaleficence, justice*, and *respect*. Medical experimentation during the Holocaust, with its systematic infliction of death and suffering, is the leviathan that haunts our collective souls and must serve as the reminder that nobody can ever be excluded from the basic set of privileges that pertain to the human status—with no exceptions and no excuses.

Nonmaleficence is a fundamental tenet of the Hippocratic Oath: First Do No Harm. The medical establishment of Nazi Germany was the best in the world. German culture, art, music, and science were second to none. Yet, despite this overwhelming intellectual excellence, the dominant fraction of German medicine found a way to escape this most fundamental of tenets of the medical profession.

The escape clause was in allowing the Oath to apply only to a subset of humankind. Once that subtly poisonous distinction is allowed to stand, once that most insidious of maladies of the soul is permitted to thrive, it gains momentum, obliterates more and more boundaries, engulfs more and more groups in a raging fire of mutual destruction, in a spiral of individualization that leaves all fighting all, to the death. In the Third Reich, it was based on a devastating principle, the desire to evolve into a genetically superior mankind through eugenics. The tragic irony is that such arguments of separation bring about exactly the opposite: They are an evolutionary *dis*advantage for all, insidiously camouflaged as an advantage for some. No differentiation into classes of citizens of humankind with different life dignities is ever acceptable. There are no exceptions, no excuses. Failure to adhere to this tenet leads to the atomic death of society.

The Hippocratic Oath contained this basic wisdom. The reaffirmation of the Oath was made necessary by the monsters created by medicine in the Third Reich and was the driving force for the Declaration of Geneva in 1948. The Hippocratic Oath had been written about 2,500 years before. As we read it today, we recognize that the Oath is far from being a "first attempt" at medical ethics, coming from a semi-primitive, ancient culture. The great instruments of medicine, the molecularly targeted pharmaceuticals, the transplants of organs, the positron imaging devices, the system biologies, the nanomedicines, the robotic surgeries, and the tissue regeneration stem cell therapies were not there at the time of Hippocrates. Yet the Odyssey was already hundreds of years old, and perhaps no other human writing since has ever matched its significance, depth, brilliance, and beauty. So perhaps it may be worth reading the totality of the Hippocratic Oath, even the passages that may cause discomfort to some—even those passages that prohibit abortions and those that prohibit materially bringing or suggesting death, with no exceptions and no excuses for doing otherwise.

Respect is perhaps the dimension that most markedly reflects the refocusing and reaffirmation of bioethics following the Shoah. Fundamental considerations within respect are autonomy and self-determination, which were horrendously violated, and the very definition of human nature. The elimination of disabled children in the Third Reich was justified by setting the threshold of human nature at certain levels of mental and physical performance, which is ethically untenable no matter what the thresholds. These levels could not be other than completely arbitrary, in that there were not then, and still there are not now, any scientifically defensible methods for determining mental "performance," other than for specific individual tasks that are so minute in their scope and relevance as to appear utterly ridiculous in front of the notion of "human-ness." At the same time, any attempt to define *race* scientifically is similarly destined to meet with even greater ridiculousness, though of an immensely tragic nature. Any decision in medicine, which is based on differences in degrees of humanness and human dignity, violates this most fundamental value of medical ethics—and so does the withholding of access to medical care on these bases, in contravention of the principle of *justice*.

We Cannot Make a Real, Positive Difference, Unless We Realize That We Are All Part-Bad Guys: It Is Not "Them Evil vs. Us Good"

I have never been a believer in the sharp distinction between good and evil in people, individually or in broad collections of humanity. Sharp lines distinguish doctrines, of course, but actions often have a composite nature, and as humans the normalcy is to have Gryffindors and Slytherins combined within ourselves. Being born in immediate postwar Italy, the rapid recognition of this duality was unavoidable in my early days, as the country turned almost overnight from nearly full-participation fascist to almost no-exception antifascist. It was the same people, but now on different sides of the fence, and in all likelihood with a very similar composition of good and evil in them. The driving forces for belonging to the two opposing sides were probably the same, before and after the turning of the coats: survival of the individual, protection of one's family, and probably, in many cases, the desire to do good even in adverse circumstances.

The horrors of fascism were followed by the systematic ethnic cleansing perpetrated against the frontier Italians in Friuli Venezia Giulia (FVG), where it is estimated that 25 percent of the population was killed by the good "liberators" under the command of Marshall Tito, with the silent permission if not the blessing of the forces of the United States and the United Kingdom. The signature killing style of the pro-communist forces was to tie two prisoners together with barbed wire, shoot one, and then throw both into the natural pits and caves of the regions, the *fojbe,* to die there. My mother has vivid memories of the fact as a young child in a frontline area that the only safety was when the German soldiers were in town— they were respectful, friendly, educated, family-minded. In the previous war, again fought quite literally in our front lawn, my mother's side of the family lived in the same home, only then it was part of Austria. My grandpa, Nonno Mario, was drafted by Kaiser Franz-Josef and fought on the Russian front wearing the Austro-Hungarian uniform, while his younger brother Zio Giulio, at the ripe old age of 16, escaped across the border into Italy and volunteered with the Italian Army, fighting against the Austrians as a *Ragazzo del '99* in the campaign on the Piave River in 1915. They were Austrians of Italian origin, they spoke Italian as their primary language, and they were both Gryffindors and Slytherins, rolled into one.

My father moved to Friuli from the deep south of Italy when he joined the Italian Army immediately after WWII, choosing his militaristic "vocation" because they were starving back home, his adoptive father was nowhere to be seen for long stretches of time, and military service was the only option to sustain himself and his illiterate adoptive mother, Nonna Teresa. My older brother Daniele followed in his footsteps and served with great distinction in the Italian Army for 40 years in Iraq, Serbia, and other dangerous places. He had the privilege to serve alongside American forces and to be honored by them with multiple medals.

One ethnicity that he and I find to have a very similar story to us Friulians, with our composite make-up, is actually the Serbs. History recognizes them (or it should, even in these revisionist times) as the ultimate stalwarts of European freedom throughout the ages, from the days of the Turkish invasions 700 years ago to the unparalleled courage in the opposition to the Nazis during WWII. The Germans and perhaps their Italian allies—though this is usually not mentioned in history books, at least those that originate from Italy—had the policy of exterminating civilians in a multiple of the number of Germans soldiers that were killed by resistance warfare. The selected multiple depended on the strength of the message that they were intending to send to the resistance movements, and it increased with the intensity of the resistors.

For instance, the multiplier was different in the South and the North of Italy. After Italy turned antifascist in September 1943, its South was basically liberated and in the hand of the liberators, the US-led Allies. The North, by contrast, was essentially without any form of government or protection for the civilian population, with large domains being controlled by the Germans, retreating with great anger against the traitor Italians, and other parts being controlled by pockets of fascist loyalists or falling under the influence of the partisan resistance movements. Some of the partisan forces were inspired by ideals of Western-style democracy, while some were communists with different degrees of radicalism, all the way to a substantially widespread allegiance to Stalin. Fratricidal wars ensued, with "white" and "red" partisans killing each other in numbers that were large (and again conveniently "forgotten" by history for a long time) but nowhere nearly as large as those suffered by essentially identical dynamics in Serbia, where over one million civilians perished, much more at the hands of different partisan factions than the forces of the Axis. In those times of mayhem and confusion, survival of self and protection of family were perhaps understandably the major driving forces, and good and evil tended to take a more relative meaning. In those days, the Italian resistance fighters suffered civilian retaliation with a multiplier of 5 to 7 executed for every German soldier that was killed. My mother's family was relocated to the small town of Nimis in FVG in 1945, after it had been completely burned to the ground by the Nazis in retaliation for an ambush on their forces.

The Serbs, on the other hand, were considered the fiercest resistors of all, and the Germans retaliated against them with a multiplier of 100: for every German soldier killed, 100 civilians were executed. I will never forget visiting the memorial park in the Serbian city of Kraguyevac, where the Germans exterminated over 2,000 civilians in retaliation for the loss of 20 of their soldiers. This savage mass murder included the extermination of the children who were taken from the local schools and executed with their teachers—it was horror beyond words, and yet a testimonial to Serbian courage and determination to fight the Nazi monster.

Yes, Serbia is the same country that perpetrated mass murder under President Milosevic and is guilty of ethnic cleansing, the rape camps in Srebernica, and other untold, unspeakable horrors against other ethnicities in former Yugoslavia in very recent times. Yes, Italy is the same country that, on the one hand, bravely fought the Germans in the mountains, but cannot ever be forgiven for sending many Jews to the death camps, essentially wiping out the historical Jewish community of Rome and other principalities. Yes, that happened with the cover of the silence of the

Vatican, which, on the other hand, was instrumental in providing free passage to life and freedom to many Jews in Italy and elsewhere. Yes, the Cossacks were aligned with the Nazis. Facing extermination by Stalin, they picked up their families and belongings and accompanied the Nazis, providing support in their attack on the Soviet Union.

When the siege of Stalingrad failed, the Germans and the Italians retreated with enormous losses and sufferings through the Russian winter. Again facing extermination, the Cossacks followed them westward—Hitler had promised them a land to settle down in peace with their families. They landed in Friuli Venezia Giulia (where else?), which was still no-man's land, awaiting liberation by the US Allies or the Soviet/Yugoslav forces. Among them, about 400 families were the first occupants of Nimis, just partially rebuilt after the retaliation burning in vindication of the anti-Nazi ambush. When the Allies finally won out, the Cossacks again feared extermination because they had sided with Hitler. So they escaped Nimis and FVG, going north to join the retreating forces of the Axis into Austria and Germany. When they got to the Drava River in Austria, they surrendered to the Allies. Neither the Americans nor the British had any desire to keep them, and it was decided that the Cossacks would be returned to the Soviet forces. Certain of extermination, at this point, the Cossacks committed mass suicide in the river—mothers drowning their children, in the self-immolation of men, women, old, and young alike. The Cossack ethnicity basically disappeared there; it perished in unspeakable pain at the combined hands of the evil, and the good. Yes, it was the same Cossacks who had been siding with Hitler against the new, radiant sunshine of … Stalinism, of all horrors! With the extermination of the Cossacks, the town of Nimis became available for my mother's family to move in as part of a new wave of refugees from communist-dominated Italy into gray-zone Italy.

Meanwhile, a major Serbian group of hundreds of families, under the leadership of one of their high priests, Orthodox Pope Djujic (who had himself been accused of heinous war crimes by his opponents), had been fighting their way from Serbia after they had lost their local war against the communist, pro-Tito partisans. They, with full families in tow while fighting, reached their promised land, breaking through forces of the Axis and communist lines in a journey that took several months. Many were lost to violence, starvation, and disease. They arrived in the town of Palmanova, in Friuli, which is maybe as many as 25 miles from Nimis, on the day of Saint George, the celebration of national pride for the Serbians. In their promised land of Palmanova, they surrendered to the Allies in an installation under the command of Australians and New Zealanders. They were requested to surrender their weapons. Rather than doing so, especially on their sacred day, many took their own lives. It was perhaps an unwarranted act of nationalism, or perhaps a commendable glorification of their people and traditions, or perhaps an unjustified act of desperation given the civility of their captors—and yet, Nimis, the Drava, the *fojbe*, Stalingrad, and Kraguyevac were still close and part of the same breath of history.

The reverberation of their desperate act remained in the stories that were told by the elders in Friuli, long before official history could do them justice (it has not yet). That is how I first learned of Serbians, when I was a small child. I was born an

Italian Army brat in the border region between the western world and communism, less than 10 miles from the border with Tito's Yugoslavia. Given my father's job, my childhood was spent more with army veterans than with kids my age. Many of them had participated on the wrong side of the siege of Stalingrad and had gone through the horrors of the retreat through the Russian winter. Nobody had volunteered for either. Nobody had had a choice to join in or not—they were drafted, basically kids, mostly from farms, with no idea of where Stalingrad was, of what democracy was, and that there even was a non-fascist world out there. At the end of the war, my mother's hometown was split into two parts. Marshall Tito wanted the train station, so in the trade he got the smaller part of town. My mother's house ended up staying on the western side, and that is basically why I only carry the embarrassment of my people of origin having worn black shirts in the darkest hours of history, and not doubling that up with those who were traded to the Soviets and ended up having to wear red shirts for the times that followed. We are ultimately caught in the wheels of history. The difference between me and those that drowned in the Drava or the Serbians in Palmanova or the many Friulian families that were executed by the Italian communist partisans and the troops of Marshall Tito is just a set of impalpable alignments of unlikely events impacting that small piece of countryside that we now call Friuli. It depended on who wanted the train station of Gorizia more, on the weather during the key week of the failed attack on Stalingrad, on the exact location of my grandparents' home—a few miles either way, and it would now be in one of four nations, and all of us in the family would have entirely different historical burdens and perspectives.

When the wall came down, and Slovenia became part of the European Union, the border fences with military guards (the *granizari*) whom I had feared so much as a child all of a sudden were removed, with free passage for everyone in both directions. I took my children across the border for lunch. As we were having a great meal in Ljubljana, I asked them to look around at all of the people and tell me what differences they could find between them and us. When they could not find any, I told them, "I thought so," and proceeded to tell them how, across those absolutely fictitious and capricious boundaries, "we" and "they" had proceeded to slaughter each other with abandon for over a thousand years.

Good and bad, you say, good and bad? Yes, there are good and bad actions; the Good Book is very clear about it. And yet, when it comes to people, the same Good Book also admonishes us about knowledge of good and evil through the narration of the actions of Adam and Eve. The original sin that caused affliction in human nature is the reaching out for the fruit of knowledge of good and evil. I suspect that is a reminder of the dangers embedded in the delusion of certitude of a world cleanly divided into the two domains. If we believe we know the boundaries with exact precision, if we delude ourselves that we have the ability to steer clear of those boundaries, then we sin, irreparably, and our hubris will lead us down the path of horror.

Yes, it was the Germany of the most sophisticated human expressions of art and culture. Yes, it was the Germany of Dietrich Bonhoffer. Yes, it was the Germany of the most advanced knowledge of medicine. And yet it was the Germany of the most atrocious crimes in the history of humankind, the Holocaust and the extermination of its own disabled children. And yes, it was the America that once again shone through as the greatest defender of freedom that humankind has ever known, sacrificing many of its own, the Greatest Generation, in the name of collective liberty in world. And yes, it was the America that would send its best doctors including Dr. Michael DeBakey to study in Germany; yes, the same America whose medical intelligentsia widely supported eugenics, especially at great institutions such as Stanford University, Cold Spring Harbor Laboratory, and the Rockefeller Foundation; and yes, the America that largely turned a blind eye on some of the tragedies and ethnic massacres of the time—and other times since.

Without the illusion of a clear separation of good and evil in real people, constant vigilance is the necessary requirement for all to remain within the bounds of ethically acceptable behavior. Our collective and individual vigilance must be extraordinarily strict, as we approach considerations that involve life and death and the dignity of the human experience from the first day to the last. In no domain of human endeavor then, is the strictest vigilance required more than in medicine. Forgetting past horrors is a condemnation to future horrors. We must acknowledge our failures, recognize how past horrors were arrived at, suffer in the recollections of the horrors themselves, and be unforgiving with ourselves, in the recognition of the elements of today's culture and actions that parallel those of the darkest of days. No discounts, no exceptions, no excuses.

There Are Major Elements of the Path to the Holocaust in Today's Medicine and Culture

Let's then face it: The path to the Holocaust is still active and is actually reflected in several, even accelerating trends in today's world of "civilization." Three aspects are particularly concerning to me: the culture of death; the culture of diversity-based exclusion; and the culture of *Herdensinn*, whereby individuals hide within the folds of mass-dominant opinions, thereby recusing themselves from responsibility of affirming their own major principles. These attitudes permeate society in the most global of manners and manifest themselves recurrently through medical policies and the practice of medicine.

Strengthened by the false dignity conferred upon them by the white coats of the medical profession, these trends loop back into society as cultural reinforcements, which in turn, lead to a spiral of ethical decay that parallels many steps that lead to the Shoah.

"Death shall be no more, Death thou shalt die," from John Donne's "Holy Sonnet X," is a concept that is inextricably linked to transcendence. The denial of transcendence is the triumph of death. It is then not surprising that in an increasingly secular society, where the separation of spiritual belief from public governance is celebrated as a *virtue*, the culture of death finds fertile grounds for uncontrolled growth. We kill those who are uncomfortable or problematic or inefficient in a utilitarian perspective of society as a whole. Death by capital execution is a medical shortcut to address a mental disability that leads to the commitment of heinous crimes—I firmly believe that the crimes that are punishable by death in western societies can only be committed by those who have an infirmity, an embedded inability to empathize or to feel pain, a frontal lobe dysfunction of some sorts, that impedes their relating to others. I specified "in western societies" simply because I understand that in certain non-Judeo-Christian cultures marital infidelity and homosexuality are examples of "crimes" deserving of the death penalty. For cases such as these, my impression is that the mental pathology is of the community administering the punishment and not of the "perpetrators."

Euthanasia is now allowed in Oregon and in certain European countries such as Belgium and the Netherlands. Giving death, or the instruments of death, or the suggestion of death to a patient is prohibited by the Hippocratic Oath, yet contemporary medical practitioners engage in these activities. Their "moral cover" can be utilitarian, as was the case in support of eugenics all the way to Auschwitz, or be masked under the pretense of autonomy. However, anyone suffering to such an extent that their will-to-live is overwhelmed cannot be considered in control of his/her mental abilities, and therefore, an argument based on autonomy is intrinsically weak. "Suffering" here is intended to cover both physical and psychiatric pain—the age when this artificial difference finally will be set aside is hopefully near with the advances in the biological basis of psychiatric disorders such as depression. Any notion of humanness and worthiness of the human experience based on levels of suffering must contend with the recognition that if those who suffered greatly throughout their lives had been eliminated by "mercy killing," the overwhelming majority of the truly superior works of art and products of human creativity never would have seen the light of day. Suffering is a component of the human experience, just as much as joy is, and just as meaningful.

Be that as it may, many instruments exist today, which were not available just a few years ago, to address mental and physical pain. The responsibility of the medical practitioner is to identify and use them, as already pointed out by Pope Pius XII and Pope John Paul II. Opting for the shortcut of hastening death by direct, enabled, or recommended action is not acceptable, though at times the very medications used to alleviate suffering may bring adverse consequences including death. All procedures in medicine carry potential risks, and medical interventions are justified when the benefits outweigh the risks. In extreme conditions, such as emergencies and in conditions where there are no alternatives, it is also common that medical intervention might present a substantial risk of death. The alleviation of extreme, otherwise intractable suffering with high-potency opioids is in this category.

By contrast, it is disheartening that some physicians find it preferable to pursue death as the primary "remedy," and to dispatch a patient with a terminal disease or a condition that we cannot control, rather than to act as necessary to accompany her/him for a longer time, admitting the failings of one's own medical science, while at the same time providing all possible comfort and relief from suffering.

While life support and resuscitation may not be warranted in some situations and must be placed at the discretion of the patient's autonomy, the discontinuation of life support actively provides death to the patient and, as such, is in violation of the Oath. By agreeing to deliver, recommend, or enable euthanasia, physicians not only violate the most fundamental of the tenets of their profession but also, by the respect accorded in society to the medical profession, arouse and reinforce in the community two extremely dangerous emotional conditions: the trivialization of death and the desensitization toward it. These conditions are signs of a deep decay in the spiritual fabric of a community, unmistakable symptoms that a profound malady has overrun the societal safeguards and protections against grievously destructive behavior, as was the case in Germany under National Socialism, in the precipitous slide all the way to Mauthausen and Buchenwald.

Third Reich eugenics allowed for the execution of those below a certain arbitrary degree of present or potential functionality. I believe that life starts at the moment of conception and, therefore, I am opposed to the performance of abortions. The ethical line is drawn by some at the notion of viability and recast in terms of age of the fetus, but it is important to realize that viability is largely a factor of circumstances such as the strength of the medical technologies available and the training of the medical personnel. With the constant advances in medical sciences and improved access to specialized care facilities, viability will draw closer and closer to conception. It appears rather capricious and arbitrary, and therefore, ethically unjustifiable to cast the moral debate on abortion and regulations thereof on a technological line that is constantly moving. Rather, society should face the question head on, recognizing its full implications and refraining from hiding behind flimsy or outright false technological arguments. It is not inconceivable that, in a near future, fetuses will be viable just a few days or weeks after conception, and if so, will we change abortion laws then? Is it prudent or acceptable to base ethics on technological advances? Especially repugnant to my conscience are the decisions to abort based on adverse genetic signatures, that is, on the same ethical bases as the Third Reich, but with the added benefit of advances in molecular biology and prenatal diagnostic medicine. Is a molecularly smarter eugenics a more justifiable eugenics?

Fertility procedures that currently enjoy widespread acceptance also raise serious concerns in my conscience. On what basis, if any, can it be acceptable, to "destroy embryos" (the artfully conceived, politically acceptable expression concealing the reality of death they embody) when too many of them are viable after a contemporary fertility procedure? I am against the destruction of embryos, period. For those who are on the fence, there are some questions to consider. Is it is ethically acceptable to destroy embryos if they carry genetic signatures of disease or imperfection (the "eugenic motivation")? An argument that is sometimes presented in the support of the decision to destroy an embryo that carries an imperfection, a disease, or a heightened risk factor is the desire to spare unnecessary suffering to the new life. Is this not a false pretense, a feel-better cover to hide from oneself the decisions of pursuing convenience over respect for life of one's own offsprings? With current technology, the embryos are normally eliminated at a time when they would not be independently viable, though they would be viable in the womb, and in this case as well, viability is moving closer and closer to fertilization. Where is the ethical argument supporting these distinctions? Is it not just a matter of economic or lifestyle convenience of the family? Let's face the truth and recognize that it is yet another slip toward the horror of the gas chambers.

Current medicine permits the killing, by withdrawal of life support, of those who are "brain dead," though in the medical sciences we have no clue as to how the brain works, and the methods we have to determine "brain death" are absolutely primitive. We do not know how to correlate electroencephalograms (EEG) to any specific activity of the brain, or how they correlate to thoughts or emotions. EEGs record some electrical activity of the brain, above a certain detection threshold. We have no idea if anything happens in the brain below that instrumental threshold. We have no idea of how the most basic functions of the brain relate to electrical activity. For instance, memory is a conformational change of the network of brain cells, a set of chemical and biological processes that is sustained in the absence of any electrical impulses.

The very notion of "self" is completely mysterious to the current state of science. I believe it is most likely related to biochemistry and the architecture of cellular networks, not the physics of electronic flow. The concept of "consciousness" relates to responses to stimuli, and some of these can be quantified by electrical measurements, but again, we have no way to recognize what happens below the instrumental threshold of EEGs, what responses are dominantly or solely owed to cell-cell contacts (such as, mechanics? cell surface interactions such as ionic exchanges? mass transport via exchange of vesicles such as exosomes?) or to gradients of metabolic activity. My expectation is that the advances of functional MRI and the advent of more sophisticated imaging techniques will reveal fully transformational discoveries about the functioning of the brain, including startling observations on the activity of brains that are officially "dead" per current EEG measurements. Despite our overwhelming ignorance of the modes of functioning of the brain, at this time the medical community has no problem dispatching into the great beyond those who fail their EEG tests, or even to perform terminal experiments on them, as was done in the search for vascular zip codes of cancer in very recent times at the MD Anderson Cancer Center in Houston. As these practices gain acceptance in medicine, society at large is desensitized to killing and to extreme human experimentation, blinded by the reflections of the white coats, without bothering to ask if they know anything about "self," consciousness, and its relation to memory and identity. It is another step toward mass graves, this time mostly in the name of practical convenience, and again enabled by some of the best scientific minds of the medical profession.

The assignment of lower "humanness" based on any discriminatory criterion is absolutely unacceptable. As society is making strong advances toward abolishing discrimination and has rejected any differences in human rights based on race and gender, still much more progress is required, and attention must also be turned toward new, insidious discriminatory criteria that are surreptitiously emerging *because* of medical "advances." Any discrimination based on skill level is unacceptable, because any individual skill or combination thereof is but a minuscule rendition of the fullness of the human experience. Most importantly, it is impossible to justify any skill-based discriminatory arguments and, at the same time, avoid sliding off the Tarpeian Rock.

The true challenge is to develop and implement methods for ensuring equitable access to health care for all, as a crucial matter of justice, since access is overwhelmingly determined by financial status, which in turn, is frequently determined by excellence in the skills that provide monetary success in today's world. These measures of excellence are, in part, natural gifts and are, in part, refined through consistent practice, sacrifice, and hard work. I am a stout believer in the importance of rewarding hard work and commitment, but at the same time, I cannot but contemplate the truth that even the ability to work long hours with dedication and passion is probably in large part owed to genetic traits as much as natural beauty or the ability to run fast or to sing wonderfully or to parallel process difficult tasks while demonstrating operational leadership and charming the world with superior emotional intelligence. The exacerbated problem of access is intertwined with the incredible speed at which medicine is advancing. A difference in access between two groups will not only mean a difference in health outcomes and life expectancy but also engender progressively increasing degrees of separation between the groups. The creation of a caste of *ubermensch* is a very feasible nightmare, which could be based on current medical practices such as the selection of favorable genetic traits for our offspring, the provision of selective enhancements of performance, the creation of entire classes of "diseases" that warrant preferential treatments for selected groups. All in all, the simplest of notions is the only tenable foundation: no differences allowed in the dignity scale for human beings. Any exception leads to the acceptability of the notion of differences based on traits, which then becomes a tool of dominance in the hands of those that have more power. Yesterday it was blond hair and blue eyes as a proxy for the industrial power of the Nazi; tomorrow there may be other traits as proxy for greater military power, or control over energy production, or natural resources. The only civilized answer is: No stratification of human dignity allowed.

I have spent most of my professional life in the hallowed halls of academe, at four truly outstanding public institutions: The University of California at Berkeley, The Ohio State University, the National Cancer Institute, and The University of Texas. I learned that at public institutions in the United States, any public profession of faith by a faculty member is considered somewhere in the continuum between inappropriate to subject to disciplinary action—a view I consider a grotesque misinterpretation of the doctrine of separation of Church and State. Over the years, and now at a faith-based institution, I have started to affirm my faith with greater ease. Actually, I now maintain that choosing not to disclose our faith is a reprehensible act of willful deception, especially in a profession such as medicine, where people trust us with their lives and thus have a right to know where we stand on fundamental ethical matters. Even with that, I always had a reluctance to declare my views on delicate issues such as abortion and euthanasia. The reluctance stemmed from a desire to respect the opinions of others, which has always been very important to me, and most certainly still is. In the midst of this reluctance came the Conference on Human Subjects Research after the Holocaust and, with it, the recognition that the silence of many enables great tragedies. Failure to speak is complicity. Thus, I have concluded that the time has come for me to speak my conscience on these controversial matters, and thus this chapter. As I am leaving my middle age, I may be finally providing signs that I am growing up, after all. My encouragement to all is to speak their truths, to give voice to their conscience, from either a religious or secular perspective, with reciprocal tolerance. Silence is not the answer.

The Future May Be Scarier Than the Past

Eugenics has evolved into the molecular science of genetics, which may prove exceptionally powerful in the fight against disease, but at the same time, carries the fearsome notion of the classification of people on the basis of their genetic profiles, with potential reductions of human dignity and basic rights based thereon. The process of selection and destruction of embryos based on genetic traits is a reduction of the dignity of life, which will become more and more apparent as technologies are perfected to bring the embryos to term in conditions that are now impossible. Genetic profiles are associated with risk factors for diseases. Thus, a number of well-studied ethical questions emerge, such as who is entitled to the knowledge of the risk profiles? Employers, before they invest in training employees who might be dead or incapacitated in a few years? Future spouses, in a manner that is reminiscent of the mandatory syphilis testing of not many years ago? Insurance companies, will they be allowed to charge higher rates or to deny coverage? Will someone with high risk for a deadly brain cancer be deprioritized in a waiting list for liver transplant?

While genomics brings risk profiles, the disciplines of post-genomic molecular medicine such as proteomics, metabolomics, and epigenetics actually carry much greater levels of information on actual, not only potential, risks. A proteomic reading in a near future may indicate the "distance" of a person from a disease, or who they had lunch with, and when and what they ate, what happened to their blood pressure as they ate, and what kind of feelings they experienced during the occasion. A connection to a smartphone could distribute the information to health care providers for automatic analysis and the potential triggering of actions in compliance with a centralized health maintenance algorithm. The opportunity for advances in health care will be extraordinary, but with those will come the "biological big brother," the fearsome notions of continuous, centralized monitoring of one's behavior and the hierarchical classification of dignities based thereon, and the ultimate tradeoff of privacy and individual freedoms in exchange for safety and convenience. This would be a highly unstable system, for sure, with a near certainty of abuse in the quest for ultimate power. With a view toward the darker side, one might easily envision molecularly targeted weapons or biological modulators designed to act selectively on those with certain traits (Race? Skill level? Age?) or on certain patterns of behavior (Smoking? Eating fat-laden foods? Following the dietary fashion of the week? Engaging in certain types of sex?)— notionally in the best interest of the community, that is, for the greater good, which is the form in which major tragedies invariably present themselves.

Perhaps even scarier, though potentially useful, is the notion of synthetic genomics. Pioneered by Craig Ventner's group, it pertains to the ability to create life forms that do not currently exist. The Ventner lab created a single-cell, bacterial life form by essentially writing its genomic sequence in a computer and building it in a machine. Perhaps this was a life form that existed in the past and was evolutionarily deselected, or perhaps it simply never existed before. The Venter lab is focusing on using this new bacterium for environmental protection against oil spills, but of course, nobody knows what could happen—they could evolve into different life forms, which might be beneficial, or not. Others may use the techniques published by Ventner's group to make more sinister creatures, or not. Who knows? Even the best of scientists are known to be carried away by enthusiasm and the best of intentions, but is anyone awake in the world as they are doing so? Could we synthesize a highly infectious bacterium that is resistant to all known drugs and that targets only the people I do not like? I suspect that, with a bit of sequencing, it will be rather easy to find a target set of genetic traits that they have in common. On the other hand, these new synthetic creatures could be used to treat diseases that are now incurable. It probably will be difficult, expensive, and time-consuming to get them approved by the Food and Drug Administration (FDA). Instead, it might prove much easier to use them for nefarious purposes. After all, terrorists and rogue nations do not require FDA approval.

In fairness to the Ventner group, I will recognize that I found myself in a similar situation in the past, with the recognition that some of the therapeutic nanoparticles I was developing to improve the efficacy of orally administered drugs also could have been used as biological weapons of mass destruction. I presume that similar occurrences may happen to many scientists who frequent the frontiers of discovery. There are different options that are available upon recognizing the possible dark sides of one's discoveries placed in the wrong hands. One option is to discontinue the research altogether, which obviously implies foregoing the potential health care benefits of the line of research. Another option is to continue to do the research, but also inform "the good guys" of the potential risks. The problem with that is obvious: Where is the line drawn that encompasses the "good guys"? Shall we then just publish public warnings with our concerns, while we continue our research? My personal decision was to discontinue nanomedical research for the oral delivery of biological pharmaceuticals and, at the same time, to publish my concerns on this type of research. It was a personal decision, not a prescription for action in all cases posing similar problems. I have no problem with the notion that different lines of decisions may be warranted, in different situations, but I do believe that the public must be fully informed in a timely and transparent fashion before any evil genie exits the bottle.

Forgiveness is the Sole Victor

Eva Kor was a small child when she and her identical twin sister were brought to Auschwitz on a cattle train in conditions that were so horrible that several people lost their lives in transit. The last time she saw her mother, father, and older sisters was on the platforms in Auschwitz when they got off the train. She and her twin sister were the subjects of extremely cruel experiments by the "Angel of Death," Dr. Josef Mengele, who was intent on studying twins to understand the genetic bases of disease and response to medical treatments. With sinister irony, because they were valuable not as human beings but as experimental animals in the mind of Mengele, she and her twin sister were helped to survive the death camps. As a father of two sets of twin girls and five children in total, it was impossible for me to sit through her presentation at the CMATH conference and not be moved to tears, many times. She showed incredible strength and acumen. And she stunned everyone with the energy and sincerity with which she announced her proclamation, her signature, her legacy: the complete forgiveness of Dr. Mengele and those who exterminated her family and so many others in the death camps and elsewhere in those dark years. Perhaps the most potent way to avoid repeating those tragic times is, indeed, forgiveness. We are all bad and we are all good. The apple from the tree of knowledge is not for us to seize and be proud of. It is not the one that seizes the apple who wins; it is the one who loves and forgives. Thanks, Eva, and God bless you.

Conclusion: For Whom Does This Bell Toll?

Medical ethics certainly pertains to the doctors who practice medicine. Obviously though, I do not see any reason why it would not apply exactly as much to nurses. Medical technicians? Hospital administrators? Medical researchers? Philanthropists? Yes, yes, yes, and yes. Politicians and government officials who determine health care policies? But of course! You and I, when we elect politicians and influence their decisions? Or, just as much, when we do not even do that, but simply sit around watching TV? Yes! You see where I am going with this: We are all deeply involved in medicine; we all have the power to bring relief from suffering and the power to cure.

We are all healers. The tenets discussed here pertain to all of us. No exceptions. No excuses.

Index

S. Rubenfeld and S. Benedict (eds.), *Human Subjects Research after the Holocaust*,
DOI 10.1007/978-3-319-05702-6, © Springer International Publishing Switzerland 2014